Phylogenomics
A Primer

Rob DeSalle • **Jeffrey A. Rosenfeld**

Garland Science
Taylor & Francis Group

NEW YORK AND LONDON

Vice President: Denise Schanck
Senior Editor: Janet Foltin
Senior Editorial Assistant: Allie Bochicchio
Production Editor: Natasha Wolfe
Typesetter and
Senior Production Editor: Georgina Lucas
Copy Editor: Heather Whirlow Cammarn
Proofreader: Chris Purdon
Illustrations: Mill Race Studio
Indexer: Indexing Specialists(UK) Ltd

ISBN 978-0-8153-4211-3

Library of Congress Cataloging-in-Publication Data

DeSalle, Rob.
    Phylogenomics : a primer / Rob DeSalle, Jeffrey A. Rosenfeld.
        pages cm
    Includes bibliographical references and index.
    ISBN 978-0-8153-4211-3
  1. Phylogeny--Molecular aspects. 2. Genomics. 3. Evolutionary genetics.
I. Rosenfeld, Jeffrey. II. Title.
    QH367.5.D47 2013
    572.8'38--dc23
                                        2012036646

Published by Garland Science, Taylor & Francis Group, LLC, an informa business,
711 Third Avenue, 8th floor, New York, NY 10017, USA,
and 3 Park Square, Milton Park, Abingdon, OX14 4RN, UK.

Printed in the United States of America

15 14 13 12 11 10 9 8 7 6 5 4 3 2 1

Visit our Website at http://www.garlandscience.com

Preface

This book is intended to serve as an introduction to a new area in comparative biology known as phylogenomics. Approximately 15 years ago, concurrent with the rapid and efficient sequencing of full genomes from living organisms, Clare Fraser and Jonathon Eisen coined the term "phylogenomics," a combination of phylogeny, which refers to the process whereby evolutionary trees are generated, and genomics, which represents the endeavor of obtaining genome-level data from organisms. Phylogenomics has developed into an important and compelling discipline in its own right. We developed this book in response to the students we have encountered over the last several years who are interested in applying genomics to comparative biology, specifically to phylogenetic, evolutionary, and population genetics problems.

Phylogenomics: A Primer is for advanced undergraduate students and graduate students in molecular biology, comparative biology, evolution, genomics, biodiversity, and informatics. Depending on their educational training, students can focus on the topics in the book that are of the most interest to them. Students who do not have strong backgrounds in evolution or phylogenomics will find the chapters that discuss evolutionary principles and the manipulation of phylogenomic-level data particularly useful. Conversely, students who are adept in ecology, taxonomy, and biodiversity will have the opportunity to learn about the evolution of genes and populations at the phylogenomic level and become familiar with applying phylogenomics to their genomics research.

We believe that there is no better way to understand the information that has been obtained about genes and genomes from life on this planet than to place it into context with the grand evolutionary experiment that has unfolded over the past 3.5 billion years. To this end, we have designed this book as a journey from the basic principles on which organic life has evolved, to the role of burgeoning databases in elucidating the function of proteins and organisms, and concluding with an interpretation of linear sequence information in the framework of organismal change.

Molecules are the currency of modern genomics and have an underlying linear arrangement of their component parts; that is, proteins and DNA can be broken down into linear sequences of amino acids and nucleotides, respectively. Chapters 1 and 2 present the essential principles underlying molecular biology and describe classical techniques used to analyze molecular sequences, including several high-throughput techniques that are known as "next generation" approaches. Chapter 3 explores evolution at the population level and introduces phylogenetic tree building. As a convenience, we make a simple demarcation between the evolutionary studies that focus on populations (microevolution) and those that focus on species relationships at higher-level systematic relationships (macroevolution).

Chapters 4 through 7 discuss the storage and manipulation of genomics-level data to enable the generation of the data sets that are used in phylogenomics. These processes include accessing databases and web programs such as PubMed, GenBank, and BLAST for downloading DNA and protein sequences; aligning linear sequences and producing matrices for evolutionary analysis; and assembling and annotating genomes.

Chapters 8 through 11 focus on the construction of evolutionary trees. Various approaches to phylogenetic analysis are presented, including distance, likelihood, parsimony, resampling, and Bayesian inference. In addition, the phenomenon of incongruence in relation to tree building is described as are the methods by which this problem is addressed.

Chapters 12 through 15 focus on the application of modern phylogenomics at the gene and population level. The transformation of population genetics by the use of DNA sequence information, the detection of natural selection on genes derived from genomic data, and the application of genome-level approaches to population genetics is essential to the understanding of natural populations in an evolutionary context.

The book concludes with a discussion of the basic applications of phylogenomics in the context of modern genome research. Chapter 16 examines the use of genome content to understand evolution. The role of phylogenomics in biodiversity studies, specifically the construction of the tree of life, DNA barcoding, and metagenomics, is explored in Chapter 17. The final chapter describes how functional genomics can be applied in a phylogenomic context, specifically transcription-based approaches and protein–protein interactions.

Working through the applications described in this book does not require an extensive computer science background beyond basic skills such as using a terminal or web browser. We have developed a set of Web Features that are linked to specific methods discussed in the book and are designed to introduce students to the websites used to obtain and analyze data. These features are designed to be accessed via a laptop or desktop computer and most are Web-based. A few stand-alone programs are referenced as well, all of which can be downloaded and installed on either a Mac or PC.

Rob DeSalle
New York, New York

Jeffrey A. Rosenfeld
Newark, New Jersey

Student and Instructor Resources Websites

Accessible from www.garlandscience.com, the Student and Instructor Resources websites provide learning and teaching tools created for *Phylogenomics: A Primer*. The Student Resources Site is open to everyone and users have the option to register in order to use book-marking and note-taking tools. The Instructor Resource Site requires registration and access is available only to qualified instructors. To access the Instructor Resource Site, please contact your local sales representative or email science@garland.com.

For Students
Web Features
Web-based exercises designed to assist students in working with the programs and databases used to analyze phylogenomic data.

For Instructors
Figures
The images from the book are available in two convenient formats: PowerPoint® and JPEG, which have been optimized for display. The resources may be browsed by individual chapter or a search engine. Figures are searchable by figure number, figure name, or by keywords used in the figure legend from the book.

Resources available for other Garland Science titles can be accessed via the Garland Science website.

PowerPoint is a registered trademark of Microsoft Corporation in the United States and/or other countries.

Contents

Detailed Contents

Why Phylogenomics Matters

Phylogenomics is a new way of looking at biological information. It refers to the intersection of several important aspects of modern biology such as molecular biology, systematics, population biology, evolutionary biology, computation, and informatics, with genome-level information as the source for testing hypotheses and for interpretation of data. Because the amount of information from genomes is orders of magnitude greater than previously available, novel approaches and new skills are needed by biologists to make sense of these data. In order to understand the biological information in a phylogenomic context, we first need to understand the nature of biological information and why and how we organize it. Understanding the nuances of computing therefore becomes an integral part of understanding phylogenomics. But we also need to have a good handle on the important molecular and evolutionary questions facing modern biology in order to formulate the right questions.

Phylogenomics and Bioinformatics

In 1976, the genome of the RNA virus MS2 (3569 nucleotides long) was sequenced by RNA sequencing. The next year, the first complete genome sequence of a DNA-based organism, φX174, was decoded. At 5386 nucleotides long, this genome opened the door for sequencing other DNA-based genomes. It took two decades to advance the technology enough that the whole genome of a living organism could be sequenced. The first living organism to be sequenced was *Haemophilus influenzae* (the bacterium that causes influenza) in 1996. In rapid succession, several bacterial genomes and eukaryotic model organism genomes were sequenced, including yeast (*Saccharomyces*), fruit fly (*Drosophila*), plant (*Arabidopsis*), mouse (*Mus musculus*), and worm (*Caenorhabditis elegans*).

As DNA sequencing technology has improved, the number of DNA fragments sequenced has risen. Recent advances in technology have resulted in an explosion of information. The trend for DNA sequencing for the three decades after genomes were first sequenced, compiled by the National Center for Biotechnology Information, is shown in **Figure 1.1**. In the years 2005–2011, advances in sequencing technology have reached what is called the "next generation" (see Chapter 2). From 2005 onward, the upswing in the amount of sequence generated by laboratories across the globe via next-generation sequencing approaches appears linear even on a logarithmic scale. With novel organisms being sequenced at extremely rapid rates, the onslaught of new gene sequences and the need to annotate, systematize, and archive them are now seen as a problem that is not solvable by simple comparative methods or simple computational approaches. The realization that billions of base pairs of sequence would soon be available to researchers studying cell biology, genetics, developmental biology, biochemistry, and evolution pushed researchers to think of the best ways to organize and interpret the data for making inferences about the functional aspects of newly sequenced genes. The first steps to achieving these goals were to use newly developed bioinformatics approaches.

such as BLAST enable searches through the large number of DNA sequences that currently exist in the public databases (see Chapter 4). Besides the human genome, additional genetic sequence information has been collected from other organisms. The set of all publicly available DNA sequences is stored in GenBank and, as of April 2012, the size of the complete database was 471 gigabytes (471 GB = 471,000,000,000 bytes). The University of California at Santa Clara (UCSC) genome browser is a very highly utilized database of annotations for the human genome. It consists of 1.5 terabytes (1.5 TB = 1,500,000,000,000 bytes) of data.

The scope of the problems addressed by bioinformatics will continue to increase in the next few years (Figure 1.1). Several large high-throughput projects (the 1000 Genomes Project and the 10K Animal Genomes project are two examples) will increase the amount of sequence in the database by several orders of magnitude. The goal of the 1000 Genomes Project is to determine the complete genome sequences of 2500 individuals from diverse ethnic groups across the world. At 3 GB per genome, it is expected that this project will produce many terabytes of data. The 10K Animal Genomes project plans to produce the whole genome sequences of over 10,000 animals. This project will generate over 60 TB of data.

The Rise of Phylogenomics

The term phylogenomics (Sidebar 1.2) was first coined by Jonathan Eisen and Claire Fraser at The Institute for Genome Research (TIGR) at the turn of the century. Phylogenomics is an updating of the term phylogenetics and refers to focus on genome-level analysis. Whereas conventional phylogenetics is based upon the analysis of a few genes, phylogenomics would investigate complete genomes of data. At first, phylogenomics was applied to the functional annotation of newly sequenced genomes. **Table 1.1** (taken from Eisen) shows the comparative approaches that can be used to assign function to a newly sequenced gene. At the genome level for higher eukaryotes, this needs to be done tens of thousands

---

Sidebar 1.2. Where does the term phylogenomics come from?

To properly understand what phylogenomics is, we need to understand the two major roots of the word: phylo and genomic. The term is really a hybrid with Greek origins and more modern twists. The first part of the term, phylo, comes from the Greek root "phylon," which means group or tribe, which has been expanded into the modern word "phylogeny," or a diagram that represents grouping. Modern-day phylogenies are, at their simplest level, branching diagrams that represent the relatedness of organisms. But, as we will see, phylogenies can also carry information about the sequence of events that have occurred over evolutionary time. The second part of the term, genomics, comes from two subroots. The root word "gene" was first coined in 1903 by Wilhelm Johannsen, a Danish botanist, to refer to a unit of heredity. The suffix "omics" has a more modern origin: it has been applied to a number of root terms to signify an entirely new way of doing biology. When this suffix is applied to a root word, it usually means the exhaustive collection of information for a particular biological level. For instance, transcriptomics is the study of the entire array of transcripts

made by a cell. Proteomics is the study of the entire array of proteins made by a cell. Similarly, genomics, a term first used in 1987 when scientists began to discuss the possibility of obtaining the DNA sequence of each base in the human genome, is the study of the entire array of DNA sequences contained in a cell. "Genome" studies proper began in 1996, when the first whole genome of a living organism (the bacterium *Haemophilus influenzae*) was produced by J. Craig Venter and his colleagues. Genomics includes the following steps:

- Obtaining sequences from the genome of an organism

- Assembly of those sequences into a single contiguous genome sequence (if the organism has a single chromosome) or sets of contiguous sequences (if the organism has more than one chromosome)

- Identification of the regions of the raw sequence that correspond to genes

- Annotation of the genes

Table 1.1. Comparative approaches to assigning gene function.

Name	Description of the approach	Example
Highest hit	The uncharacterized gene is assigned the function (or frequently, the annotated function) of the gene that is identified as the highest hit by a similarity search program.	Tomb et al., 1997
Top hits	Identify top 10+ hits for the uncharacterized gene. Depending on the degree of consensus of the functions of the top hits, the query sequence is assigned a specific function, a general activity with unknown specificity, or no function.	Blattner et al., 1997
Clusters of orthologous groups	Genes are divided into groups of orthologs based on a cluster analysis of pairwise similarity scores between genes from different species. Uncharacterized genes are assigned the function of characterized orthologs.	Tatusov et al., 1997

of times because the genomes of eukaryotes contain 10,000 to 30,000 genes. To date over a dozen species of *Drosophila* have had their genomes sequenced. The main reason for all of this fly sequencing was not because scientists were specifically interested in these other species, but rather because the sequences of these species gave scientists better tools to understand the function of the genome of *Drosophila melanogaster*, the model organism. In other words, these other species were sequenced simply because they would help with the annotation of a model organism genome. These kinds of approaches are called functional phylogenomics, because they attempt to get at the processes involved in the function of gene products. As time progressed, scientists realized the power of reconstructing phylogenetic relationships by use of genome-level information. So after about 5 years of usage of the term with its original meaning, other aspects of the use of genome-level sequences were assigned to the umbrella of phylogenomics. These include using an evolutionary approach to understand the function of genes and using whole genome sequences to interpret the relationships of organisms.

To give the student a sense of the power of a phylogenetic evolutionary approach to genomics, we present two examples. The first example concerns understanding the functional nature of protein products from genes (known as functional phylogenomics) and the second concerns the use of whole genome sequences to infer the pattern of relationships of organisms (known as pattern phylogenomics).

Functional phylogenomics employs common ancestry to infer protein function

Phylogenomic analysis allows for a way to use common ancestry to infer the function of an unknown protein. Brown and Sjölander have used the example of G protein coupled receptors to demonstrate how a phylogenetic approach can lead to annotation of function in a large group of proteins that might seem unrelated in the beginning. A branching diagram of protein sequences, derived from the opioid/galanin/somatostatin gene family that allows for two important inferences about assigning function to unknown proteins, is shown in **Figure 1.2**. Diamonds represent fully annotated and well-understood proteins, and ovals represent unannotated proteins. The structure of the tree allows researchers to focus on three subtrees—the opioid, galanin, and somatostatin receptors—and to assign a function for the unannotated proteins in the study. Thus unknown proteins can

Figure 1.2 Phylogenetic tree of opioid, somatostatin, and galanin genes in mammals and an unknown opioid gene subtype. The dark green ovals and diamonds represent the opioid family of genes. The gray diamonds represent the somatostatin genes, and the black diamonds represent the galanin genes. The unknown subtype (starburst) is not a relative of any of the specific types (opioid, galanin, or somatostatin) and so must be considered related to the progenitor of all three kinds. (Adapted with permission from D. Brown & K. Sjölander. *PLoS Comput. Biol.* 2(6):e77, 2006.)

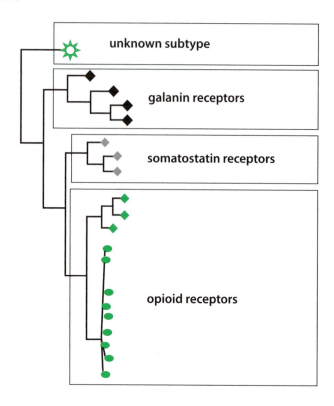

be annotated through their relationships to known proteins, without requiring costly and time-consuming experimentation. Additionally, an unknown protein (see starburst in Figure 1.2) that does not fit into one of the three subtrees can be classified as having an unknown function and targeted for further analysis for function.

Pattern phylogenomics gathers information from branching patterns of a group of organisms

The relationships of bacteria in the family *Pasteurellaceae* are an important subject because of the role these bacteria play in pathogenesis. A robust understanding of the relationships between bacteria in this family will help researchers understand better the pathogenicity of the members of this group. If an ancestral bacterium becomes pathogenic, its descendant species should also be pathogenic. By reconstructing evolutionary relationships, the history of pathogenicity can be traced and interpreted better. Fortunately, whole genomes exist for several species in the family. Di Bonaventura et al. compiled these sequences and constructed a branching diagram (**Figure 1.3**). The overall branching pattern of this group of organisms is clear when all of the 2000 or so genes in the genomes of these bacteria are combined into a single analysis (called a concatenated or simultaneous analysis; see Chapter 9). The inferences made for the relationships in this tree are robust and give researchers a tool for understanding the role of pathogenicity in the group. Interestingly, no single gene recovers the concatenated analysis tree (Figure 1.3). Rather, several genes need to be combined in order to get the simultaneous analysis or concatenated tree. Even more importantly, given the tree in Figure 1.3 and knowing the distribution of sequences in the species in the tree, we can easily trace the sequence changes that occur in common ancestors. This approach can also allow for the analysis of horizontal exchange of genes, as bacteria are very prone to this process. Horizontal exchange has been suggested as a problem in phylogenetic reconstruction of microbial species, and we will address this problem in more detail in Chapter 11.

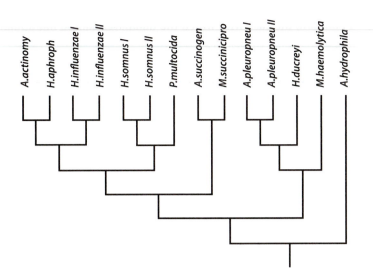

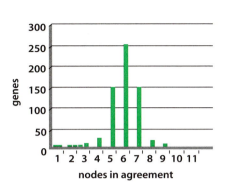

The Phylogenomic Toolbox

The science of phylogenomics requires numerous and varied tools. First and foremost are the molecular biology techniques used for the fundamental experiments, such as generating whole genome sequences for organisms, determining the expression level of genes in a particular cell type, and determining the functions of proteins, all of which we will discuss in Chapter 2. The second most important tool in the phylogenomics toolbox is a thorough understanding of the evolutionary process. We provide an overview of the subject in Chapter 3 and point out that many of the approaches and important questions of phylogenomics are based in evolutionary theory. The third tool for phylogenomics is computing.

Sequence alignment programs and databases are essential computational tools in phylogenomics

Computing is at the heart of phylogenomics. While the reader need not be proficient in programming or systems management, we do feel it necessary that the student understand the role of computing in bioinformatics and phylogenomics. Sequence alignment tools to generate matrices of characters are also extremely important, as these matrices are the basis for generating the phylogenetic hypotheses that are central to phylogenomics studies. We discuss the various approaches to sequence alignment in Chapter 5. As we will see in Chapter 7, the sequences used in phylogenomics can be partitioned from whole genomes into smaller functional units like genes, or untranscribed regions, or exons and introns, or even positions within codons in coding regions (that is, first positions versus second positions versus third positions). This partitioning is essential when attempting to establish function to regions of the genome. The Phylogeny-generating programs and other programs that assist in the interpretation of trees are extremely relevant to phylogenomics. These approaches and programs implementing the approaches will be discussed in Chapter 8. As we will detail in subsequent chapters, there are a wide array of programs and functional Websites that perform much of the data manipulation in phylogenomics, and these are at the heart of how phylogenomics is carried out on a practical level. This book will not require students to know programming, and instead we will emphasize the existing programs and Websites that are available to students.

Databases are computational tools that have become an extremely important aspect of phylogenomics. The ease with which scientists can obtain sequences for phylogenomic studies is stunning, and the programs and Websites that accomplish this important function are also central to phylogenomics (see Chapters

Figure 1.3 Phylogenetic analysis of *Pasteurellaceae* species. Left: The best tree when all the data are concatenated; exactly what we mean by "best" will be explored in subsequent chapters. Species represented in the tree are *Aggregatibacter actinomycetemcomitans*, *Haemophilus aphrophilus*, *Haemophilus influenzae*, *Haemophilus somnus*, *Pasteurella multocida*, *Actinobacillus succinogenes*, *Mannheimia succiniciproducens*, *Actinobacillus pleuropneumoniae*, *Haemophilus ducreyi*, *Mannheimia haemolytica*, and *Aeromonas hydrophila*. Right: Number of single-gene trees ("genes") that agree with the concatenated tree. For instance, there are about 256 genes that have exactly six nodes in common with the concatenated tree. Numbers on the x-axis represent the number of nodes in the trees that agree with the concatenated tree. There are 11 nodes in the tree; hence a tree that completely agrees with the concatenated tree will have 11 nodes in agreement. However, note that the most nodes for which any single-gene tree agrees with the concatenated tree is 9. (Adapted with permission from M. Di Bonaventura, E. Lee, R. DeSalle, & P. Planet. *Mol. Phylogenet. Evol.* 54(3):950–956, 2010. Courtesy of Elsevier.)

the other. These include microevolution, macroevolution, and the area between them that entails studies of speciation. Microevolution includes population genetics and studies of natural selection, and most microevolutionary studies will focus on populations within a single species. Macroevolutionary studies include phylogenetics and the interpretation of other major evolutionary patterns that occur over longer evolutionary periods, like biogeography or adaptive radiation. Speciation studies are at a unique level that uses both macroevolutionary and microevolutionary approaches. Hence, the phylogenomicist has to have a broad command of the basics of evolutionary theory. A phylogenomicist should also have a working command of the databases that exist and how these databases are constructed (Chapter 4), how to extract information from the databases, and how to manipulate the data. Finally, the phylogenomicist should have ideas about genes, genomes, species, and phylogeny. In essence, a phylogenomicist is only as good as his/her ideas. These ideas can then be tested by use of the phylogenomics toolbox, or if need be, new tools can be developed to address new problems.

Summary

- Bioinformatics, or the use of computational tools to answer biological questions and manage biological data, has become increasingly important in handling the large amounts of data generated by DNA sequencing.

- Two examples of modern high-throughput biology that have required a shift in the way we think about biological information are microarrays and the Human Genome Project.

- Phylogenomics is the study of complete genomes of data. Functional phylogenomics employs common ancestry to infer protein function, while pattern phylogenomics gathers information from branching patterns of a group of organisms

- Computing and statistical analysis are essential to phylogenomics. Both parametric and nonparametric statistical techniques are used. Maximum likelihood and Bayesian analysis are additional statistical methods that are important in phylogenomics.

- A good phylogenomicist should understand the molecular biology of genes and genomes; have a broad command of the many aspects of evolutionary theory; possess a working knowledge of existing databases, how to extract information, and how to manipulate the data; and have ideas about genes, genomes, species, and phylogeny that can be tested and explored.

Discussion Questions

1. List the various levels of biological organization that can be used as entities in bioinformatics and phylogenetic analysis. Arrange these from most inclusive to least inclusive. Give biological examples of each level. Hint: start with molecules.

2. Discuss the interplay of objectivity, operationality, and repeatability. How do the following scientific endeavors stack up with respect to these three hallmarks of scientific research?

   - Microscopy of nerve cells
   - Discovery of new chemical elements
   - Discovery of new planetary systems
   - Determination of ecological diversity

3. List four different kinds of statistical distributions and the kinds of statistical tests that are appropriate for each.

4. Discuss the similarities and differences between Bayesian and likelihood approaches.

5. Discuss the difference between parametric and nonparametric statistics.

Further Reading

Brown D & Sjölander K (2006) Functional classification using phylogenomic inference. *PLoS Comput. Biol.* 2, e77. DOI: 10.1371/journal.pcbi.0020077.

Delsuc F, Brinkmann H, & Philippe H (2005) Phylogenomics and the reconstruction of the tree of life. *Nat. Rev. Genet.* 6, 361–375.

Di Bonaventura MP, Lee EK, DeSalle R, & Planet PJ (2010) A whole-genome phylogeny of the family Pasteurellaceae. *Mol. Phylogenet. Evol.* 54, 950–956.

Eisen JA (1998) Phylogenomics: improving functional predictions for uncharacterized genes by evolutionary analysis. *Genome Res.* 8, 163–167.

Eisen JA & Hanawalt PC (1999) A phylogenomic study of DNA repair genes, proteins, and processes. *Mutat. Res.* 435, 171–213.

Eisen JA & Fraser CM (2003) Phylogenomics: intersection of evolution and genomics. *Science* 300, 1706–1707.

Fiers W, Contreras R, Duerinck F et al. (1976) Complete nucleotide sequence of bacteriophage MS2 RNA: primary and secondary structure of the replicase gene. *Nature* 260, 500–507.

Fleischmann R, Adams M, White O et al. (1995) Whole-genome random sequencing and assembly of Haemophilus influenzae Rd. *Science* 269, 496-512.

Nixon KC & Wheeler QD (1990) An amplification of the phylogenetic species concept. *Cladistics* 6, 211–223.

Philippe H, Snell EA, Bapteste E et al. (2004) Phylogenomics of eukaryotes: impact of missing data on large alignments. *Mol. Biol. Evol.* 21, 1740–1752.

Tatusov RL, Koonin EV & Lipman DJ (1997) A genomic perspective on protein families. *Science* 278, 631–637.

Tomb JF, White O, Kerlavage AR et al. (1997) The complete genome sequence of the gastric pathogen *Helicobacter pylori*. *Nature* 388, 539–547.

Yoon CK (2009) Naming Nature: The Clash between Instinct and Science. W. W. Norton.

The Biology of Linear Molecules: DNA and Proteins

In this chapter we will examine the molecular biology of the molecules used in phylogenomics. We start by describing the beautiful symmetry of nucleic acids known as DNA (deoxyribonucleic acid) and RNA (ribonucleic acid). According to the "central dogma of molecular biology," DNA is transcribed into RNA and then translated into protein, so we next turn to proteins. The genetic code that implements the translation of RNA into protein is an integral part of understanding phylogenomics. With a good understanding of DNA, RNA, proteins, and the genetic code, we can examine the procedures that are used to generate the immense amounts of data in the nucleic acid and protein databases. To accomplish this end, we discuss current methods of DNA sequencing and how whole genomes are generated. In some cases entire genomes are not necessary, and the focus is on detecting variants. Ingenious techniques have been developed for characterizing variation in the natural world. Advanced techniques have also allowed us to examine entire collections of RNA transcripts and to analyze the results of differential protein expression. All of these approaches allow an organism or a population of organisms to be characterized with linear information from DNA, RNA, and proteins. The significance of the linear arrangement of this variation and how it is interpreted is an important part of the analysis of biological information.

Nucleic Acids

DNA is a perfect molecule for transmitting information

Like most biological molecules, DNA is made up of simple atoms of carbon (C), hydrogen (H), oxygen (O), phosphorus (P), and nitrogen (N). These atoms are combined in specific ways to make the building blocks of DNA, nucleotides. The nucleotide building blocks in DNA are cytosine (C), guanine (G), thymidine (T), and adenine (A). These nucleotides have an upstream and a downstream end due to their structure (**Figure 2.1**). In the terminology of the molecular biologist, the sugar rings that make up part of these nucleotides have a 5′ ("5-prime") end and a 3′ ("3-prime") end (**Figure 2.2**). These 5′ and 3′ designations refer to the locations of carbon atoms in the sugar–phosphate backbone of the DNA molecule.

DNA is synthesized by specific pairing

Due to the double-helical structure of DNA, if the sequence of one strand is known, the sequence of the other strand can be deduced. Watson and Crick, in their initial publication describing the double-helix structure of DNA, referred to this phenomenon as "specific pairing," which "immediately suggests a possible copying mechanism for the genetic material." The process is a molecular one, and it requires an understanding that the nucleotide building blocks of DNA are not just pieces that stick to each other but rather resemble interlocking jigsaw pieces with directionality. When a DNA molecule is synthesized, the assembly can proceed only in the downstream direction of 5′ to 3′ because of the morphology of the DNA polymerase that copies DNA. This means that, given a particular strand

Figure 2.1 Diagrammatic representation of deoxyribonucleic acid (DNA) showing the molecular structures of the four bases.

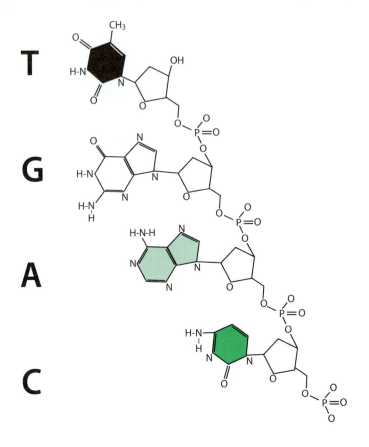

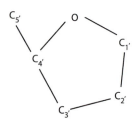

Figure 2.2 Diagram of the ribose ring with the carbons in the ring numbered. The 3′ (3-prime) and 5′ (5-prime) carbons are important, as these are the carbons that participate in the phosphate bonds along a single strand of DNA and impart a sense of direction to the strand.

of DNA, bases can only be added to one end of the molecule. As the name DNA polymerase implies, it polymerizes, or elongates, DNA molecules.

Now, imagine a DNA strand with the 5′ to 3′ orientation and the nucleotide sequence of

$$5′G3′\text{-}5′A3′\text{-}5′A3′\text{-}5′T3′\text{-}5′C3′ \rightarrow \qquad (2.1)$$

Any nucleotides added to this strand will be added in the direction of the arrow after the C. There is no mechanism for DNA polymerase to add a nucleotide before the first G. In addition to working from the 5′ to 3′ direction, DNA polymerase requires a template to tell it what base to add. This template is the other strand of DNA, and the new bases are added according to the rules of "specific pairing," as outlined by Watson and Crick but initially predicted by the discoveries of Erwin Chargaff in the 1940s. Here are the rules followed by DNA polymerase:

- If a G appears on the first strand, then put a C in the other
- If a C appears on the first strand, then put a G in the other
- If an A appears on the first strand, then put a T in the other
- If a T appears on the first strand, then put an A in the other

These rules will now be used to illustrate the production of a new DNA strand.

```
Template:    5′G3′-5′A3′-5′A3′-5′T3′-5′C3′

Step 1:      5′G3′-5′A3′-5′A3′-5′T3′-5′C3′→
                                    ←3′G5′

Step 2:      5′G3′-5′A3′-5′A3′-5′T3′-5′C3′→
                               ←3′A5′-3′G5′
```

```
Step 3:        5′G3′-5′A3′-5′A3′-5′T3′-5′C3′→
                     ←3′T5′-3′A5′-3′G5′

Step 4:        5′G3′-5′A3′-5′A3′-5′T3′-5′C3′→
                ←3′T5′-3′T5′-3′A5′-3′G5′

Step 5:        5′G3′-5′A3′-5′A3′-5′T3′-5′C3′→
          ←3′C5′-3′T5′-3′T5′-3′A5′-3′G5′
```

The first line shows the template sequence introduced in Equation 2.1. DNA polymerase reads the sequence of the template strand and adds nucleotides one at a time, according to the rules of specific pairing, to create the complementary strand. Note that the new strand is created in the 5′ to 3′ direction, antiparallel to the template strand.

A key point to understand is that the two strands of DNA in a double helix are created and read in opposite directions (**Figure 2.3**). This directionality is illustrated by arrows.

DNA can mutate and impart heritable information important in understanding descent with modification

Although DNA polymerase almost always follows the rules of specific base pairing in synthesizing DNA, it does (but rarely) make mistakes. In human cells, the rate of mistakes is about 1 in every 50 million (5×10^7) nucleotides, which is enough to make about 100 mistakes in every new egg or sperm cell of a sexually reproducing organism. These mistakes made by a polymerase are called mutations.

Mutations can occur both in regions of the genome that code for protein and regions that are noncoding. Changes in noncoding regions of the genome can be very useful for understanding descent with modification because they are transferred from generation to generation. In fact, the grand majority of DNA sequence change that occurs in nature is in noncoding regions. The impact of a change in coding regions is discussed below.

Polymerase chain reaction is a milestone development

One of the most important developments in molecular biology in the last century takes advantage of this simple yet elegant way of synthesizing DNA. This approach is called the polymerase chain reaction (PCR), and it has been adapted for use in many of the techniques that we will discuss throughout this book (Sidebar 2.1). This technique was developed by Kary Mullis, and it is so significant a contribution that it earned Dr. Mullis a share of the 1993 Nobel Prize in Chemistry.

Proteins

Proteins are linear polymers of amino acids

As with DNA, proteins are composed of simple atoms, almost the same atoms as DNA with one exception: instead of C, O, H, N, and P, proteins use C, O, H, N, and sulfur (S). These atoms are combined into basic protein building blocks called amino acids. The backbone of an amino acid is a simple chemical structure that has an amino (NH_3) group on one end and a carboxy (COOH) group on the other (**Figure 2.4**). Between these two groups is a central carbon with a group called a side chain connected to it. The chemical structure of the side chain is what gives the amino acid its shape and dictates how it will interact with the environment, with other amino acids, and with other molecules. There are 20 most common amino acids (**Figure 2.5**). The side chains in these 20 amino acids vary in their

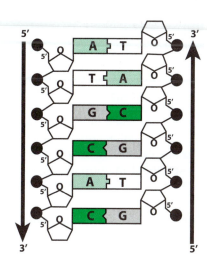

Figure 2.3 Antiparallel nature of DNA. Arrows indicate the direction of synthesis of each strand in the double helix. Note that guanine (G) pairs only with cytosine (C) and adenine (A) pairs only with thymine (T), as shown by the specific hydrogen bonds between them. (Adapted with permission of Madeleine Price Ball.)

Sidebar 2.1. Polymerase chain reaction: the linchpin of the data explosion.

The PCR protocol is the first step in a myriad of applications such as DNA sequencing, microarrays, DNA fingerprinting, and DNA bar-coding, and in fields and endeavors such as the human genome project, forensics, law enforcement, medicine, pharmacology, and combating bioterrorism. Due to the need to synthesize substantial amounts of a gene or DNA fragment in these endeavors, PCR is the one step that is essential to all of them.

PCR is designed to mirror the process by which a cell will copy its DNA prior to cell division. When cells make DNA, they need certain components. Those components include a short primer sequence, the individual nucleotide building blocks of DNA (dATP, dCTP, dTTP, and dGTP; a generic name for any one of these four is dNTP, deoxynucleoside triphosphate), and a polymerase enzyme. The dNTP building blocks are discussed later in this chapter. For a PCR reaction, these reagents are provided and the repeated raising and lowering of temperature provides the optimal conditions for the reactions to occur that will copy a DNA molecule. Here is a basic overview of PCR:

Step 1. A researcher decides that he or she needs to use PCR to make copies of a particular sequence that we will call sequence X. This sequence is usually a small part of the full genomic DNA sample that we have. This full genomic DNA is known as the template.

Step 2. A small pair of single-stranded DNA fragments (known as primers) that are complementary to opposite ends of the positive and negative strands of sequence X are designed.

Step 3. The source template DNA, the primers designed for sequence X, dNTPs, DNA polymerase, and other reagents are all added to the same test tube.

Step 4. The tube is heated to 100°C to cause the double-stranded DNA in the template to separate into single strands.

Step 5. The mixture is cooled, and this lower temperature allows for binding of the primer sequences and initiation of the copying of sequence X by DNA polymerase. This cooling is essential for the primer sequence to bind to the template (known as annealing), and the proper temperature is dependent upon the nucleotide composition of the template.

Step 6. The DNA polymerase produces two copies of sequence X by adding nucleotides to the primer sequence.

This reaction is repeated in order for the chain reaction to work. There is no need to add either new dNTPs, polymerase, or target genomic DNA. The temperature is raised and lowered, which causes the DNA, primer sequences, and dNTPs to react. Raising the temperature of the reaction to 100°C, lowering it to, say, 55°C, and raising it again to 72°C is called a cycle.

Since the temperature of the reaction is initially raised to 100°C, it is essential that the protein does not denature, or become inactivated. An enzyme, which is found in the bacterium *Thermus aquaticus*, is used as a polymerase enzyme. *Taq* polymerase remains stable at very high temperatures that are significantly higher than human body temperature.

Every 1 minute, the sample of DNA doubles again, resulting in many copies of the sample in a short amount of time. Today, scientists use machines called thermal cyclers, which automatically cycle the reaction temperature. Each temperature cycle doubles the amount of copies of sequence X that are produced, so that the number of sequences grows in a binary fashion. The number of copies increases as 2, 4, 8, 16, 32.... For most purposes, a PCR reaction proceeds through 30 cycles, which results in 2^{30} or ~1 billion copies of the desired DNA sequence.

See Web Feature 2 for more information on the Polymerase Chain Reaction (PCR).

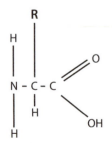

Figure 2.4 Structure of an amino acid. The large R in the figure designates the position of the side chain, the composition of which varies among different amino acids.

chemical makeup and arrangement. For instance, glycine is a very simple amino acid with a simple hydrogen side chain. On the other hand, lysine is a more complex amino acid with a large side chain composed of four carbons, 10 hydrogens, and one nitrogen atom. Each side chain imparts physical–chemical characteristics to the specific amino acid: acidic or basic, hydrophobic (water-hating) or hydrophilic (water-loving), positive or negatively charged, and so on. Proteins are made up of amino acids strung together (**Figure 2.6**).

Proteins have multiple levels of structure

The specific arrangement of amino acids gives the protein its structure and dictates its function in a hierarchical fashion (**Figure 2.7**). The sequence of amino acids in a protein is known as its primary structure. This primary structure folds

Amino acid	Three-letter code	One-letter code	Polarity	Charge
Alanine	Ala	A	NP	N
Arginine	Arg	R	P	P
Asparagine	Asn	N	P	N
Aspartic acid	Asp	D	P	G
Cysteine	Cys	C	P	N
Glutamic acid	Glu	E	P	G
Glutamine	Gln	Q	P	N
Glycine	Gly	G	NP	N
Histidine	His	H	P	P/G
Isoleucine	Ile	I	NP	N
Leucine	Leu	L	NP	N
Lysine	Lys	K	P	P
Methionine	Met	M	NP	N
Phenylalanine	Phe	F	NP	N
Proline	Pro	P	NP	N
Serine	Ser	S	P	N
Threonine	Thr	T	P	N
Tryptophan	Trp	W	NP	N
Tyrosine	Tyr	Y	P	N
Valine	Val	V	NP	N

P= polar   N= netural
NP= nonpolar   P= positive
                G= negative

Figure 2.5 The 20 most common amino acids. (Courtesy of the National Institutes of Health.)

into secondary structures. Secondary structure is produced by hydrogen bonding between amino acids in different parts of the same protein. This leads to the formation of three-dimensional structures in proteins called β-pleated sheets and α-helices, which are important for maintaining part of the three-dimensional shape of the protein. The next level of complexity in proteins is called tertiary structure. The tertiary structure of a protein is implemented by a variety of chemical reactions, such as disulfide bridges (between cysteines, which have sulfur atoms in their side chains), hydrogen bonds, hydrophobic interactions, and ionic bonds, and gives further three-dimensional structure to the protein. The final level of structure for some proteins is called quaternary structure. This level of structure involves the interaction of different discrete amino acid chains, known as protein subunits. For instance, collagen proteins will wind around each other into triple-helical structures. In this case, three collagen subunits make up a higher-order quaternary protein structure.

The broad diversity of dimensional shapes that proteins can take as a result of their primary amino acid sequences is daunting. But this broad array of shapes

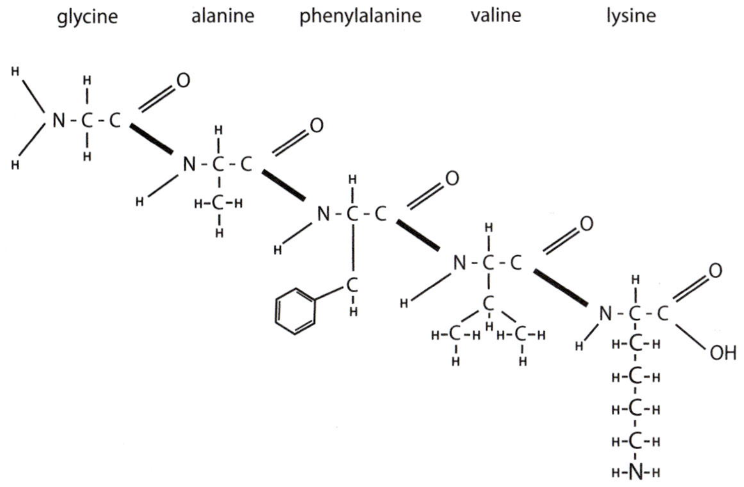

Figure 2.6 Molecular structure of a five amino acid protein, Gly-Ala-Phe-Val-Lys. Amino acids are joined together by the formation of an amide bond between the carboxy group of one amino acid and the amino group of the next.

Figure 2.7 Hierarchical levels of protein structure. Primary structure leads to secondary structure, which leads to tertiary structure. For proteins with multiple subunits, tertiary structure leads to quaternary structure. (Courtesy of the National Institutes of Health.)

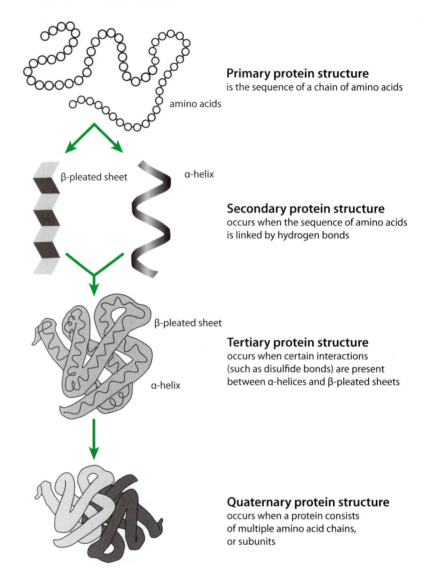

Primary protein structure
is the sequence of a chain of amino acids

amino acids

β-pleated sheet   α-helix

Secondary protein structure
occurs when the sequence of amino acids is linked by hydrogen bonds

β-pleated sheet

α-helix

Tertiary protein structure
occurs when certain interactions (such as disulfide bonds) are present between α-helices and β-pleated sheets

Quaternary protein structure
occurs when a protein consists of multiple amino acid chains, or subunits

and distribution of chemical attributes in proteins, dictated ultimately by the sequence of the protein, is what produces the broad array of functions that proteins can accomplish. We will address this diversity in Chapters 12 and 13, but for now we return to how nucleic acids are involved in the translation of proteins. This task requires us to examine how nucleic acids, with just four bases (G, A, T, and C), can code for the 20 amino acids that make up proteins.

Translation of the information in DNA is accomplished by the genetic code

The genetic code is a set of rules by which information encoded in mRNA is synthesized using DNA in the genome. To understand this process, scientists needed to discover how DNA's four-letter alphabet could be translated into a 20-letter alphabet, with one letter for each of the amino acids in proteins. Discovering this code was one of the most exciting and important discoveries in biology of the 1960s, and many scientists—from mathematicians to biologists to physicists—attempted to decipher it. In pursuit of this endeavor, several beautiful experiments were undertaken, but in the end, as with most problems in molecular biology, its solution turned out to be quite simple.

The solution was not a straight "one to one" translation process. In other words, one nucleotide cannot code for one amino acid. The four nucleotides are simply not enough to code for the 20 amino acids; if they were, we would only have four amino acids to correspond with the four nucleotides. Nature is simple in its design most of the time, and so the "one to one" solution was set aside. The "two for one" solution, proposing that two nucleotides code for a single amino acid, also does not provide enough combinations for 20 amino acids. Here are all the possible two-letter combinations of the four nucleotides: GG, GA, GC, GT, AA, AG, AC, AT, TT, TA, TC, TG, CC, CA, CT, and CG. There are only 16 possible combinations. One could think of ways to modify the two-letter codes, but these involve complications that are not as plausible as a direct translation system. If one considers the "three for one" solution, there are 64 possible three-letter combinations of GATC. Now we have too many three-letter combinations for the 20 amino acids: 44 too many, to be precise. Does this mean that the three-letter system is not possible? Either DNA sequences containing some of the three-letter combinations do not occur, or multiple three-letter combinations (called codons) can code for the same amino acid. After extensive study, it was concluded that the three-letter code was correct.

How would such a code work specifically? Are there some three-letter GATC words that do not exist in DNA sequences? Or are some of the amino acids coded for by more than one codon? To understand this concept, scientists synthesized all the possible nucleotide triplets (in chains of repeated triplets), converted each double-stranded DNA sequence into a synthesized single-stranded messenger RNA (mRNA), and put these synthesized RNAs into cell extracts to identify which amino acids these repeated triplet RNA chains would string together. The first experiment was stunning. Marshall Nirenberg at the National Institutes of Health made an RNA nucleotide chain that was composed of all Us (in RNA, the T from DNA is simply replaced by a U with a slightly different structure; both T and U pair specifically with A). When he put this mRNA chain into a cell-free system that he devised, an all-phenylalanine amino acid chain was produced. This meant that RNA consisting only of Us made an amino acid chain consisting only of phenylalanines. Remember that DNA → RNA → protein. In the step RNA → protein, a molecule called transfer RNA (tRNA) is used to match the code on the mRNA and "drop off" the correct amino acid. The tRNA molecule is made up of RNA with an amino acid attached at one end and has a region called an anticodon in one of its loops (**Figure 2.8**). An anticodon is a three-base sequence that is complementary to a codon in the mRNA transcript. The anticodon for the codon UUU is AAA. The tRNA with an AAA anticodon has a phenylalanine attached at the other end. The codon UUU is recognized by the anticodon AAA, and phenylalanine is delivered to the protein chain; thus UUU codes for phenylalanine.

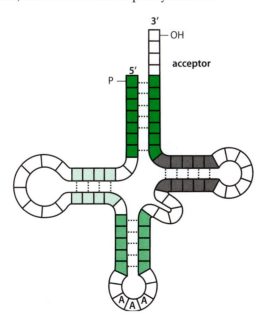

Figure 2.8 Diagram of a transfer RNA. The structure is made of "stems" and "loops." Shaded segments represent nucleotides that pair in a complementary fashion to form a stem, and open segments are those that remain unpaired, forming a loop. The position of the anticodon is indicated. The attachment site of the amino acid on the tRNA is marked "acceptor."

This discovery did not answer the question of whether the code had missing words or was redundant. In order to investigate this, all possible three-letter RNA sequence combinations needed to be examined. For instance, an RNA string of ACCACCACC..., or $(ACC)_n$, was synthesized and placed in the cell-free extract. This resulted in an amino acid string of threonines. The RNA string of ACGACGACG, or $(ACG)_n$, was then placed in the same extract and a string of threonines was also produced. In fact, there are four different nucleotide codons for threonine: ACA, ACC, ACG, and ACU. Note that it is the third position in the triplet that seems to be able to vary. This is known as the codon "wobble" factor, and it is found for other amino acids. In many cases, the third nucleotide can be either A,C, G, or U and the same amino acid will be produced. This is not the case for the first two nucleotides, where a change is very likely to produce an altered amino acid.

After testing all 64 possible triplet combinations of G, A, U, and C, scientists were able to write out the genetic code (**Figure 2.9**). All 20 amino acids are accommodated by the code, and there are also three triplets that tell a protein where to end. These three "punctuation marks" are called terminators, and they act like periods at the end of a sentence. Further study determined that one amino acid, methionine or M, was at the beginning of almost every protein. The single triplet that codes for M is ATG, known as the initiation codon. It acts like the capitalization at the beginning of a sentence.

The genetic code allows translation of the RNA sequence into protein and identification of the start and the end of a protein. By knowing what these 64 triplets mean, we can look at any DNA or RNA sequence and determine what protein sequence it would produce. With an understanding of the genetic code and the occurrence of mutations, we can look at how DNA sequences can be used to examine descent with modification and the change of genes over time at the population level (see Chapter 3).

A single nucleic acid sequence has multiple reading frames

One way to identify whether a protein is made from a particular DNA sequence is to examine how it would translate into a protein by using the genetic code. This procedure gets a little tricky, because for every one DNA sequence there are actually six different proteins that could possibly be made from the DNA sequence. Because the genetic code uses triplets, there are three ways a protein can be translated from one strand of a DNA sequence, and since there are two strands of DNA, there are 2 × 3 ways a protein can be made. Consider the following short DNA sequence:

Figure 2.9 The genetic code. All 64 codons are represented. The codons are grouped first by common bases in the first position and second by common bases in the second position. Within each two-base combination for the first and second positions, all four possible third positions are listed. The three-letter representations for the amino acids are given as well as the single-letter code. (Courtesy of the National Institutes of Health.)

Second letter

First letter	U		C		A		G		Third letter
U	UUU	Phe F	UCU	Ser S	UAU	Tyr Y	UGU	Cys C	U
	UUC		UCC		UAC		UGC		C
	UUA	Leu L	UCA		UAA	Stop	UGA	Stop	A
	UUG		UCG		UAG	Stop	UGG	Trp W	G
C	CUU	Leu L	CCU	Pro P	CAU	His H	CGU	Arg R	U
	CUC		CCC		CAC		CGC		C
	CUA		CCA		CAA	Gln Q	CGA		A
	CUG		CCG		CAG		CGG		G
A	AUU	Ile I	ACU	Thr T	AAU	Asn N	AGU	Ser S	U
	AUC		ACC		AAC		AGC		C
	AUA		ACA		AAA	Lys K	AGA	Arg R	A
	AUG	Met M	ACG		AAG		AGG		G
G	GUU	Val V	GCU	Ala A	GAU	Asp D	GGU	Gly G	U
	GUC		GCC		GAC		GGC		C
	GUA		GCA		GAA	Glu E	GGA		A
	GUG		GCG		GAG		GGG		G

```
           123
5´-CAGATGCAGATGCAGATG-3´
3´-GTCTACGTCTACGTCTAC-5´
```

A translation could start at position 1, 2, or 3 on the top strand as labeled. If the translation began at the 5´ end of the bottom strand, it could begin with C, A, or T. These are the six possible reading frames for the nucleotide sequence. Note that starting at the fourth position would produce the same triplet groupings (or reading frame) as starting at the first position; likewise, starting at the fifth or sixth position would produce the same reading frame as starting at the second or third position, respectively. Changes in nucleic acids do not have to be involved in regions that code for proteins. Web Feature 2 demonstrates the nuances of translating DNA sequences into protein sequences.

The DNA Data Explosion

The deluge of information in biology has been phenomenal in the past two decades. In fact, this book might not be necessary if it were not for the amazing advances in technology. There would be little use for the development of methods that process massive amounts of information without the information. Methods for obtaining biological information have greatly expanded our knowledge but have also swollen the databases. To understand the explosion of data now available for genome analysis, it is important to consider the history of the developments that have occurred in DNA sequencing technology.

Sequencing methods for linear molecules have become more powerful

DNA is a linear polymer of nucleotides, and proteins are linear polymers of amino acids. Determining the order of the nucleotides or amino acids is the key to understanding the information encoded in the molecule. Frederick Sanger was a pioneer in developing methods to sequence linear molecules. He used innovative methods to manipulate proteins so as to obtain their primary sequence, which earned him his first Nobel Prize in Chemistry in 1958. He later turned his attention to sequencing DNA, for which he was awarded a share of a second Nobel Prize in Chemistry in 1980. Both processes were laborious. For example, Sanger and his colleagues worked for over a year to obtain the sequence of a simple bacteriophage of approximately 3000 nucleotides. Obtaining the same number of base pairs of sequence can now be accomplished in seconds on a modern sequencing machine.

Major intellectual leaps in our knowledge of DNA have been accompanied by 10-fold increases in our capacity to obtain DNA sequences. Powers of 10 are used to calculate results because they explain leaps of technology. From experience, the senior author of this book can say that the average number of base pairs a single research laboratory could generate in a year in 1980 was about 10,000, or 10^4 in scientific notation. This number jumped to 10^5 base pairs per year by 1990. By the year 2000, the throughput in most laboratories was well over a million (10^6), and for large multilaboratory sequencing projects, the amount of sequence generated was into the billions (10^9) per year. Most of the increases in efficiency and throughput were implemented by automating the process of DNA sequencing (Sidebar 2.2).

By 2005, spectacular increases in sequencing capacity were starting to be realized. Using a combination of robotics and miniaturization of the DNA sequencing approach, scientists were able to produce 10^7–10^9 base pairs of sequence per day. The sequencing process had entered what genomicists call "the next generation."

Sidebar 2.2. The "old-school" recipe for DNA sequencing.

This approach may be outdated soon, so one might wonder: why learn about it? The reason is simple. Like other elegant techniques, certain aspects of this method (the Sanger method) are used in the newer techniques discussed later in this chapter, and it is necessary to know how this old-school method works in order to understand the new-school methods.

Step 1. Obtain tissue from an organism.

Step 2. Isolate the DNA in the cells.

Step 3. Chop the DNA into short pieces of around 2000 base pairs either by use of enzymes or by mechanical techniques.

Step 4. Copy DNA by PCR, clone PCR fragments into cloning vectors and pick clones for step 5.

Step 5. Treat the copied DNA with chemicals that have four different fluorescent dyes in them, a different dye for each of the four bases in DNA.

Step 6. Place the samples on an automated sequencer that simulates stretching the insert out like a string so that the fluorescently labeled bases can be detected by a highly focused laser beam.

Step 7. Via computer analysis, compile the fluorescence data and convert into a read-out that reflects the actual sequence in the insert DNA.

Step 8. For a sequence of a bacterial genome, repeat steps 4–7 tens of thousands of times. For a genome the size of the human genome, repeat steps 4–7 several million times.

Step 9. Take all of the sequences generated in step 8 and reassemble them. Because the inserts were generated randomly, there should be overlap on the ends of the insert, which act like interlocking puzzle pieces. Since most humans have trouble with even 1000-piece puzzles, a computer is used to do this last step.

Step 10. Identify the genes and label them (genome annotation).

Web Feature 2 explains this process in detail.

Next-generation sequencing allows small genomes to be sequenced in a day

With next-generation sequencing methods such as 454 and Illumina sequencing (Sidebar 2.3), several small bacterial genomes (about 4 million bases) could be sequenced in one day. Recently the entire genomes of Dr. Craig Venter and Dr. James Watson have been sequenced by using a fraction of the time and cost expended in the initial human genome projects. Scientists anticipate several more leaps of powers of 10 in the efficiency of obtaining DNA sequence information in the near future. In fact, these leaps have started to be realized with the advent of next-generation sequencing methods, as there was a 1000-fold increase in efficiency of DNA sequencing between 2005 and 2010. More developments like these will eventually result in the generation of an individual human genome sequence (3 billion bases) in a very short period of time on a single next-generation sequencing machine. There are also spectacular applications of the next-generation sequencing approaches that will enhance our understanding of biology. Some of these applications will be described briefly here and discussed in more detail in later chapters of this book.

Next-generation sequencing leads to practical applications

Two remarkable applications of next-generation sequencing have recently been developed. The first is the use of this technique to characterize micro-RNAs. These molecules are produced as transcripts from the genome, and immense numbers of them exist in a typical genome like *Arabidopsis* or *Caenorhabditis*. Next-generation sequencing techniques have been adapted to assay and characterize the kinds and distribution of these potentially important regulatory elements.

The basic protocol for characterizing small RNAs is as follows. Total RNA is first isolated from tissue or cells, and the small RNAs are separated from other RNAs by urea gel purification. The isolated small RNAs are first ligated to a cloning adaptor oligonucleotide. A second adaptor oligonucleotide is ligated to the other end of the small RNAs. Next, a reverse transcription reaction is accomplished with primers complementary to the ligated 5′ adaptor oligonucleotide to produce a pool of cDNA products with both adaptor sequences abutting the small RNA sequences. The cDNA is then amplified via PCR with primers complementary to the two adaptor primers, producing an amplified pool of DNA products. These PCR products are sequenced by next-generation technology, and in some cases

Sidebar 2.3. Next-generation sequencing methods: 454 and Ilumina sequencing.

454 sequencing was the first commercially available advanced sequencing technique. It was introduced in 2005 by the 454 Corporation, originally a subsidiary of CuraGen Corporation but now a subsidiary of Roche Diagnostics, known as 454 Life Sciences.

Step 1. DNA is first denatured into single strands and attached to microscopic beads.

Step 2. The DNA on the beads is amplified by PCR.

Step 3. The beads are placed and sorted in wells on a fiber optic chip.

Step 4. The chip (along with the DNA-attached beads) is then treated with enzymes that produce light when exposed to adenosine triphosphate (ATP). Free nucleotides (G, A, T, and C) are then washed over the chip one at a time. When a nucleotide matches a base on one of the DNA strands, ATP is generated and this results in a reaction that generates a light signal.

Step 5. While these reactions are occurring, a charge-coupled device (CCD) camera in the 454 instrument records the light emission, the well in which the light emissions occur, and the particular nucleotide that was washed over the chip at the time that the light emission occurred.

Step 6. A computer then interprets this information for each well on the microchip and compiles it into a DNA sequence.

Illumina sequencing was initially developed by Solexa Inc. and introduced in 2006; shortly after its introduction, Solexa was purchased by Illumina Inc. This technique uses a flow cell surface that allows for the detection of over 1 million individual reads. This flow surface has eight channels on it, so eight separate sequencing experiments can be run per cell. The process uses the four nucleotides labeled with four different fluorescent molecules and involves the following steps:

Step 1. Target DNA is fragmented and two small linkers are attached to the ends of each randomly fragmented piece of DNA, so that each fragment has one type of linker on one end and the other type of linker on the other end.

Step 2. The DNA is then bound to the surface of a channel via small oligonucleotides on the wall of the channel that are complementary to the linkers.

Step 3. PCR is performed on all the samples. This step will create a large number of paired end repeats for DNA sequencing, all of which are bound to the channels.

Step 4. The DNA on the flow cell surface is denatured. This step will create a large number of single-stranded paired end repeats that are suitable for another round of PCR.

Step 5. Another round of PCR is performed. This step produces several million templates for DNA sequencing.

Step 6. The first base in each fragment is sequenced by using special nucleotides that are labeled with specific fluorescent dyes for each of the four bases. These special nucleotides are also blocked at the 3′ end of the nucleotide so that no new base can be added until the block is removed. This structure of the nucleotides ensures that only one base at a time is added during the sequencing reaction.

Step 7. The base added to each of the millions of fragments is identified and recorded by exciting the cell with a laser. In this way, the first base of each fragment is recorded.

Step 8. The blocked ends of the newly added first base are unblocked and step 6 is repeated. In this way the second base of each fragment can be obtained and recorded (as in step 7).

Step 9. By reiterating step 8, each base of each of the millions of fragments can be identified.

Currently the length of fragments that can be sequenced in this way is up to 70 or so bases. This number of bases can also be enhanced by noting that the paired ends of fragments can be easily identified. By combining sequences from paired ends, the total length of fragments can be pushed to 150 bases. Most Illumina sequencing done prior to 2009 had an upper read length of 36 bases for single reads and about 70 bases for paired ends. The length of the reads will only get longer, perhaps up to 400 or 500 bases.

Web Feature 2 will guide the student to several excellent animations for next generation sequencing (NGS) approaches.

column with no DNA in it, UV absorption will be low or absent. As DNA starts to come out of the column, it is detected as a spike of UV absorption (**Figure 2.11**). If the target DNA is identical to the normal DNA, the UV absorption pattern shows only a single peak for the homoduplex. However, if the target DNA has a different allele from the normal DNA, then two homoduplex peaks and two heteroduplex peaks are observed.

Temperature-gradient gel electrophoresis (TGGE) or temperature-gradient capillary electrophoresis (TGCE) techniques similarly use target and normal DNA fragments that are denatured together and allowed to re-anneal. In this case, the products are separated by electrophoresis instead of HPLC. Electrophoresis involves running a current through a gel made up of a polymer mixture. The DNA fragments or proteins that need to be sized are loaded into the gel via a slot or loading lane. These molecules are given an electrical charge that makes them move when an electrical current is applied to the gel. The gel acts as a sieve in which smaller fragments move through easily and travel longer distances, while longer fragments are retarded and travel shorter distances. Instead of appearing as UV peaks, as in DHPLC, the homo- and heteroduplex products are detected as size shifts on a gel. In this way, a target or unknown sample can be characterized as having the same or a different allele compared with the normal DNA.

Figure 2.11 Diagrammatic representation of the DHPLC procedure. The top panel shows the annealing patterns of a reaction where the queried allele matches the normal allele. Because the sequences match, only a single peak is obtained after chromatography. The bottom panel shows what happens when the normal allele is different from the queried allele. In this case four products are annealed, two heteroduplexes and two homoduplexes, and so four peaks appear after chromatography.

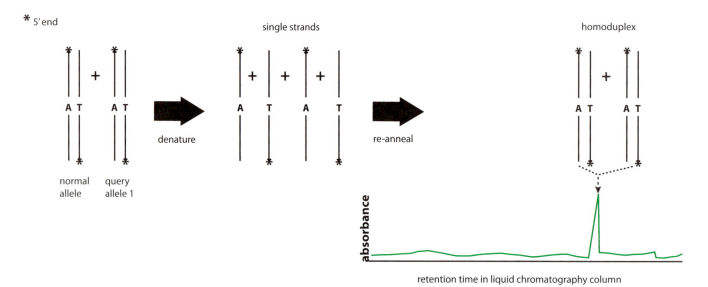

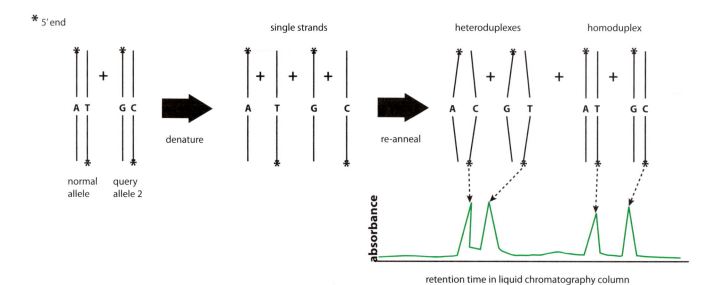

called a northern
its cells. The RNA
ferred from the gel
membrane is hybr
the gene being ex
the specific mRNA
placed over the n
position of the mF
film. This approac
was being made fr
of the mRNA was
was assumed not
intensive and is no
as the transcripto

Microarrays al

The major techno
miniaturizing the
miniaturized and
the process of det
described above.
affixed in a rectan
is recorded using
and fluorescently
hybridize with its
the slide matches
because of the flu
duces a specific a
array can be tallie
the location of eac
back to the associ
turned on and wh

One of the most e
nology to examin

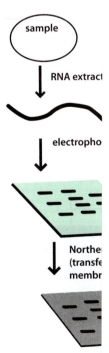

Microarrays, which are discussed in more detail below, can also be used to scan a genome for the state of thousands of single-nucleotide polymorphisms, in a procedure known as DNA resequencing. (Resequencing means that a member of the species has already been sequenced in full and its sequence information, known as the reference genome, is available for comparison.) As with all microarray approaches, this technique is based on DNA hybridization. In this application, genomic DNA is used as the template. Because the array can be spotted with millions of oligonucleotides to act as tools to detect polymorphisms, these approaches can assess the state of thousands of SNPs in the genome. There are several ways that the determination can be done. First, it should be pointed out that the specific position of the SNP where the variation occurs is called the "interrogated site," because the researcher is asking or interrogating what base is present at the variable SNP site. One approach involves using the "gain of signal" characteristics of the hybridized array. The relative hybridization of the target DNA labeled with a fluorescent dye is assessed with each of the four possible nucleotides that could be present at an interrogated site. Four oligonucleotides are synthesized, each with a different base inserted in the interrogated site for each potential SNP. The oligonucleotide spot on the array that lights up with the specific fluorescent dye is the base called the SNP polymorphic site (**Figure 2.12**). Another method is called "loss of hybridization signal." In this technique, the target DNA is labeled with a particular color of fluorescent dye, and a reference or normal DNA is labeled with a second differently colored dye. The two labeled DNAs are then hybridized to an array with potential SNP oligonucleotides. The decrease of target-labeled fluorescence relative to normal-labeled fluorescence indicates the presence of a sequence change at an interrogated site.

Figure 2.12 Diagram of a microarray resequencing chip. A: The first panel shows a DNA strand with the interrogated site as a question mark. The second panel shows a region of the microarray with four oligonucleotides attached to the chip. The only differences among the four oligonucleotides is that each one has a different base in the interrogated site: from left to right, T, C, A, and G. When fluorescently labeled query DNA with an A in the interrogated site is added to the slide, it will hybridize to the oligonucleotide with a T at that site attached to the slide. After fluorescence detection methods are employed, a spot corresponding to the location of the attached complementary oligonucleotide will appear (continued overleaf).

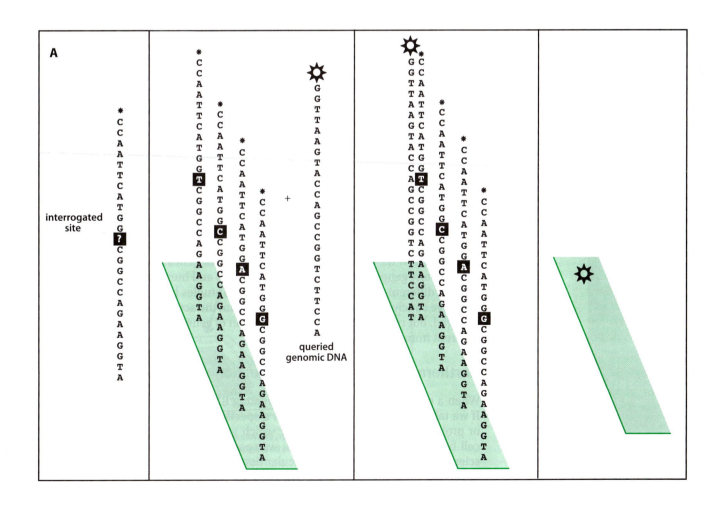

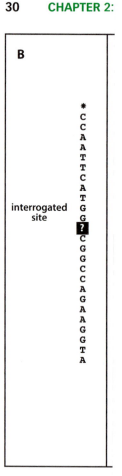

B

interrogated
site

Figure 2.12 Diagram
resequencing chip (
B: Here the DNA being
in the interrogated site
oligonucleotide with a
fluorescently labeled q
a different location on
a fluorescent spot.

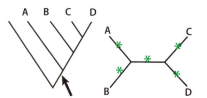

Figure 3.3 Two ways of
representing evolutionary
relationships. The drawing on the left
represents an evolutionary tree. The
arrow indicates a root that gives the
tree direction. The drawing on the right
is a network with four taxa, from which
relatedness cannot be inferred. The
asterisks indicate points at which the
network can potentially be rooted.

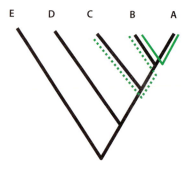

Figure 3.4 Relationships of three
taxa on an evolutionary tree. A
and B share a common ancestor to
the exclusion of all other taxa in the
tree (solid green line). B and C share a
common ancestor as indicated by the
dotted green line.

like bacteria in order to obtain objective phylogenetic hypotheses about groups. Parsimony methods, developed in the 1950s and 1960s, were championed as a valid phylogenetic approach and became entrenched in morphological systematics in the 1970s. Likelihood approaches were introduced in the 1960s and were a promising way to view systematization of molecular attributes for phylogenetics. All three approaches described above have been, and continue to be, used to accomplish phylogenetic or systematic analysis for the past four decades. Due to the different underpinnings of each approach and the philosophical tenor of the revolution in systematics, the scientific worth of each has been judged by the philosophical rigor each one brings to the table. The assumptions and methods of each approach will be covered (Chapters 8, 9, and 10), but first the products of phylogenetic analysis (trees) and their meaning will be described.

Modern phylogenetics is dominated by trees and tree thinking

Since the universal element of modern evolution is the phylogenetic tree, a discussion of trees and how they are interpreted is necessary. This tool is not new to evolution. Naturalists before Darwin suggested that the relationships of organisms in nature could be represented by branching diagrams or, as they called them, trees. Darwin included a single figure in *On the Origin of Species*, which is a tree. Ever since an understanding of descent with modification arose from Darwin and Wallace's work, trees have been used to detail the relationships of organisms. Currently, one of the most extensive projects in evolutionary biology is the Assembly of the Tree of Life (AToL), an endeavor in which the relationships of the majority of the named species (1.7 million of them) on our planet will be displayed by 2015. In microevolution, the construction of branching diagrams that represent the genealogy of various kinds of molecules has become common fare.

A branching diagram may be presented as a tree or as a network (**Figure 3.3**). The tips of the branches on both tree and network represent observed entities that are being studied. The tips of the branches are also called **terminals**, and any level of biological organization can occupy a terminal. For instance, individuals within a specific population (as in population biology studies), species, genera, and even higher categories can all be placed at the terminals of the tree or network. Although it appears that a tree and a network have the same content, there are fundamental differences between the two systems. Like real trees, the phylogenetic tree has a **root**. The arrow indicates a root that allows direction to be imposed on the tree. In contrast, the network is not rooted; thus, overall statements of direction of change cannot be inferred from this system. However, a network with four taxa can potentially be rooted in any one of five places. Each potential root of the network would mean something different with respect to the relationships of the taxa. The phylogenetic tree is rooted via a terminal that is not part of the group to which the four taxa belong; this is called an **outgroup**.

Trees are rich with information, in that they can show which terminals share a most recent common ancestor (**Figure 3.4**). As an example, trace terminals A and B to where the two branches meet, a point that is called a **node** or **fork**. Note that there are no other terminal branches leading to where the branches from A and B meet. This observation means that A and B share a most recent common ancestor. Now look at terminals B and C and trace the branches back until they meet. Note that terminal A is now included in the trace of B and C, meaning that B and C do not have a most recent common ancestor, a fact already known from tracing A and B. What is interesting with the BC trace is that it tells us that C and the node connecting B and A have a most recent common ancestor to the exclusion of all other terminals in the tree. The node representing the trace of A and B is important in that it represents a hypothetical ancestor. The nature of this ancestor as represented by trees is the reason why fossils do not necessarily have to be, nor are they usually, ancestors. Fossils are observed entities; they usually reside in a tree at a terminal at which we analyze them and are not usually found at nodes in trees; thus, they cannot be ancestors.

The rest of the tree with respect to relationships of other terminals can be read in the same way. Additional aspects related to trees concern the shape and complexity of trees. There are several types of trees. A strictly **bifurcating** tree has two branches connected to each node in the tree. When a node has three branches emanating from it, the node is called a **polytomy** in general and a **trichotomy** in particular. Polytomies can represent either true divergence (a "hard" polytomy) or a resolution that is weakly supported (a "soft" polytomy). In general, for computational and intellectual convenience, bifurcating trees are primarily used in evolutionary study.

A problem with numbers demonstrates the issue of complexity in systematics. For three terminals, there are exactly three strictly bifurcating rooted trees. If there are three terminals (A, B, and C), we can put A with B, or B with C, or finally C with A. For four, five, and six terminals, there are 15, 105, and 945 strictly bifurcating rooted trees, respectively. Note the almost exponential increase in the number of trees with the addition of each terminal. The number of trees needed to accommodate a certain number of terminals will become important when we discuss tree searches and bioinformatics strategies for generating trees in subsequent chapters.

Evolutionary trees are extremely flexible. Two trees that seem to be different at first glance can, in fact, express the same set of relationships (**Figure 3.5**). The reason for this somewhat disconcerting observation is that the branches coming out of a node leading to terminals can be rotated without disrupting the implied relationships of the terminals.

The complexities involved with this process lead to difficulty in reading trees. Sidebar 3.1 presents a short test of tree reading. The idea for this test originated from an article in *Science*. Once the ability to read trees has been mastered, the construction of trees can be understood.

Modern phylogenetics aims to establish homology

As discussed in Chapter 1, homology is perhaps the most important aspect of comparative biology (see also Sidebar 3.2) and is often considered one of Darwin's most impressive contributions to evolutionary thinking. Darwin was one of the first to discern the differences between homology and analogy, thus making an important distinction between sameness because of common ancestry (homology) and similarity due to convergence (analogy). Homology then becomes a term of absoluteness and must be discovered via hypothesis testing, while analogy becomes a term of measurement and is calculated from observation of two entities. The literature on homology is rich with debate. Why has there been so much confusion about the term? Basically, because there are so many characteristics about organisms that can be used to establish homology, things become confusing. In addition, an initial confusion with the independence of attributes also causes scientists to misconstrue analogy and homology and hence misconstrue whether they are offering up a hypothesis to be tested or whether they have demonstrated homology.

Despite problems with distances, many researchers rely on distance-based approaches. In bioinformatics, distances between sequences of DNA or proteins are estimated by use of a program such as BLAST, which we will discuss in great

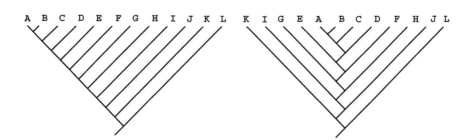

Figure 3.5 Two trees that look very different but in fact represent the same set of phylogenetic relationships. If the branches are rotated at the nodes several times, it can be seen that the two trees represent the same set of relationships.

Sidebar 3.1. Testing your tree thinking.

Answer each of the following. Some knowledge of the organisms in question will be helpful.

1. Which of the trees shown (a, b, c, or d in **Figure 3.6**) does not imply the same set of relationships as the others? You can also answer (e) they are all the same. The organisms depicted are fungus, crustacean, fish, human, and starfish.

2. If you were to add a marsupial to the tree shown, where would it attach (a, b, c, d, or e in **Figure 3.7**)? You can also answer (f) none of the above.

3. If you were to add a fruit fly to the tree shown, where would it attach (a, b , c, d, or e in Figure 3.7)? You can also answer (f) none of the above.

4. Which statement is true for the tree shown (Figure 3.7)?

   (a) The starfish is more closely related to the human than to the fish.

   (b) The fish is more derived from the outgroup than the starfish.

   (c) The fish is as closely related to the crustacean as the fish is to the starfish.

   (d) The fish shares a more common ancestor with the crustacean than with the starfish.

   (e) All of the above.

   (f) None of the above.

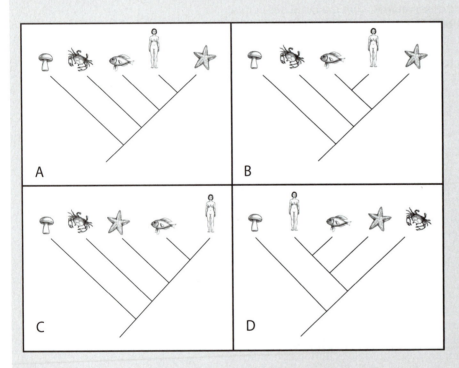

Figure 3.6 Four evolutionary trees for fungus, crustacean, fish, human, and starfish. See text for further explanation.

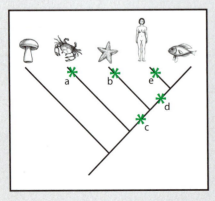

Figure 3.7 An evolutionary tree with marked positions. See text for further explanation.

Sidebar 3.1. Testing your tree thinking (continued).

5. Which of the trees shown (a, b, c, or d in **Figure 3.8**) is different from the others? You can also answer (e) they are all different or (f) they are all the same.

Extra Credit. A good root or outgroup for the five taxa (fungus, crustacean, fish, human, and starfish) in this exercise would be

(a) an orangutan

(b) a tiger

(c) a snake

(d) a turtle

(e) a plant

Answers: 1 (a), 2 (e), 3 (a), 4 (b), 5 (f), EC (e).

A web link to a more detailed "tree test" is provided in Web Feature 3.

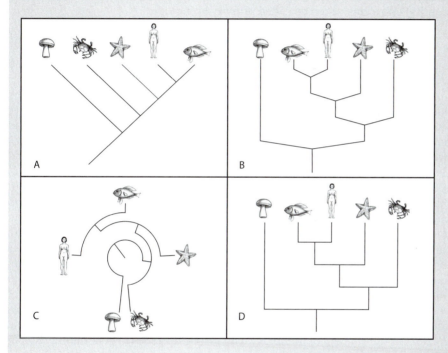

Figure 3.8 Four evolutionary trees of diverse appearance. See text for further explanation.

detail in Chapter 5. On the other hand, several authors have discussed the assessment of homology by use of characters and phylogenetic principles. de Pinna suggested that homology assessment in systematics consists of two steps. The first step consists of "getting in the ballpark." In other words, the first step is simply setting up the hypothesis that two things are homologous. This is usually done by assessing measures of similarity of characteristics of organisms, and if things appear similar enough, then they are said to have primary homology. This concept should sound familiar, as primary homology is simply the observation that a characteristic in different organisms appears similar. The hypothesis of homology, or the primary homology statement, is then tested by phylogenetic analysis. If the characteristics that are under examination and are thought to be similar enough to be homologies are indeed shared and derived in a phylogeny, then they pass the second test, called secondary homology. In other words, if the secondary homology statement is a shared derived character or a synapomorphy, then true homology of the characteristics can be inferred. If they do not pass the second test, then the characteristics are analogies.

de Pinna was mostly interested in morphological homology, and his hypothesis testing scheme for homology makes great sense. But molecular data are a bit different. Brower and Schawaroch, thinking about molecular data, refined de Pinna's homology scheme by adding a third step. They begin the assessment of homology with a method known as topographic similarity assessment. Once topographic similarity is established, the next step involves establishing character state identity. The final step is testing of the hypotheses established in the first two steps. This scheme was proposed because of the basic difference between molecular and morphological approaches to systematics; namely, morphological systematists often take topographic similarity for granted. Systematics is concerned with establishing character state identity through interpretation of the detailed anatomical aspects of organisms. In molecular systematics, the establishment of topographic similarity involves the alignment of molecular sequences that are assumed to be orthologous across all the organisms under study. Since molecular sequence studies utilize character states (G, A T, C, and gaps for DNA) that are relatively simple, the character state identity step is attainable. In essence, de Pinna compressed the two steps (topographic similarity assessment and character state identity) into a single step he called primary homology.

Despite differences in opinion, the distinctions between homology and analogy clarify the issues inherent in each aspect. This clarity began to emerge when scientists from both classical systematics and the new field of molecular evolution collaborated in an effort to define homology more precisely. Fitch and colleagues published an important statement about the terms homology and similarity that were then being used in the early comparisons of molecular sequences (see Sidebar 3.2). They suggested that scientists use the term "homology" with care and pointed out that most molecular biologists in the early 1980s were misusing and/or overusing the term. A typical sequence comparison paper from that period would claim that two sequences of 100 amino acids long had 70% homology if 70 out of 100 residues in the two proteins were identical. Fitch and colleagues pointed out that this was a misuse of the term. Homology, as stated above, is an absolute term about common ancestry. Similarity, while sometimes correlated with common ancestry, does not necessarily address it. When two sequences are compared, there is necessarily a lack of investigation of common ancestry because it requires at least three target sequences and an outgroup sequence to discover common ancestry. This reality is best summed up by Walter Fitch's famous punchline about homology: "Homology is like pregnancy. Someone is either pregnant or not. A person cannot be 70% pregnant." In this context, two sequences can be homologous, but they cannot be 70% homologous; rather, two sequences are 70% similar. We discuss the concept of homology with respect to genes in genomes in Sidebar 3.2 and address this problem throughout the book.

Species

In this chapter, we have discussed the two significant branches of evolutionary biology: population genetics and systematics, or microevolution and macroevolution. However, one of the most important subjects in this field is the study of species and speciation. Consider the fact that Darwin entitled his treatise *On the Origin of Species* and not *On the Origin of Higher Categories* or *On the Origin of Populations*. Studying species is complicated. If homology is considered a contentious subject, the controversy surrounding species is even more difficult. What a species is, and how to study species and speciation, has been at the heart of many arguments in modern biology and, consequently, is an important consideration in conducting bioinformatics.

The definition of species is a subject of debate

Scientists use the attributes of species in their day-to-day work. Taxonomists need a clear definition of species because they are responsible for naming new species and organizing and revising old ones. Systematists are interested in the patterns of speciation and how these patterns affect the phylogeny of groups of organisms.

Sidebar 3.2. A digression in gene families.

The process of homology assessment in molecular sequence studies is somewhat different from establishing homology of gene family members. Fitch and colleagues established a framework for examining genes in multigene families by proposing that the term orthology refer to genes that are identical by descent as a result of speciation of two entities. The term paralogy then refers to a pair or set of genes related to each other but not through a speciation event. In this case, paralogy refers to members of a gene family that have arisen through duplication and not followed by a speciation event. While both terms refer to types of homologous relationships, orthology is what an anatomist would refer to as homology and paralogy would be akin to serial homology (as the fourth vertebra is serially homologous to the fifth vertebra). A third term, xenology, was coined by scientists who study gene transfer and refers to the similarity of entities as a result of horizontal transfer. It should be obvious from these differences in the terminology that any departure from orthology for members of gene families negates sound evolutionary or biological analysis. The old adage of comparing apples to oranges also applies to genes in gene families.

Establishing homology or orthology of genes in gene families can be viewed slightly differently from the classical scheme in morphological homology assessment as follows. The potential members of a gene family are identified by topographical similarity. The use of this method in orthology studies is very similar to that used in morphological homology assessment, except that the determination of topographical similarity is complicated by potential paralogy problems. Instead of going straight to alignment to determine topographical similarity, a preliminary step of determining group membership, by use of similarity as a means to assess membership, is implemented. Once this first step is accomplished (via similarity comparisons of sequences such as BLAST, BLAT, or COG), the alignment step can be undertaken. The establishment of character state identity is then made, and the hypothesis can be tested by phylogenetic approaches. The inclusion of genes into a particular gene family is often accomplished by setting a similarity cutoff and including all genes in an ortholog group that conform to the predetermined cutoff.

Population geneticists study the process of speciation and attempt to understand the genetic, ecological, or physiological changes that accompany speciation. All of these scientists need to be aware of an important distinction in this process: biological phenomena that are the result of speciation versus phenomena that result in speciation. These are two very different occurrences. Phenomena that are the result of speciation include branching patterns and divergence. While these phenomena are interesting and important to systematists and taxonomists, the more interesting phenomena are those that result in speciation.

The study of species can be approached from one of two directions: population genetics or systematics. The approach to understanding species from the population level is called the "bottom-up" approach, meaning that population genetics is used to examine the problem and lower categories are defined at the population level. Systematics or phylogenetics is called a "top-down" approach because systematics deals with both higher taxa and higher levels of organization.

Before the potential characteristics of a species are discussed, it is necessary to settle on an appropriate definition of the term. In science, objectivity is an important aspect of definition. The more objective a concept, the better science can be conducted (operationality) around the concept. However, there is a tradeoff between objectivity and operationality, where operationality is understood as whether or not, or to what degree, a particular concept can be applied and utilized. For instance, with respect to tree building, it is necessary to objectively assess which tree is the best in terms of studying all possible trees. This same principle is invoked to facilitate operationality: it is necessary to choose a tree in which three taxa are involved (as described in Chapters 8 and 9), a process that can be calculated without a computer. However, when large data sets with 100 taxa are assessed, the operationality of an exhaustive search is very small because of the huge number of trees needed to exhaustively assess the most suitable tree. Finally, any approach to the definition of species needs to be repeatable. Repeatability and objectivity usually go hand in hand.

One definition of species that most students remember is the one generated from the biological species concept: "A species is set of actually or potentially interbreeding populations." This definition is highly objective in that if one observes individuals of two different populations having sex, which results in the production of offspring, then they are the same species. Likewise, any two individuals that cannot have sex and produce viable offspring would be considered to be members of different species. The act of mating is an objective marker for whether individuals of two populations can be considered the same species. But how operational is this definition? In real terms, it fails this standard because it is tedious and impractical to establish a research program that is designed to observe every instance of individuals from populations attempting to mate.

Other problems with the conditions of operationality and objectivity are related to the problems of being either indiscriminate or too selective regarding sexual partners. A suitable definition of species would be objective if it applied to all organisms on the planet. However, in some cases, two apparently adequately separated species can cross-breed. An example is the Hawaiian silversword alliance, in which silversword plants from different species and even from different genera produce viable hybrid offspring. Species definitions that ignore this potential problem are considered inferior because they do not accommodate these cases. On the other hand, this definition cannot be easily applied to organisms such as bacteria that reproduce asexually.

Another interesting question is how to classify organisms such as mules. A mule is produced from the union of a donkey and a horse, and it is sterile. Should the phenomenon of a mule require that horses and donkeys be grouped as one species, or is the production of sterile offspring excluded from the species definition? A similar situation is a coydog, which is a hybrid between a coyote and a dog. Evolutionary biologists have attempted to accommodate these problems of objectivity and operationality with theories such as the recognition species concept and the cohesion species concept. After examining the 25 or so species concepts that exist or have existed, the authors have done what most biologists do when faced with the species problem: we chose the one we prefer. We have several reasons to prefer the definition that is presented below, based on its operationality and objectivity and because of its attraction to bioinformatics. This concept is identified as a phylogenetic species concept because it uses the principles of phylogenetics or a top-down approach to understanding species.

A simple solution is to define species phylogenetically

There are two ways in which a phylogenetic species can be defined. The first is tree-based and suggests that species are monophyletic assemblages in a tree. Monophyly is a simple concept in systematics and in viewing trees. A group in a phylogenetic tree is monophyletic if every member of the group has the same common ancestor to the exclusion of all other taxa in the tree.

The second way in which species can be defined in a phylogenetic context has been championed by Joel Cracraft. He defines a species as "an irreducible (basal) cluster of organisms, diagnosably distinct from other such clusters, and within which there is a parental pattern of ancestry and descent." In other words, a species is the smallest cluster of organisms that have a common ancestor. This definition does not require the aid of a phylogenetic tree. Rather, organisms are aggregated according to a hypothesis and diagnostics are identified by which the hypothesis can be tested. Rather than compare or contrast the different species definitions, this book will focus on the one that is more defensible as objective and operational: the Cracraft version.

Davis and Nixon developed an operational and objective framework for this approach, called population aggregation analysis (PAA). This approach is very simple but requires identification of the original populations in order to erect hypotheses to test. Consider four hypothetical cases (**Figure 3.9**). In each panel,

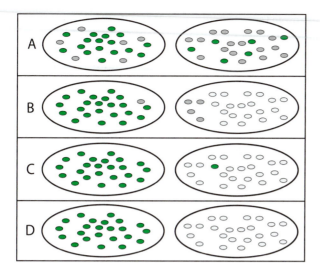

Figure 3.9 Population aggregation analysis to determine diagnostics. The figure describes four situations (A–D). In case A, there are two populations with several individuals represented with small circles. There are two attributes in the populations: "green" and "gray." In this first case there is no diagnosability. In case B, there is still no diagnosability because of the single gray individual in the left population. Likewise for case C, there is no diagnosability because of the single green individual in the right population. Finally, case D shows pure diagnosability.

the populations circumscribe a group of individuals with attributes that are either green or gray. The trick to PAA is to find attributes that can diagnose one population as distinct from the other. In case A, a situation is depicted in which both populations are polymorphic for green and gray. There is no diagnostic, so the hypothesis that these two populations are from the same species cannot be rejected. In cases B and C, the two populations are close to being fixed for the different colors. But we cannot reject the hypothesis that these two populations are from the same species. In case D, if green and gray attributes are fixed for the two populations, then we are left with an unrejected hypothesis and we can conclude the two populations are different species. Now consider a second set of hypothetical cases (**Figure 3.10**). Instead of colors of the individuals in the populations, DNA SNPs are examined as the attributes. The SNP distribution in case A in Figure 3.10 is similar to the color distribution in case A in Figure 3.9. The SNPs are polymorphic and not fixed for either of the populations. The SNP distribution in case B in Figure 3.10 is similar to the color distribution in case B in Figure 3.9, where the fixation is close but not complete. The SNP distribution in case C in Figure 3.10 demonstrates a case where the SNPs are diagnostic. Finally the SNPs in case D in Figure 3.10, while polymorphic within the two populations, are actually diagnostic for these two populations. The population (now species) on the left is diagnosed by either an A or a T, while the population on the right is diagnosed by a G.

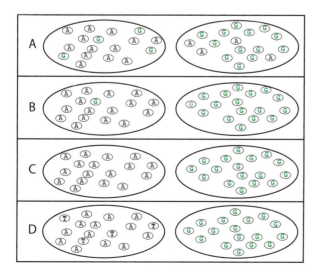

Figure 3.10 Four additional cases where population aggregation analysis is used to find diagnostics. In these cases, hypothetical DNA single-nucleotide polymorphisms (SNPs) are used. Case A has no diagnosability. Case B is also not diagnosable because of the presence of the "G" in one individual in the population on the left. Cases C and D are diagnosable.

In the color example (case D in Figure 3.9) and the SNP example (cases C and D in Figure 3.10), the populations are "diagnosably" distinct from each other. The one question that we are always asked about this approach is "How many characters are needed to call something a species in this system?" The most objective answer is "one."

A common misconception in the analysis of individual entities within a species is that these entities have a demonstrable hierarchy or phylogeny. In sexually reproducing species, hierarchy within the species is destroyed by the reproduction of individuals with each other. In the context of individuals within a species, any phylogeny has no meaning. Only when a single marker is used will the hierarchy resulting from systematic analysis have meaning, and in these cases, the trees produced indicate genealogies of the genes themselves that exist in the species' genomes. For instance, mitochondrial DNA (mtDNA) sequences are commonly used in population-level analysis of organisms. The mtDNA marker (usually D-Loop or control region sequences), because of its maternal inheritance and lack of recombination, will be an excellent tracer of females in a population. Likewise, the Y chromosome of many organisms experiences a similar clonal inheritance and is an effective tracer of males in a population. Even single-gene genealogies of organisms or observed taxonomic units (OTUs) can be useful in showing the pattern of descent of alleles in the gene genealogy, and if the gene is involved in pathology, the genealogy could be of importance in interpreting disease. But none of these (mtDNA, Y chromosomes, or single nuclear genes) in and of itself is suitable to establish a "phylogeny" of individual entities within a species. This problem makes dealing with single genes or single sources of characters an issue when examining species. We will return to this problem, called the "gene tree versus species tree problem," in Chapter 11.

Modern Challenges to Darwinian Evolution

Modern evolutionary biology is a vibrant discipline. This vibrancy and the controversies that arise as a result of the expansion of evolutionary biology have made some outside observers suggest that evolutionary theory can be supplanted by other nonscientific ideas like intelligent design. In fact, evolutionary theory has stood the test of time and has been tested over and over, again and again, and our understanding of the basic processes of evolutionary biology has only been strengthened through the process. We liken the establishment of gravitational theory in physics to the development of evolutionary theory. While ideas about gravitation have changed over the past four centuries, from no theory to Newtonian physics to relativity to quantum mechanics, at no time during the debates about how gravity works was gravity doubted as a natural force. While evolutionary biology has a shorter historical stretch, the same can be said for evolution; at no time in the past 150 years of evolutionary theory, at no time during the debates about evolution and how it worked, should there have been doubts that evolution is a natural force. There have been challenges to the idea of evolution, though, and we discuss two of them below.

Punctuated equilibrium suggests that not all evolution is gradual

One tenet of Darwinian evolutionary theory is that evolution proceeds in a gradual fashion. Darwin made this clear, suggesting that evolutionary change was imperceptible. Several challenges to this way of thinking have emerged in the expansion of evolutionary thought. Some of the arguments have been resurrected as a result of new genomic and developmental biology advances made in the later part of the twentieth century. In the 1970s, two young paleontologists at the American Museum of Natural History noticed that some of their fossil trilobites were not evolving gradually. Niles Eldredge and Stephen Gould suggested that evolution proceeded by jerks and jolts, and they termed this idea punctuated equilibrium. Punctuated equilibrium and gradualism can be compared diagrammatically

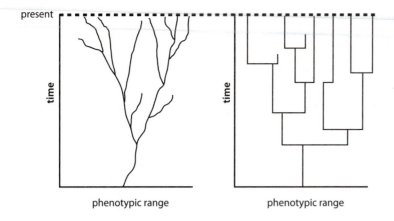

Figure 3.11 Graphs comparing gradual evolutionary change (left) with punctuated equilibrium (right).

(**Figure 3.11**). While not all evolution on the planet follows the jerky pattern of punctuated equilibrium, Gould and Eldredge certainly showed that not all evolution was gradual. Arguments could be made as to which (gradualism or punctuated equilibrium) is the predominant form of evolutionary change, but the bottom line is that there is room for competing processes of change in evolutionary theory.

Epigenetic changes caused by outside influences can be inherited

One of the burning questions of evolutionary biology concerns the phenomenon of epigenetics. One category of epigenetics involves the alteration of gene expression patterns through the addition of small molecules (methyl or acetyl groups) to DNA or to the nucleosome proteins that bind to DNA. If a protein is being expressed normally in a cell, and its methylation pattern changes, then it could become expressed at either a higher or a lower level. This change is controlled by proteins whose actions can be influenced by the external environment and the nutrients that a person consumes. One important example of this phenomenon is the consumption of folate by pregnant women. Folate provides the substrate for histone methylation, which is critical for development. A lack of methyl groups leads to altered gene expression and is responsible for neural tube defects in fetuses, which can be extremely tragic. For this reason, pregnant women in the United States are recommended to consume a high level of folate.

The reason that epigenetics is important for a discussion of evolutionary theory is that epigenetic changes are thought to be heritable between generations to some extent. Thus, to use traditional Lamarckian language, this would be the "inheritance of acquired characteristics." If a parent were to do something that altered the epigenetic profile of their gametes, then this change would be carried on to their offspring. This contrasts with the Darwinian idea that all change comes from genetic mutations, which are not subject to outside influence. The field of epigenetics and how it is incorporated into modern evolutionary theory is a young and exciting one. Certainly, with the development of modern genetic approaches such as high-throughput sequencing, we will learn more about the process, which will allow us to fold the impacts of this interesting force in nature into evolutionary theory. We conclude this chapter by describing a simple mathematical model that accommodates change through time, that has become an important tool in modern evolutionary biology.

Markov chain models

The Markov chain model (MCM) is a method for describing relationships between objects or conditions over time and hence has been a favored tool of evolutionary biologists and phylogenomicists. The model's objective is to compress large amounts of data into a simple predictor. It is used extensively in many areas of quantitative science, including meteorology, economics, and bioinformatics and phylogenomics.

The MCM depends on the process of transformation, which includes a property of change known as the Markov property. This property accommodates the transformation of one state (for example, a G or a C in a specific position in a sequence) to another state. There are a finite number of possible states in the system (in the case of DNA these are G, A, T, C, and gap) and the transformations are random, meaning the process is what mathematicians refer to as "memoryless." Therefore, an MCM is formally defined as a discrete random process with the Markov property described above.

How the MCM works is important for understanding concepts discussed later in the book. There are many examples that illustrate the basic parameters of the MCM, one of which is drawn from a business case, as follows. *The New York Times* conducts an analysis to determine if an advertising campaign for the Sunday edition is effective given certain parameters of customer preference that have been previously measured. At the time the survey is initiated, 60% of the customers who purchase Sunday newspapers buy the *Sunday Times* and 40% do not. An advertising agency recruited by the newspaper estimates that, if people are currently reading the *Sunday Times*, they will continue to buy and read this edition 90% of the time, while 10% will shift to reading *The New York Post* Sunday edition. On the other hand, 80% of the people who do not currently buy and read the *Sunday Times* will convert to the *Sunday Times* after seeing the advertisement. This also means that 20% of the initial non-*Sunday Times* readers will not switch to the *Times*.

To demonstrate a MCM using this example, an initial state distribution (probability) matrix is created. Matrices are mathematical statements that contain columns and rows in which the values in the matrix correspond to specific states. They can be added and multiplied by each other using simple rules. For the above example, the probability matrix is depicted below, where column T indicates *Sunday Times* readers and column N indicates non-*Sunday Times* readers.

$$\begin{array}{cc} \text{T} & \text{N} \\ I_0 = [\ 0.6 & 0.4\ ] \end{array}$$

A new matrix is then generated that represents the preferences of readers who are identified the week following the advertising campaign. This transition matrix assesses the probabilities that readers will switch according to the findings established by the advertising company. These preferences and their probabilities are represented in **Figure 3.12**.

Most MCMs begin with a clear statement of the transitions involved, which can be illustrated with a transition diagram as shown in Figure 3.12. This diagram can easily be transformed into a mathematical statement, again in matrix form. For this matrix, we have two rows and two columns. One row represents people who are currently reading the *Sunday Times* (T) and the other row represents people who are not currently reading the *Sunday Times* (N). The first column summarizes the percentage (P) of readers who purchase and read the *Sunday Times* and the percentage of readers who decline to buy the *Times* following a week of advertising. The transition matrix for this example is shown below.

$$P = \begin{array}{c} \\ \text{T} \\ \text{N} \end{array} \begin{array}{cc} \text{T} & \text{N} \\ \left[\begin{array}{cc} 0.9 & 0.1 \\ 0.2 & 0.8 \end{array} \right] \end{array}$$

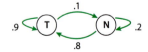

Figure 3.12 Example of a Markov Chain Model. Transition diagram. T, readers of the *Sunday Times*; N, readers who avoid the *Sunday Times*.

Now we can easily estimate the probabilities for the percentage of people who will buy the *Sunday Times* and those who will not after a week of advertising by constructing a probability tree. If we start out with the initial proportions in the probability matrix, we see that there are two decisions that someone who originally bought the *Times* can make after receiving the advertisement; that is, to either buy it again (0.9) or not to buy it (0.1). Therefore, if there is a 60% probability that a person will purchase the *Times* initially and a 90% probability that this person will continue to buy the newspaper based on the advertising campaign, then it can be estimated that 0.54 percent (0.6×0.9) will buy the *Times* (T) based on the promotion. On the other hand, of the 60% of the people who buy the *Times* initially, 10% will switch to the *Post* (N), meaning that 0.06 percent (0.1×0.6) will *not* (N) buy the *Times*.

There are two additional decisions that someone who initially did not buy the *Times* can make after receiving the advertisement. Of the 40% who initially did not buy the newspaper, 80% will switch to the *Times* as a result of the promotional campaign, so 0.32 of the total (0.4×0.8) will now buy the *Times* as well. Conversely, of the 40% who did not buy the *Times* at first, 20% will not be persuaded by advertising; therefore, 0.08 of the total (0.4×0.2) will *not* buy the *Times* as a result of the advertising campaign.

The probability of people buying the *Sunday Times* after a week of advertising is the sum of the two probabilities estimated above for people who read the times after receiving the advertisement, or 0.86 ($0.54 + 0.32$). This is about a 30% increase in readership. The number of people not reading the *Sunday Times* can be computed by totaling $0.06 + 0.08$ (or, alternatively, by subtracting 0.86 from 1.0), which indicates that 0.14 of people will *not* buy the *Sunday Times* edition after the first week of advertising. These calculations yield new starting probabilities of 0.86 (buying the *Times*) and 0.14 (not buying the *Times*).

The *New York Times* also wants to know how much longer they should continue the advertising, because at a certain point they realize that they will not see an increase in readership. Using the same approaches outlined above, we can compute the probabilities using the MCM for two, three, four, and five weeks of advertising. The results are shown below.

Week→	0	1	2	3	4	5
T	0.6	0.86	0.88	0.89	0.89	0.89
N	0.4	0.14	0.12	0.11	0.11	0.11

Note that after week 3 there is no increase in readership. The *Times* should therefore conduct its advertising campaign for a maximum of two to three weeks.

MCMs come in a wide variety of forms and have discrete properties, such as being irreducible or aperiodic. The first property relates to being able to use the transition matrix to move from a specific value in the matrix to any other value in the matrix via the rules of the matrix irrespective of the number of steps. Aperiodicity is characterized by a return to an initial value at irregular time intervals.

In the example above, the prediction of the next value is only dependent on the value of the previous week. This result is known as first-order MCM and is the simplest type of the model. In many cases, the MCM is more complex, and the output is dependent on the previous number of values. This effect is intuitive for many examples since one can presume that the movement of a value can be better predicted based on data from more points in the past than simply from one measure.

In the world of phylogenomics, there are many applications of MCMs. One of the most important uses is the evaluation of the DNA substitution process over evolutionary time. For example, if a particular nucleotide position is an A at the current time, we can query the probability if it was an A or a T previously. This probability can be estimated by analyzing large amounts of aligned nucleotide sequence data from the database, calculating the percentage of times that an A remained an A between two divergent species, and calculating the number of times that position

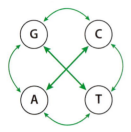

Figure 3.13 DNA transition as depicted in a Markov Chain Model. Transition diagram for DNA sequences.

changed from a T to an A in two divergent species. Such nucleotide transitions would be modeled by a 4 × 4 or 16-cell transition matrix.

The four base pairs in DNA sequences (G, A, T, C) can be arranged in a transition diagram (**Figure 3.13**). In this example, it is equally probable to change from one base to any other base but the bases cannot stay the same, a property known as transitional probability, which is shown below.

$$P = \begin{array}{c} \\ G \\ A \\ T \\ C \end{array} \begin{array}{cccc} G & A & T & C \\ \left[\begin{array}{cccc} 0 & \frac{1}{3} & \frac{1}{3} & \frac{1}{3} \\ \frac{1}{3} & 0 & \frac{1}{3} & \frac{1}{3} \\ \frac{1}{3} & \frac{1}{3} & 0 & \frac{1}{3} \\ \frac{1}{3} & \frac{1}{3} & \frac{1}{3} & 0 \end{array} \right] \end{array}$$

A probability matrix for a model where it is equally probable to change from one base to any other including itself is shown below. A wide range of state transitions can be accommodated by transition matrices.

$$P = \begin{array}{c} \\ G \\ A \\ T \\ C \end{array} \begin{array}{cccc} G & A & T & C \\ \left[\begin{array}{cccc} \frac{1}{4} & \frac{1}{4} & \frac{1}{4} & \frac{1}{4} \\ \frac{1}{4} & \frac{1}{4} & \frac{1}{4} & \frac{1}{4} \\ \frac{1}{4} & \frac{1}{4} & \frac{1}{4} & \frac{1}{4} \\ \frac{1}{4} & \frac{1}{4} & \frac{1}{4} & \frac{1}{4} \end{array} \right] \end{array}$$

A more applicable case is that of amino acid codons. Each pair of the 20 amino acid and stop codons has a transition probability. Due to the redundancy of the genetic code and the chemical properties of amino acids, there is a hierarchy of transitions that will be tolerated for a given codon (see Chapter 13). For instance, the most tolerated change would be for a different codon producing the same amino acid; the next tolerated change would be a different amino acid with the same hydrophobicity; the next would be an amino acid with a similarly sized side-chain; and the last tolerated change would be a switch from a coding codon to a stop codon. By taking alignments of homologous sequences in the database, an MCM for amino acid substitutions can be determined.

The MCM is extremely useful in Bayesian phylogenetics, as discussed in Chapter 10, as well as in DNA sequence alignment algorithms (see Chapters 5, 6, and 7). It is also used in human genetics to understand linkage and distribution of site changes in population genetics. The general approach used in these kinds of studies is to use the MCM to simulate a field of values and then to use it as a null hypothesis to test. The model tested depends on the initial transition matrix.

Summary

- Evolutionary biology can be split into macro- and microevolutionary domains.
- The goal of microevolutionary studies, implemented by population genetics approaches, is to understand variation.
- The goals of macroevolutionary studies, from a systematics approach, are twofold: first, to understand the relationships of organisms using trees and tree thinking, and second, to understand the characters that have evolved as homologies.

This latter goal may seem a bit disconnected from phylogenomics until one realizes that molecules like DNA and proteins are nothing more than characters.
- Speciation lies between macro- and microevolution, and the process of speciation can be approached from both macro- and microevolutionary perspectives.
- The Markov chain model (MCM) can be used as a tool to simulate changes through time and hence has become an important tool in phylogenomics.

Discussion Questions

1. Draw the 15 possible trees when there are four taxa in the ingroup. Use A, B, C, and D as labels for the terminals.

2. Discuss the merits of tree building in determining homology and analogy. What added value do phylogenetic trees give to the discussion? Can approaches other than tree-based ones be used to determine homology?

3. Draw a Venn diagram that best hierarchically groups the icons (**Figure 3.14**). What character did you use to establish the hierarchy?

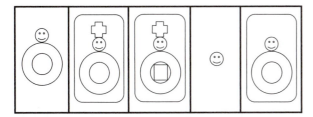

Figure 3.14 Five "icon organisms" to arrange hierarchically.

4. Several populations with particular attributes are compared (**Figure 3.15**). Which of the scenarios in A–F have diagnosable entities in them? Consider scenarios G and H. Just by looking at G, are these two "populations" diagnosable? Now in H more individuals have been examined for the right population. Are these "populations" still diagnosable?

5. The number of rooted trees, N_r, given the number of ingroup taxa, n, is

$$N_r = (2n - 3)!/[2^{n-2}(n-2)!] \tag{3.1}$$

where ! is the notation for factorial [$n!$ is the product of all positive integers less than or equal to n].

As an example, consider $n = 3$. If we fill in the equation above, we get

$$N_r = [(2 \times 3) - 3]!/([2^{3-2} \times (3-2)!])$$

$$N_r = (3)!/[2^1 \times (1!)]$$

$$N_r = (3 \times 2 \times 1)/[2 \times 1]$$

$$N_r = 6/2 = 3$$

Now compute the exact number of trees for $n = 4$, 5, 6, and 100 taxa.

6. Confirm the values for weeks 2, 3, 4, and 5 using the MCM for the *New York Times* example. Show your calculations. The values for these weeks are shown below.

Week→	0	1	2	3	4	5
T	0.6	0.86	0.88	0.89	0.89	0.89
N	0.4	0.14	0.12	0.11	0.11	0.11

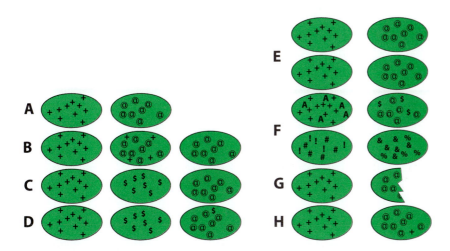

Figure 3.15 Population aggregation analysis tests.

Further Reading

Brower AVZ & Schawaroch V (1996) Three steps of homology assessment. *Cladistics* 12, 265–272.

Baum DA, Smith SD, & Donovan SS (2005) Evolution: the tree-thinking challenge. *Science* 310, 979–980.

Cracraft J (1989) Speciation and its ontology: the empirical consequences of alternative species concepts for understanding patterns and processes of differentiation. In Speciation and its Consequences (D. Otte, J. Endler eds.), pp 28–59. Sinauer Associates.

Darwin C (1859) On the origin of species by means of natural selection. J. Murray.

Davis J & Nixon K (1992) Populations, genetic variation , and the delimitation of phylogenetic species. *Syst. Biol.* 41, 421–435.

de Pinna MGG (1991) Concepts and tests of homology in the cladistic paradigm. *Cladistics* 7, 367–394.

Dobzhansky Th (1970) Genetics of the Evolutionary Process. Columbia University Press.

Edwards AWF (1972) Likelihood. Cambridge University Press

Gould SJ & Eldredge N (1977) Punctuated equilibria: the tempo and mode of evolution reconsidered. *Paleobiology* 3, 115–151.

Haldane JBS (1964) A defense of beanbag genetics. *Perspect. Biol. Med.* 7, 343–359.

Hennig W (1966) Phylogenetic Systematics. University of Illinois Press.

Hudson RR (1991) Gene genealogies and the coalescent process. *Oxford Surv. Evol. Biol. 7*, 1–44.

Kimura M (1983) The Neutral Theory of Molecular Evolution. Cambridge University Press..

Kingman JFC (1982) On the genealogy of large populations. *J. Appl. Probab.* 19A, 27–43.

Lewontin RC (1974) The Genetic Basis of Evolutionary Change. Columbia University Press.

Marjoram P & Tavaré S (2006) Modern computational approaches for analysing molecular genetic variation data. *Nat. Rev. Genet.* 7, 759–770.

Mayr E (1963) Animal Species and Evolution. Belknap Press of Harvard University Press.

Paterson H (1985) The recognition concept of species. In Species and Speciation (E Vrba ed.), pp 21–29. Transvaal Museum.

Platnick NI & Gaffney ES (1977) Systematics: A Popperian perspective. *Syst. Zool.* 26, 360–365.

Reeck GR, de Haën C, Teller DC et al. (1987) "Homology" in proteins and nucleic acids: a terminology muddle and a way out of it. *Cell* 50, 667.

Simpson GG (1944) Tempo and Mode in Evolution. Columbia University Press.

Sneath PHA & Sokal RR (1973) Numerical Taxonomy: The Principles and Practice of Numerical Classification. Freeman, San Francisco.

Templeton A (1989) The meaning of species and speciation: a genetic perspective. In Speciation and its Consequences (D Otte, J Endler eds), pp 3–27. Sinauer Associates.

Databases

In order to do genome-level analyses, we need to learn how to obtain genomic data and manipulate them. This chapter is an introduction to the databases that exist for phylogenomics research. It will read somewhat like a "how-to" manual, but this is unavoidable for the beginning student in phylogenomics. We start by introducing the historical aspects of databases. Next, we delve directly into GenBank, the premier repository of trillions of base pairs of DNA that have been sequenced by scientists. We take a detailed journey through this database, showing how to locate, identify, and access information. There are two major kinds of searches that can be accomplished in most databases. The first uses keywords such as the gene name or species name to find suitable sequences. In this chapter we detail this method for searching the database. The second kind of search that can be accomplished is by using a sequence as the query. These searches will be discussed in detail in the next chapter. Because whole genomes are very taxon-specific, we consider this kind of a search and manipulation of the data much like the keyword-based searches we described above. Since the early 2000s, the National Center for Biotechnology Information has developed a facility that focuses specifically on genome biology, and in this chapter we also demonstrate how these genomic biology functions are utilized.

Databases and Phylogenomics

Prior to the onslaught of DNA sequencing that began in the 1980s, data storage and retrieval was conducted on a laboratory to laboratory basis, where the providers and seekers of sequence data communicated and transferred it directly. This approach was sufficient for the small amounts of sequence that were being utilized at that time. But in the 1980s the growing amount of DNA and protein sequence data being produced reached a point at which there was a need for a shared centralized repository. On November 4, 1988, the Congress of the United States, prompted by Senator Claude Pepper, established the beginnings of a large database that would focus on the collection of molecular data. The center that was established was named the National Center for Biotechnology Information (NCBI) and was placed under the aegis of the National Library of Medicine (NLM) at the National Institutes of Health (NIH). The National Library of Medicine was originally established in 1836 as a repository for the Surgeon General's library and has since developed into the largest collection of medical literature in the world. The National Institutes of Health was established in 1930 as an arm of the federal government of the United States; it administers research and health advances for the country. The placement of the informatics center for biotechnology at a library might at first glance appear odd, but because libraries store information for retrieval—such as journals and books—the placement is not so odd after all. Add to this that the NLM has been a repository for all kinds of medical information other than literature since the 1960s, and the transition to the treatment of biotechnology information is a natural move.

DNA sequences are stored in large international databases

The DNA sequence portions of the NCBI data took the form of a database known as GenBank. This database is linked to and coordinated with similar databases in Europe at the European Molecular Biology Laboratories (EMBL), and in Japan at the DNA Databank of Japan (DDBJ). The resulting DDBJ/EMBL/GenBank collaboration is based upon the principles of international cooperation and freedom of access. These three centers exchange data daily in order to provide the most up-to-date sequence information for researchers worldwide.

As with any large database, GenBank needed a method to organize all of the sequence data that was being produced. This led the creation of the GenBank feature table definition: http://www.insdc.org/documents/feature_table.html. This document outlines the various pieces of information that need to be included when a genetic sequence is added to GenBank. For instance, a person submitting data needs to include information such as the name of the organism that the sequence came from, along with the gene name, the name and institution of the submitter, what sequencing technique was used, and whether this sequence was produced as part of a scientific publication. In addition, a GenBank record generally includes links to other online databases having records on the same sequence. The key to the usefulness of GenBank is its search capabilities, which are similar to Google except they are specialized for the analysis of sequence information. As will be seen below, when a researcher needs the sequences of a gene or the complete genome of an organism, a query can simply be typed into the search box.

As shown in Figure 1.1, populating the database was gradual at first because sequencing was painstakingly slow in the first 10 years of GenBank's existence. There was a large upswing in the number of sequences stored beginning in 2000. This development is a result of the use of robotic high-throughput sequencers that were developed to sequence the human genome. The information in Figure 1.1 is recorded only up to 2008. Since that time, next-generation sequencing approaches have been introduced that have rapidly increased the amount of sequence produced. If total sequence output were plotted graphically, a sharp increase in the slope of the graph would be evident. As discussed, these machines have hundreds to thousands of times the throughput of the previous Sanger sequencing machines.

DNA sequence data consists of a long string of characters such as ACTCCATTAACA. In general, the data storage requirements of sequence are 1 byte per base pair (bp). This means that a 1000 bp sequence would take 1 kilobyte (kB) of storage, 1 million base pairs would take 1 megabyte (MB) of storage, and 1 billion base pairs would require 1 gigabyte (GB) of storage. Therefore, a human genome of approximately 3 billion base pairs would require 3 GB of storage. Currently, this is not considered a huge amount of storage, but in the 1980s, when sequencing was beginning, data were measured in megabytes, not gigabytes, and this was considered a huge amount of data. One important innovation that programmers have utilized is the fact that DNA uses an extremely reduced alphabet of only four nucleotides, plus a few additional characters to code for uncertainty. This reduced alphabet has allowed for great compression of DNA sequence data and has allowed for a human genome to be compressed to less than 800 MB in size.

Specific data sets may be held in special repositories

In 1996, the first genome of a whole organism was sequenced. At 2 million or so nucleotides, the genome information from *Haemophilus influenzae* did not pose a particularly large problem for storage and information retrieval. However, as we discussed in Chapter 2, the genome sequences from other organisms started to roll in, in droves, by the year 2000. Since the genome sequence of a single organism is a database in and of itself, novel ways of archiving genomes have been developed. Currently, there are hundreds of repositories for sequences that individual institutions or consortiums maintain for specific organisms (for example, FlyBase compiles sequences for the fruit fly *Drosophila*, and WormBase stores information from the roundworm *Caenorhabditis elegans*). While other databases offer excellent search and retrieval services, in general they are all modeled after GenBank. Thus, this chapter discusses GenBank in detail. There is no better way to show how this database is structured, how it works, and how to use it than to explore it yourself. All of the examples in this chapter are based upon the state of GenBank in the spring of 2012, but GenBank's content is constantly increasing (nothing is ever deleted!). Therefore, performing the identical queries at a later date will most likely yield greater numbers than we mention. This does not mean that anything is wrong, just that science has been progressing.

Microarray experiments were first performed in the late 1990s. As we will see when we discuss microarrays in subsequent chapters, these experiments are extremely data-rich, and they require archiving and curating. Not surprisingly, one of the main resources for microarray data is managed by the NCBI; it is known as the Gene Expression Omnibus (GEO). This database contains all of the raw microarray data from thousands of experiments, and at the time of this writing, contained over 500,000 records.

Besides the sequencing of DNA, the understanding of protein structures has led to tremendous increases in biological knowledge. Most of these structures are obtained by either X-ray crystallography or NMR spectroscopy and are full of information. These data are stored at NCBI in the Structure database.

Databases offer free access and availability for scientific inquiry

There are two important guiding principles for all of these databases that have greatly enhanced their value. First, all of the databases are completely open for scientists to download without paying any fees. When there is a need for confidentiality to protect either pre-publication or patient data, appropriate provisions are made, but overall, the vast majority of the data can be accessed without restrictions. Even the restricted data can be easily accessed by researchers who present a compelling justification for obtaining it.

The second principle behind these databases is of the required submission of unprocessed data, which allows for other researchers to completely reanalyze it. This means that, for microarray data, the raw intensity calls for each probe from each sample need to be submitted before any corrections or distillations of the data have been performed. An easy alternative would have been to just require the submission of the final data reports, which are much smaller in size and easier to interpret, though this would have hindered opportunities for scientific inquiry. By using data from GEO, a researcher can download the complete data from a publication and perform his or her own analyses to determine the veracity of the conclusions made by the authors. Indeed, there are whole groups of computational researchers who are completely focused on reanalyzing and reinterpreting the data produced by others.

In order to make sure that these databases contain as many experimental results as possible, many funding agencies and publications require submission of the underlying data to an NCBI database. Therefore, researchers who want to publish their work or obtain continued funding will share their data.

Information Retrieval from the NCBI Database

Sequence data are maintained in different formats that are designed to be human- or machine-readable. All sequences are available in the simple FASTA format and in the GenBank format with the feature table containing extensive annotation. In addition, data are available in the XML format that is readily accessible to

Table 4.2 Typical GenBank accession fields

LOCUS
    Locus name
    Sequence length
    Molecule type
    GenBank division
    Modification date

DEFINITION
ACCESSION
VERSION "GenInfo identifier"
KEYWORDS
SOURCE (organism name)

REFERENCE (or direct submission)
    AUTHORS
    TITLE (title of published work or tentative title of unpublished work)
    JOURNAL
    PUBMED

FEATURES

Figure 4.3 Screenshot of top half of an NCBI accession page. The result shown is for the search "Mouse Fox F2." (Courtesy of the National Center for Biotechnology Information.)

line gives the original accession number that this sequence is identified with in GenBank (in this case NP_043555). The VERSION line lists the up-to-date accession number information and can have two numbers on the line. These two numbering systems are an artifact of the NCBI numbering system and refer to the

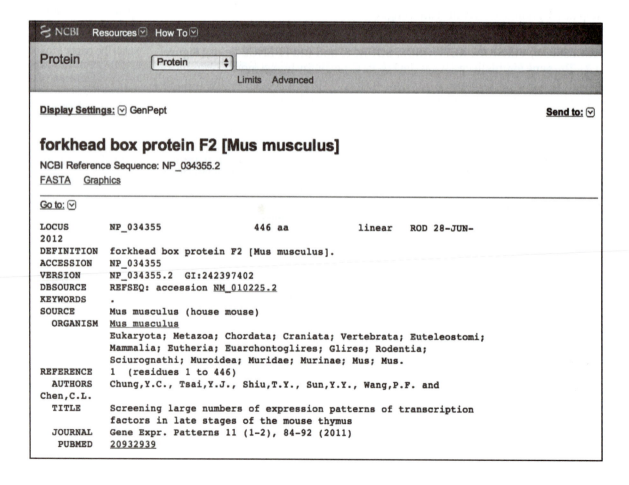

same sequence. Initially sequences accessioned into GenBank were numbered consecutively as they were received by GenBank. This numbering system is called the GenInfo (GI) system and was augmented by what is called the accession.version system. The major difference between the two systems is that the GI system would have very different sequences from different taxa in successive numbers. The accession.version system, established in 1999, uses a numbering system that better organizes the sequences along taxonomic and genetic lines. Hence the ACCESSION line has these two accession numbers, the first of which is the up-to-date accession.version number and should have the same root as the ACCESSION number. If the up-to-date accession.version number has changed since the first entry of the sequence into the database, this number increases with each change such as NP_04355 to NP_04355.1 to NP_04355.2, etc. The second number on the VERSION line is the GI number discussed above. This number will increase in parallel with the accession.version number. The next line is a KEYWORDS line, in which in many cases will be blank since it is not a required field in the accession process. The next lines carry the SOURCE information. The source is simply the organism from which the tissue was taken to do the sequencing. In this case the source organism is *Mus musculus* or the common house mouse. Below the brief description of the source is a detailed taxonomy for the organism. This information is extremely important and flexible and has its own browser called the NCBI Taxonomy Browser. We will address this taxonomy browser in a later chapter when we discuss taxonomy. The next several lines detail the literature mentioning the sequence. The REFERENCE line itself mentions the number of publications in which the sequence is referred to and the specific positions of the sequence referenced. Below this first line for the REFERENCE information are listed the authors, the title of the article, the journal, and the PUBMED reference number of the article. PubMed is the largest repository of medical reference material in the world, and as such, each journal article, book, and technical report is given an accession number. One can access the publication referencing this Fox F2 sequence simply by clicking on the PUBMED accession number on the web page. Because some sequences are deposited directly into GenBank without being mentioned in the literature, these accessions will have "direct submission" status, where the direct information for who submitted the sequences will be given.

The bottom half of the accession for mouse Fox F2 gene contains additional fields (**Figure 4.4**). There is only one field in this figure that we need to discuss, and that is the FEATURES field (see Table 4.2). This field is rather complex, but easy to follow.

Additional information on the tissue source of the sequence is given on the "source" line. The accession we have been examining has the following source information:

```
source       1..2366
             /organism="Mus musculus"
             /mol_type="mRNA"
             /strain="C57BL/6"
             /db_xref="taxon:10090"
             /chromosome="13"
             /map="13"
```

The first line gives the number of nucleotides in the accession that are relevant. The second line reiterates the organism source, and the third line reiterates the kind of molecule (DNA, RNA, or protein for instance) that the sequence was generated from. If a specific strain of the species was used to generate the sequence, then that is listed on the fourth line. The fifth line lists the taxonomic reference number, which we will explore in more detail later. The chromosomal location and map locations of the gene are given in the sixth and seventh lines. These last

```
                               /gene_synonym="Fkh20; FREAC2; LUN"
                               /note="forkhead box F2"
                               /db_xref="GeneID:14238"
                               /db_xref="MGI:1347479"
          exon                 1..1441
                               /gene="Foxf2"
                               /gene_synonym="Fkh20; FREAC2; LUN"
                               /inference="alignment:Splign"
                               /number=1
          CDS                  265..1605
                               /gene="Foxf2"
                               /gene_synonym="Fkh20; FREAC2; LUN"
                               /note="forkhead homolog 20"
                               /codon_start=1
                               /product="forkhead box protein F2"
                               /protein_id="NP_034355.2"
                               /db_xref="GI:242397402"
                               /db_xref="CCDS:CCDS26424.1"
                               /db_xref="GeneID:14238"
                               /db_xref="MGI:1347479"
                               /translation="MSTEGGPPPPPPRPPPAPLRRACSPAPGALQAALMSPPPAATLE
                               STSSSSSSSASCASSSSNSVSASAGACKSAASSGGAGAGSGGTKKATSGLRRPEKPP
                               YSYIALIVMAIQSSPSKRLTLSEIYQFLQARFPFFRGAYQGWKNSVRHNLSLNECFIK
                               LPKGLGRPGKGHYWTIDPASEFMFEEGSFRRRPRGFRRKCQALKPMYHRVVSGLGFGA
                               SLLPQGFDFQAPPSAPLGCHGQGGYYGGLDMMPAGYDTGAGAPGHAHPHHLHHHHVPHM
                               SPNPGSTYMASCPVPAGPAGVGAAAGGGGGGDYGPDSSSSPVPSSPAMASAIECHSP
                               YTSPAAHWSSPGASPYLKQPPALTPSSNPAASAGLHPSMSSYSLEQSYLHQNAREDLS
                               VGLPRYQHHSTPVCDRKDFVLNFNGISSFHPSASGSYYHHHHQSVCQDIKPCVM"
          exon                 1442..2366
                               /gene="Foxf2"
                               /gene_synonym="Fkh20; FREAC2; LUN"
                               /inference="alignment:Splign"
                               /number=2
          STS                  1826..1916
                               /gene="Foxf2"
                               /gene_synonym="Fkh20; FREAC2; LUN"
                               /standard_name="Foxf2"
                               /db_xref="UniSTS:515964"
ORIGIN
        1 aggagacggt tgcgcaagga gccggccgga ggctcggaag agggatgcgc ggggcgttgc
       61 ctccgacccg ccgccgccgc cgccgcccga agccccagag gagctgaggg aggcgacgcc
      121 gaagcgctgg cccgcagtgg cccgggctgc agcgcggccg cgcgcagtagg gcactcgccc
```

two identifiers will be common for model organisms, as the chromosomal locations in non-models are not well known.

Following the source listing is information on the gene itself. The coordinates of the sequence are given first, followed by the gene name, synonyms, and other information that will help in identifying the gene. In the Fox F2 case, the gene section looks like this:

```
gene        1..2366
            /gene="Foxf2"
            /gene_synonym="Fkh20; FREAC2; LUN"
            /note="forkhead box F2"
            /db_xref="GeneID:14238"
            /db_xref="MGI:1347479"
```

The next section, "exon," contains information about the structure of the gene and its relevance to the accessioned sequence. Many eukaryotic genes contain coding sequences that are translated into protein, known as exons, interspersed with non-coding regions that are removed, known as introns. The accession will list the base pair positions that conceptually code for exons:

```
exon        1..1441
            /gene="Foxf2"
            /gene_synonym="Fkh20; FREAC2; LUN"
            /inference="alignment:Splign"
            /number=1
```

In this case, there was a single potential exon recognized. Note that the accession sequence length is 2366 base pairs and that the exon identified is from 1 to 1441 nucleotides. Below the exon details are listed more gene information and the criterion by which the exon was established.

For some accessions the intron/exon structure is much more complex, as in this accession (Z31725.1 GI 469799) for the trithorax gene (a developmental gene in *Drosophila*):

```
FEATURES      Location/Qualifiers
   source     1..14546
              /organism="Drosophila melanogaster"
              /mol_type="genomic DNA"
              /strain="Oregon"
              /db_xref="taxon:7227"
              /dev_stage="Embryo"
   mRNA       join(1..3553,3624..3830,3895..12121,12192..12325,
              12403..12578,12639..14546)
   mRNA       join(1..1139,2380..3553,3624..3830,3895..12121,
              12192..12325,12403..12578,12639..14546)
   mRNA       join(1..690,1140..3553,3624..3830,3895..12121,
              12192..12325,12403..12578,12639..14546)
   mRNA       join(1..690,2380..3553,3624..3830,3895..12121,
              12192..12325,12403..12578,12639..14546)
   mRNA       join(1..690,2380..3553,3624..3830,3895..12121,
              12192..12325,12403..12578,12639..12968)
   exon       1..690
   exon       691..1139
   exon       1140..2379
```

In this example, there are several alternative ways to delete the introns and splice the exons together. These are shown by the five "mRNA" lines. The numbers on the "mRNA" lines refer to positions in the 14,546 nucleotide long DNA sequence accession for this gene. To get the first mRNA, for instance, the sequence from positions 1–3553 is joined with the sequence from positions 3624–3830, and then joined with the sequence from positions 3895–12121, etc. These are the coding regions of the first mRNA. Introns in the DNA sequences, then, are the sequences that are excluded from the joining process to make the mRNA. For instance, the DNA sequences from positions 3553–3624 is the first intron in the first mRNA sequence from above. The "exon" lines at the bottom indicate the positions in the DNA sequence that correspond to potential protein coding regions or exons. In the first 2400 bases of the gene for this protein, there are three such potential exons that correspond to DNA positions 1–690, 691–1139, and 1140–2379. In this way, with the DNA sequence in the accession, the mRNA joining information, and the exon information, the structure of the gene can be reconstructed from this GenBank information.

Returning to the Fox F2 example, the codons for the gene that can be translated into the Fox F2 protein are listed on the "CDS" line. In this case the Fox F2 protein is coded for by nucleotides 265–1605. Note that this stretch of mRNA is 1305 nucleotides long, and this number divided by three (the number of bases in a codon) is 435, indicating that the Fox F2 protein is 435 amino acids long. Information about the protein that could be generated from this 1305 nucleotide-long mRNA is given below, and the amino acid translation of the mRNA is given on the "translation" line.

```
CDS   265..1605
      /gene="Foxf2"
      /gene_synonym="Fkh20; FREAC2; LUN"
      /note="forkhead homolog 20"
      /codon_start=1
      /product="forkhead box protein F2"
      /protein_id="NP_034355.2"
      /db_xref="GI:242397402"
      /db_xref="CCDS:CCDS26424.1"
      /db_xref="GeneID:14238"
      /db_xref="MGI:1347479"
      /translation="MSTEGGPPPPPPRPPPAPLRRACSPAPGALQAALMSPPPAATLE
      STSSSSSSSASCASSSSNSVSASAGACKSAASSGGAGAGSGGTKKATSGLRRPEKPP...
```

In all of the lines we mention above, some of the information is hyperlinked to lead the user to more information about the listed feature. For instance, clicking on CDS leads the user to a web page (**Figure 4.5**) with just the mRNA sequence that codes for the protein, whereas the original accession lists all of the nucleotides in both 5′ and 3′ untranslated regions. In other words, the CDS button leads to the stripping away of any nucleotide that does not code for an amino acid in the protein.

Whole genomes are collected on the Genome Page

Figure 4.5 Screenshot of the FASTA-formatted Mouse Fox F2 mRNA sequence. (Courtesy of the National Center for Biotechnology Information.)

Whole genome information can also be accessed via the Genome Page of the NCBI Website (**Figure 4.6**). A genome of a specific organism can be accessed via the query box at the top. Genomes that have been sequenced and accessioned into

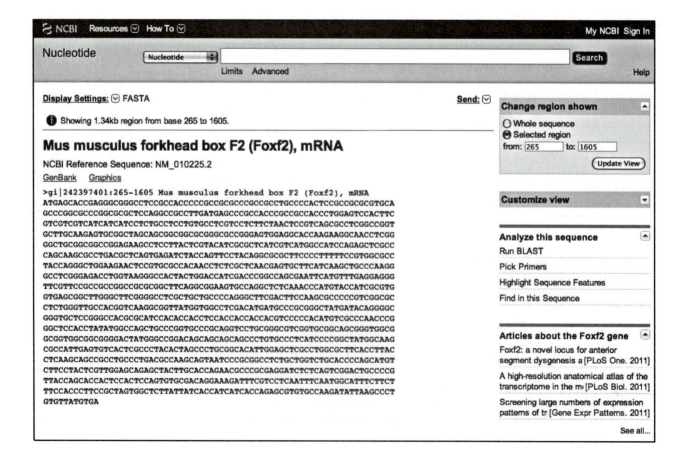

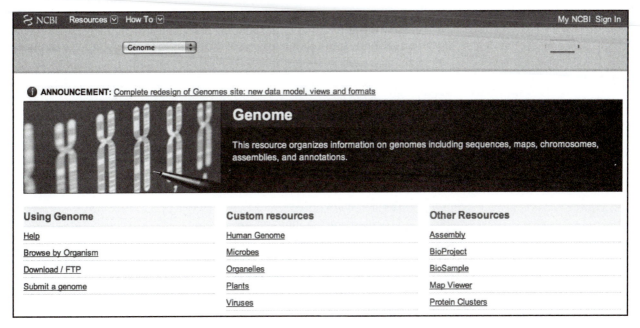

Figure 4.6 Screenshot of the Genome webpage. Types of functions offered can be accessed by clicking on the entries. One can browse the data base by organism name, download full genomes using ftp, and submit sequenced genomes using the functions on this webpage.

GenBank can be obtained using the Browse by Organism buttons. One can search a genome for a specific sequence by using the Genome Tools buttons. The human genome, or any one of a thousand or so microbial and eukaryotic genomes, can be accessed this way. Organelle genomes can be accessed via the button under the Custom resources heading. Web Feature 4 demonstrates how to access the "Genome Tools" on this site.

An example of the web page for a typical genome accession, for the Ames strain of *Bacillus anthracis*, is shown (**Figure 4.7**) The page is a large one with information

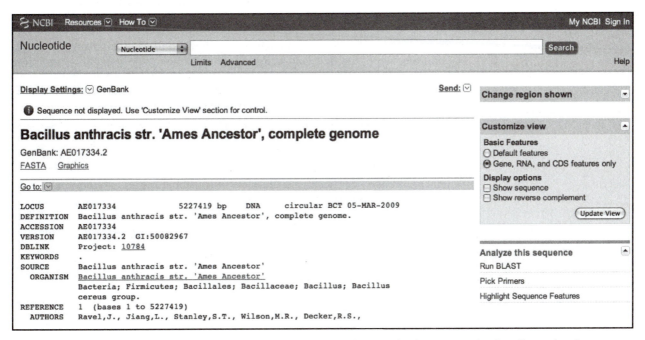

Figure 4.7 Screenshot of the Genome accession page. The accession page for the Ames strain of *Bacillus anthracis*. (Courtesy of the National Center for Biotechnology Information.)

on over 1500 genes that exist in this species genome. One can scroll down the page to view the annotation of the genome and the list of all the genes in the *B. anthracis* genome. Information for one of the genes in the *B. anthracis* genome is shown here:

```
gene     407..1747
         /gene="dnaA"
         /locus_tag="BA_0001"
         /old_locus_tag="BA0001"
         /db_xref="Pathema:BA_0001"
CDS      407..1747
         /gene="dnaA"
         /locus_tag="BA_0001"
         /old_locus_tag="BA0001"
         /note="identified by similarity to EGAD:14548; match to
          protein family HMM PF00308; match to protein family HMM
          PF08299; match to protein family HMM TIGR00362"
         /codon_start=1
         /transl_table=11
         /product="chromosomal replication initiator protein DnaA"
         /protein_id="AAP24059.1"
         /db_xref="GI:30253517"
         /db_xref="Pathema:BA_0001"
```

To obtain the information for this particular gene, which happens to be a chromosomal replication protein called DnaA, the "gene" (for the amino acid sequence) or the "CDS" hyperlinks can be clicked. The methods discussed above for downloading and saving such sequences from GenBank are the same for accessions from the Genome pages.

Other databases such as Online Mendelian Inheritance in Man (OMIM) can be used to obtain information

So far we have discussed sequence- and publication-based databases. In much genomics work, the end goal is to understand phenotype too. There are a few Web-based approaches to this problem. One is the Online Mendelian Inheritance in Man database (OMIM; http://www.ncbi.nlm.nih.gov/omim/). This database has existed since the 1960s and contains a collection of genetic disorders in man. A keyword for a disease or genetic disorder such as "hemophilia" can be typed into the query box on the database query page, which will return information on this disorder and ones with similar names. In the case of "hemophilia," several hits are obtained that include hemophilia, hemophilia B, and other hemophiliac-related diseases. By clicking on the accession numbers obtained from the search, the profile for the disease can be obtained as well as the list of items available for display. Alternatively, the entire report for the disorder can be downloaded. It is best to explore this site by reading through the very detailed descriptions and the tutorial for the site.

Summary

- Genome-level sequence information is of great importance to modern phylogenomics and is stored in large international databases.
- Specific data sets may be held in special repositories, such as the FlyBase for *Drosophila* sequences.
- Open access and availability of unprocessed data for further scientific inquiry are crucial for sequence databases.

- Publication information is archived in the PubMed database.
- Extensive sequence information may be retrieved from GenBank.
- Whole genomes may be accessed through the Genome Page of NCBI

Discussion Questions

1. Find the oldest publication listed in PubMed for the keyword "phenylketonuria" (PKU, a metabolic genetic disorder).

2. Use OMIM to do a search for the keyword "phenylketonuria". What is the oldest publication listed in this database for the keyword?

3. Download the Fox F2 sequences for human, chimpanzee, and orangutan. Write down the first 20 or so amino acids in each sequence. Do this for the mouse sequence too. How would you compare these sequences? What might be the first step in making the sequences usable for comparison?

4. Search in PubMed for "alcohol dehydrogenase," followed by a separate search for ADH (the gene abbreviation for alcohol dehydrogenase). How do the results of the searches differ? Which search results in more "hits"?

Further Reading

Boguski MS (1992) A molecular biologist visits Jurassic Park. *BioTechniques* 12, 668–669.

Boguski M & McEntyre J (1994) I think therefore I publish. *Trends Biochem. Sci.* 19, 71.

Edhlund BM (2007) PubMed Essentials. Form & Kunskap.

van Etten-Jamaludin F & Deurenberg R (2009) A Practical Guide to PubMed. Springer.

PubMed Tutorial. http://www.nlm.nih.gov/bsd/disted/pubmedtutorial/

Homology and Pairwise Alignment

Homology is the cornerstone of phylogenomics, if not the cornerstone of all biology. In Chapter 3 we discussed the basics of homology, but in this chapter we delve deeper into homology and how it is used in phylogenomics. We defined homology as a character that arises as a result of common ancestry. In other words, if a character is shared and derived, it is considered a homology. In this chapter, we first look at the processes that result in diverging genes and noncoding genomic regions. This examination reveals that there are two ways that DNA and protein sequences can diverge. The first is through speciation, or splitting of lineages. Once speciation occurs, two copies of the gene exist and go on to separate evolutionary trajectories. The second process of divergence can be caused by gene duplication: whether it by single gene, segmental, or whole genome duplication, these events produce diverging genes that can also be homologized. One of the major tools of gene comparison and homology assessment is called alignment. In this chapter we also take a detailed look at pairwise alignment and how it results in statements about sequence homology that we can use in phylogenomics. Finally we return to the NCBI databases and explore the use of alignment tools to search the database. In this last part of the chapter, we explore the use of the Basic Local Alignment Sequence Tool, or BLAST, to accomplish efficient database searches. We also detail how BLAST has become one of the basic tools in the phylogenomics toolkit.

Homology of Genes, Genomic Regions, and Proteins

Our discussion of homology requires that we look at the phenomenon of sequence change at two levels. The first level is the level of genes in genomes, where the goal is to determine which genes or genomic regions are homologous. The second level is the level of sequences in genes, where the goal is to determine which residue positions in genes, proteins, and noncoding regions are homologous. Determining homology at these two levels is the cornerstone of phylogenomics.

Genomes can diverge by speciation and by duplication

There are two processes by which genomes can diverge. The first is through **speciation**, or splitting of lineages during the evolutionary process. As we have discussed in Chapter 3, speciation is generally considered to be the forming of two groups that are reproductively isolated from each other. Speciation-based divergence can be thought of as vertical divergence, where divergence directly occurs along a pedigree (**Figure 5.1**). If this kind of divergence were the only way genomes varied, then all organisms would have the same number of genes and these genes would have basic characteristics that (one would hope) would allow researchers to determine that they were all related. In addition, the determination of homologous genome regions would be relatively easy. But genomes also diverge through **duplication** and loss of genetic material. These duplications and deletions can involve nucleotides, parts or domains of genes, full genes, segmental

Figure 5.1 Genes diverge as a result of speciation and gene duplication. Left: three-dimensional view of the process. Right: tree-based relationships of the same genes.

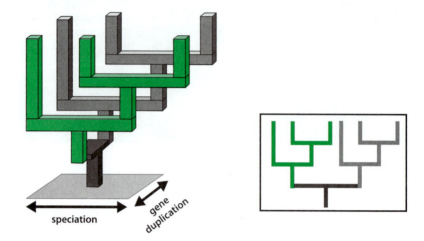

speciation

gene duplication

duplications, chromosomal duplications, and whole genome duplications. This divergence through duplication and loss can be thought of as nonvertical, and needs to be taken into consideration when doing phylogenomics.

Both speciation and duplication can impact the homology of genes (Figure 5.1). In this case, all of the genes are related to each other through the single starting gene in the common ancestor of all the genes in the figure. Gene duplication produces a situation where the genes that existed before the duplication are related to each other through a different common ancestor than the genes that are the product of duplication. Since common ancestry is one of the arbiters of homology, we could say that all of the genes in the figure are homologous at some level. They all come from the same common ancestral gene. But some genes are homologous through a more recent common ancestral gene than others. To accommodate this interesting evolutionary phenomenon, two terms for homology of molecular sequences were developed: orthology and paralogy. **Orthology** refers to the relationships of genes that are separated by speciation. **Paralogy** refers to all other relationships of genes in genomes that occur as the result of processes other than speciation. Paralogous genes are most commonly generated by duplication and unequal crossing over during recombination. A third term for homology of genes has also been coined to accommodate genes that have transposed or jumped from one species to another, disrupting common ancestry. This kind of relationship is called **xenology**.

It is important to realize that orthology is not the same thing as similarity (see the discussion of homology in Chapter 3). In the context of sequences and genes, two genes can be very dissimilar to each other at the sequence level and still be orthologous. On the other hand, two genes might show the highest similarity to each other and not be orthologous.

Alignment is a proxy for homology at the sequence level

Because of the importance of homology in phylogenomics, there is a need for a method to compare sequences to determine whether they appear homologous, the extent of their homology, and where the changes in the sequences have occurred. In order to assess this homology, the techniques of sequence alignment have been developed. In essence, the alignment of two sequences is a form of the classic computer science problem of string alignment. Research in this area has produced techniques such as regular expressions and underlies Internet and database searching. But because of the specialized nature of DNA sequences and the great importance of sequence alignment, large amounts of intellectual might have been directed to the string alignment problem in a biological context.

Several kinds of changes can occur at the nucleotide level (**Figure 5.2**). These include changing from one base to another, losing a base (deletion), and gaining

gaps can occur

· **before the first character of a string**

    e.g. CTGCGGG---GGTAAT
          | | | | | | | |
          --GCGG-AGAGG-AA-

· **inside a string**

    e.g. CTGCGGG---GGTAAT
          | | | | | | | |
          --GCGG-AGAGG-AA-

· **after the last character of a string**

    e.g. CTGCGGG---GGTAAT
          | | | | | | | |
          --GCGG-AGAGG-AA-

Figure 5.2 Three main positions where insertions can be accrued during the alignment process. (Courtesy of the National Institutes of Health.)

a base (insertion). If insertion and loss did not occur at the level of nucleotides within genes, then alignment of genes would be trivial. Genes would not change length and one would simply need to align base number 1 of a specific gene from species 1 and base number 1 of a gene from species 2. The rest of the bases would be perfectly aligned without any tinkering, all the way to the last base in the sequence. But with the possibility of insertions and deletions (collectively termed **indels**), this task becomes much more complicated.

Alignment is the cornerstone of sequence-based database searches, and it is the foundation for sequence-based phylogenetic tree building (discussed in detail in Chapters 8, 9, and 10). There are two basic types: pairwise alignment and multiple sequence alignment. **Pairwise alignment** is a method that analyzes two sequences to find the best way to arrange the sequences, given a set of rules for optimizing their arrangement. Because most sequence-based database searches use pairwise approaches, we discuss this kind of alignment in detail here. **Multiple sequence alignment** is more computationally intense than pairwise alignment and involves the arrangement of several sequences at one time, given a set of rules. Multiple sequence alignment is examined in detail in Chapter 6.

Alignment of sequences is controlled by imposing a set of rules or optimality criteria on the process. Given the cost of either inserting a gap into a given sequence or placing two different bases or amino acids in the same position in an alignment, the costs for any arrangement are calculated, and the least costly option wins. Computing the cost of inserting an open space in sequences to obtain residues "in register" is complex. The cost of gap insertion is critical in deciding the optimality of an alignment. **Figure 5.3** demonstrates two trivial ways to align the same two sequences. In this example, the cost of placing gaps in the sequences in the top example is so extreme that it prohibits putting gaps in the sequence. The bottom example shows a case where changes in base identity cost are so extreme that it is less costly to place gaps anywhere in the sequences. There are three ways to insert gaps into a string of amino acids or base pairs. The first is to simply place a gap or gaps inside a string of bases or amino acids. The other two ways involve placing gaps at the end or at the beginning of a string. Depending on how the gaps are introduced, different costs are assigned. For instance, an alignment can be penalized for introducing too many gaps in a row. All of these costs are given defaults by alignment programs, but the costs can be altered by a user. In general, the default values parameters should be appropriate for the vast majority of cases.

Nucleic acid sequence alignments can be evaluated manually

Consider two nucleic acid sequences, sequence 1 and sequence 2 (**Figure 5.4**) and two extreme ways to align them—align 1 and align 2. If the cost of a base pair change is set at 1 and the cost of inserting a gap in a sequence is set at 2 (a change to gap ratio of 1/2), then the cost of align 1 is calculated as $C_{aln1} = c_1 + c_2 + c_3 + c_4 + c_5 + c_6 + c_7 + c_8 + c_9$, because it has nine positions. This is just a simple accounting problem. The uppercase C represents the cost of the entire alignment, and the lowercase c stands for costs at specific positions in the alignment. In the first alignment shown (align 1), positions 1, 3, 5–7, and 9 have zero cost because there is no change or gap in any of these positions. Positions 2, 4, and 8 each incur a cost of 1 because they have base changes, and the entire cost of the alignment

```
sequence 1    GGGAATGCGCTAGCATCGA

sequence 2    GGCACTGATCGATGCTACG
```

```
sequence 1   GGG-----AAT-------GCGCT-----AGC----ATCGA

sequence 2   ---GGCAC---TGATCGA-----TGCTA-----CG-----
```

Figure 5.3 Trivial alignment of two sequences. In the upper panel, gaps are "expensive," and in the lower panel, changes are "expensive."

Figure 5.4 Counting alignment costs in simple DNA sequences.
Top: two DNA sequences to be aligned. Bottom: two different alignments and how the costs are summed in pairwise alignment.

```
sequence 1    CAACGATGA
sequence 2    CCAAGATCA


align 1
CAACGATGA
CCAAGATCA
```
$C_{\text{aln 1}} = c_1 + c_2 + c_3 + c_4 + c_5 + c_6 + c_7 + c_8 + c_9$

```
align 2
-CAACGATGA
CCAA-GATCA
```
$C_{\text{aln 1}} = c_1 + c_2 + c_3 + c_4 + c_5 + c_6 + c_7 + c_8 + c_9 + c_{10}$

is $0 + 1 + 0 + 1 + 0 + 0 + 0 + 1 + 0 = 3$. The second alignment (align 2) is calculated as $C_{\text{aln2}} = c_1 + c_2 + c_3 + c_4 + c_5 + c_6 + c_7 + c_8 + c_9 + c_{10}$, because it has 10 positions. In this alignment, positions 2–4, 6–8, and 10 have zero cost because there are neither gaps nor changes from one base to another in these positions. There are gaps at positions 1 and 5, though, and each of these has a cost of 2, and position 9 has a base change with a cost of 1, for a total cost of $C_{\text{aln2}} = 2 + 0 + 0 + 0 + 2 + 0 + 0 + 0 + 1 + 0 = 5$. Comparison of the costs clearly shows that alignment 1 is the better of these two options (however, an even better alignment may still exist). Suppose that the costs are reversed, so that base changes have a cost of 2 and gaps have a cost of 1. In that case, $C_{\text{aln1}} = 0 + 2 + 0 + 2 + 0 + 0 + 0 + 2 + 0 = 6$ and $C_{\text{aln2}} = 1 + 0 + 0 + 0 + 1 + 0 + 0 + 0 + 2 + 0 = 4$. In this scenario, alignment 2 is the better outcome. Finally, if the costs are set so that gaps and changes are equally costly, then both alignments would have the same calculated cost, and either one would be considered an optimal match.

In considering the appropriate change to gap ratios in alignments, it is assumed that indels should cost more than changes because they are known to be rarer and they cause a change in reading frames. From various studies, including the recent 1000 Genomes Project, it has been found that sequence variations are 10 times more common than indels. This is presumably because of the strong effects of an indel. If a single nucleotide is changed in a gene, only the codon overlapping that position would be changed, and the rest of the protein would be the same. In the case of an indel (unless its size is a multiple of 3), the reading frame of the gene will be altered and every downstream residue in the protein will be affected.

With respect to costs of changes within a sequence, one of the best ways to estimate these costs is via empirical analysis. For instance, if it is known how often a particular amino acid changes to another, a frequency can be estimated and used in the accounting process. There are two methods of calculating frequency. The first method uses the genetic code (see Chapter 2) to estimate the number of steps it would take to move from the codon of one amino acid to the codon of another. This approach is similar to those used to detect selection in protein coding sequences (see Chapters 13 and 14). This method produces a matrix for calculating the costs of changes in the amino acids. The second method employs a database of proteins from different species and empirically calculates the frequencies of all possible kinds of changes and translates these results to costs. The changes that occur frequently in the observed database have a low cost, and the changes that are rare have a high cost.

The same process can be used for estimating the costs of gaps, but this calculation is less quantifiable. To avoid this problem, scientists employ a procedure called "exploring the alignment space," meaning that the change costs are established with a predefined cost matrix and several independent alignments are conducted in which the gap cost is altered. In this way, the behavior of alignment can be examined as a function of the gap cost. Usually, three or four gap costs are used. If the actual alignments produced are the same for various costs, then the alignment is said to be "stable to gap cost," thus resulting in a particular acceptable alignment. If the alignments look very different when gap cost is altered, then the alignment is said to be "sensitive to gap cost," and it is more difficult to choose an alignment.

	A	C	D	E	F	G	H	I	K	L	M	N	P	Q	R	S	T	V	W	Y
[A]	0	2	1	1	2	1	2	2	2	2	2	2	1	2	2	1	1	1	2	2
[C]	2	0	2	3	1	1	2	2	3	2	3	2	2	3	1	1	2	2	1	1
[D]	1	2	0	1	2	1	1	2	2	2	3	1	2	2	2	2	2	1	3	1
[E]	1	3	1	0	3	1	2	2	1	2	2	2	1	2	2	2	1	2	2
[F]	2	1	2	3	0	2	2	1	3	1	2	2	3	2	1	2	1	2	1
[G]	1	1	1	1	2	0	2	2	2	2	2	2	2	2	1	1	2	1	1	2
[H]	2	2	1	2	2	2	0	2	2	1	3	1	1	1	1	2	2	2	3	1
[I]	2	2	2	2	1	2	2	0	1	1	1	1	2	2	1	1	1	1	3	2
[K]	2	3	2	1	3	2	2	1	0	2	1	1	2	1	1	2	1	2	2	2
[L]	2	2	2	2	1	2	1	1	2	0	1	2	1	1	1	1	2	1	1	2
[M]	2	3	3	2	2	2	3	1	1	1	0	2	2	2	1	2	1	1	2	3
[N]	2	2	1	2	2	2	1	1	1	2	2	0	2	2	2	1	1	2	3	1
[P]	1	2	2	2	2	2	1	2	2	1	2	2	0	1	1	1	1	2	2	2
[Q]	2	3	2	1	3	2	1	2	1	1	2	2	1	0	1	2	2	2	2	2
[R]	2	1	2	2	2	1	1	1	1	1	1	2	1	1	0	1	1	2	1	2
[S]	1	1	2	2	1	1	2	1	2	1	2	1	1	2	1	0	1	2	1	1
[T]	1	2	2	2	2	2	2	1	1	2	1	1	1	2	1	1	0	2	2	2
[V]	1	2	1	1	1	1	2	1	2	1	1	2	2	2	2	2	2	0	2	2
[W]	2	1	3	2	2	1	3	3	2	1	2	3	2	2	1	1	2	2	0	2
[Y]	2	1	1	2	1	2	1	2	2	2	3	1	2	2	2	1	2	2	2	0

Figure 5.5 Cost matrix for amino acid substitutions. The costs are based on the number of steps inferred from the genetic code. See text for further explanation.

A paired protein alignment can be evaluated manually

An amino acid cost matrix can be calculated by use of the genetic code (**Figure 5.5**). The letters across the top and down the left-hand side of the matrix refer to the 20 amino acids found in proteins. Numbers in the matrix refer to the number of changes in a codon required to implement the transformation of one amino acid to the other. For instance, the underscored 2 in the matrix for a change from A to C indicates that two nucleotide changes are required to get from the codon for alanine to the codon for cysteine. Note that the costs in the diagonal are all zero. These zero costs imply that there is no cost to maintain the same amino acid and suggest that changes in codons that are redundant (see Chapter 2) are ignored in estimating the costs in this matrix. Note also that the matrix is symmetrical; that is, the top half (above diagonal) is identical to the bottom half (below diagonal). This is because the same number of nucleotide changes would be needed in either direction (note the symmetrical placement of the underscored 2s, from A to C or from C to A).

Consider two short amino acid sequences, two different ways of aligning these sequences, and how the costs are calculated using the cost matrix above (**Figure 5.6**). In alignment 1, the cost of a gap is set to be twice the value of the most costly sequence change in the matrix (Figure 5.5). The most costly change in the matrix is 3, so the gap cost would be 6. For alignment 1, the cost would be $C_{aln1} = c_1 + c_2 +$

alignment 1

cost

| sequence 1 | MCHHHVVIVD-DL |
| sequence 2 | MCHH-VVIVDEDL |

$$C_{ain\ 1} = c_1 + c_2 + c_3 + c_4 + c_5 + c_6 + c_7 + c_8 + c_9 + c_{10} + c_{11} + c_{12} + c_{13}$$

alignment 2

cost

| sequence 1 | MCHHHVVIVDDL |
| sequence 2 | MCHHVVIVDEDL |

$$C_{ain\ 1} = c_1 + c_2 + c_3 + c_4 + c_5 + c_6 + c_7 + c_8 + c_9 + c_{10} + c_{11} + c_{12}$$

Figure 5.6 Counting alignment costs in simple protein sequences. Top: two protein sequences to be aligned. Bottom: two different alignments and how the costs are summed in pairwise alignment.

Traceback uses the filled matrix to obtain the alignment

The last step in the dynamic programming method involves tracing the matrix back to obtain the alignment. Beginning in the cell with the highest score, a move is made in one of three directions (up, left, or diagonal) to a cell from which the score in the current cell could have been obtained. Equivalent scores in the up, left or diagonal directions simply means that all three pathways could have arisen from the previous cell. When all three options have the same score in their cells, then the diagonal option is chosen. In the example where we have filled the matrix, the best score for the alignment is 5 (boxed value in row 7, column 9). The cell that is the most reasonable for this value (score = 5) is then selected. Since all three positions (indicated by arrows in the diagram) are equivalent (all have a score of 4), the diagonal position is the optimal cell. The diagonal cell is chosen, and the rest of row 7 and column 9 are grayed out:

```
        T  G  C  T  C  G  T  A
     0  0  0  0  0  0  0  0  0
  T  0  1  1  1  1  1  1  1  1
  T  0  1  1  1  2  2  2  2  2
  C  0  1  1  2  2  3  3  3  3
  A  0  1  1  2  2  3  3  3  4
  T  0  1  1  2  3  3  3  4  4
  A  0  1  1  2  3  3  3  4  5
```

Next, go to the cell on the diagonal, at row 6, column 8, and move to the cell with a score that can best be explained as coming from a score of 4. Since all three cells have the same score (3), the diagonal cell is chosen. The remaining cells in row 6 and in column 8 can then be grayed out:

```
        T  G  C  T  C  G  T  A
     0  0  0  0  0  0  0  0  0
  T  0  1  1  1  1  1  1  1  1
  T  0  1  1  1  2  2  2  2  2
  C  0  1  1  2  2  3  3  3  3
  A  0  1  1  2  2  3  3  3  4
  T  0  1  1  2  3  3  3  4  4
  A  0  1  1  2  3  3  3  4  5
```

Next, the value in row 5, column 7, is compared to the cells in the up, left, and diagonal positions. Note that all of these cells contain a score of 3. Again, the diagonal cell is selected because all three choices are equally likely. The rest of the column and row are then grayed out:

```
        T  G  C  T  C  G  T  A
     0  0  0  0  0  0  0  0  0
  T  0  1  1  1  1  1  1  1  1
  T  0  1  1  1  2  2  2  2  2
  C  0  1  1  2  2  3  3  3  3
  A  0  1  1  2  2  3  3  3  4
  T  0  1  1  2  3  3  3  4  4
  A  0  1  1  2  3  3  3  4  5
```

Repeating this process one more time, we note that up, left, and diagonal values are all 2s. So we choose the diagonal cell and we also gray out the remaining cells in row 4 and in column 6:

```
      T G C T C G T A
    0 0 0 0 0 0 0 0 0
T   0 1 1 1 1 1 1 1 1
T   0 1 1 1 2 2 2 2 2
C   0 1 1 2 2 3 3 3 3
A   0 1 1 2 2 3 3 3 4
T   0 1 1 2 3 3 3 4 4
A   0 1 1 2 3 3 3 4 5
```

The cell in row 3, column 5 is chosen next. Since the values in the up, left, and diagonal positions have scores of 1, the diagonal cell is chosen and the remainder of row 3 and column 5 are then grayed out:

```
      T G C T C G T A
    0 0 0 0 0 0 0 0 0
T   0 1 1 1 1 1 1 1 1
T   0 1 1 1 2 2 2 2 2
C   0 1 1 2 2 3 3 3 3
A   0 1 1 2 2 3 3 3 4
T   0 1 1 2 3 3 3 4 4
A   0 1 1 2 3 3 3 4 5
```

The next diagonal cell is a 1 in row 2, column 4. From this cell, the up and diagonal cells both have scores of 0, while the left cell has a value of 1. Since no change in score (1 → 1) is more likely than a change (1 → 0), the left cell is the next cell to which we move, while the rest of the cells in the column and the diagonal are grayed out:

```
      T G C T C G T A
    0 0 0 0 0 0 0 0 0
T   0 1 1 1 1 1 1 1 1
T   0 1 1 1 2 2 2 2 2
C   0 1 1 2 2 3 3 3 3
A   0 1 1 2 2 3 3 3 4
T   0 1 1 2 3 3 3 4 4
A   0 1 1 2 3 3 3 4 5
```

The next cell selected for the traceback is the cell to the left (row 2, column 3) with a score of 1. The cell to the left contains a 1, so it is chosen and the diagonal would then be grayed out:

```
      T G C T C G T A
    0 0 0 0 0 0 0 0 0
T   0 1 1 1 1 1 1 1 1
T   0 1 1 1 2 2 2 2 2
C   0 1 1 2 2 3 3 3 3
A   0 1 1 2 2 3 3 3 4
T   0 1 1 2 3 3 3 4 4
A   0 1 1 2 3 3 3 4 5
```

The last cell has only two choices: the diagonal cell and the cell to the left of it. Since these contain the same score, the diagonal cell is selected, yielding the final traceback matrix shown:

```
      T G C T C G T A
    0 0 0 0 0 0 0 0 0
T   0 1 1 1 1 1 1 1 1
T   0 1 1 1 2 2 2 2 2
C   0 1 1 2 2 3 3 3 3
A   0 1 1 2 2 3 3 3 4
T   0 1 1 2 3 3 3 4 4
A   0 1 1 2 3 3 3 4 5
```

In order to interpret the results of the dynamic programming method illustrated above, the following guidelines are applied. Whenever the diagonal is not traversed, a gap is placed into one of the sequences. If the diagonal is violated in a row, then the gap is placed in sequence 2. If the diagonal is violated in a column, a gap is inserted in sequence 1. In our example, sequence 1 can be read as TGCTCGTA (there are no violations of the diagonal in any column). Sequence 2 can be read as T--TCATA (there are two violations of the diagonal after the first cell in the diagonal and none afterward). The following alignment, with a score of 5, is thus obtained:

```
TGCTCGTA
T--TCATA
```

As can be seen from the preceding section, this type of analysis is tedious to perform manually. Fortunately, computer programs have been created to perform this task. This Smith–Waterman dynamic programming algorithm is fast when two sequences are being aligned; however, it is too slow for the similar tasks of multiple sequence alignment and database searching.

Database Searching via Pairwise Alignments: The Basic Local Alignment Search Tool

Just as the standard Smith–Waterman algorithm is too slow for multiple sequence alignment, it is also not suitable for database searching. In a database search, a researcher will take a particular sequence and compare it to a very large set of known and annotated sequences. This task is important for genomic annotation as well as for obtaining homologous sequences for a phylogenomic analysis. The nucleotide and protein databases, such as GenBank, that are generally used are very large and would be impossible to query by use of the Smith–Waterman algorithm. A search of the full GenBank database for even a short sequence with the Smith–Waterman algorithm could easily require days of computation.

In order to allow for rapid searching of genetic databases, the Basic Local Alignment Search Tool (BLAST) was developed. This program and its derivatives are probably the most commonly used piece of bioinformatic software in all of phylogenomics, regularly used not only by bioinformaticians but also by regular wet-lab biologists. Its simplicity of concept and interface has given it such a high level of usage.

The BLAST program answers the following question rapidly: what are the annotated sequences that most closely match my query sequence? Since it is so important, we will walk through the steps of doing a BLAST search. While the objective of Entrez searches in GenBank is to obtain accessions based on a name or a keyword, the object of a BLAST search is to use a sequence (DNA or protein) to obtain related accessions. While the Entrez approach described in Chapter 4 is useful, the name of the gene in question may not be available, or the purpose of the search may be aimed at identifying as many accessions that are similar to a query sequence as possible. In terms of these scenarios, searching the database with a known sequence is often the most efficient way to obtain sequences from a large number of species for comparative purposes.

BLAST can identify inexact matches of high similarity

Here are two real-life examples of BLAST usage. A researcher has sequenced an unknown gene from a newly discovered bacterium and they want to know what gene they have sequenced. They would input the sequence into BLAST and the program would return the names of all of the matching genes from bacterial species which would allow the researcher to identify their gene. For another example, assume that a researcher is trying to obtain the α-globin sequences for many different species of mammals. They tried using the Entrez keyword-based system, but they ran into the same difficulties that were outlined above. So instead, they find the α-globin sequence for one species and then they perform a BLAST search to identify similar sequences. This BLAST search will not be hampered by the difficulties of a keyword search, as it would act directly on the protein sequence and attempt to identify similar sequences.

One potential question may be asked: if keywords, DNA sequences, and protein sequences are all really just text, why does one need an algorithm like BLAST to search the sequence databases? Couldn't they just use the same search tool that is used for the keywords? The answer is that for sequence, most of the time, the search is not for an exact match but for an inexact match with high similarity. This is because DNA sequences are subject to mutation and therefore are rarely identical between individuals. But even though two individuals have a different genetic sequence in their microsatellite loci (one of the most variable parts of the human genome), a BLAST search would easily identify both of them as being microsatellite loci. This is because BLAST looks for pairwise similarity of sequences and not just identical matches.

BLAST is optimized for searching large databases

The question that BLAST is trying to answer is the same question that the Needleman–Wunsch and the Smith–Waterman algorithms were developed to solve: namely, what is the pairwise alignment score between nucleotide or protein sequences? The main difference is that Needleman–Wunsch and Smith–Waterman algorithms are used to compare individual pairs of sequences, whereas BLAST is used to compare a particular sequence against a large database of sequences. With the incredibly large size of genetic sequence databases, optimizations were needed to the search algorithms to enable a search to be completed in a reasonable amount of time. BLAST addresses this issue by optimizing the search algorithm. More recently, the BLAT program (BLAST-Like Alignment Tool) was developed that is even faster than BLAST.

The BLAST program was developed to balance computational speed with sensitivity of the search. To access BLAST, go to http://blast.ncbi.nlm.nih.gov/Blast.cgi (**Figure 5.7**). The program compares two sequences and finds regions of similarity based on local comparisons in the sequences. It can compare a nucleotide or amino acid sequence to a collection of sequences in the database and will calculate a statistical significance for the comparison. When a BLAST search report is generated, it contains what are called "hits." Each report can contain a large number of hits, arranged from best to worst. Each hit includes an accession number, the name of the accession, the quality of the hit (given as an *E*-value, which is discussed below), selected metrics on the hit, and the alignment of the "query" sequence with the hit sequence, called the "subject" sequence. The coordinates within the gene or genome of the organism of the hit region are given for both the query and the subject sequence.

There are two key components to the increased speed that BLAST provides. First, BLAST does not directly align the query sequence to every database sequence; rather, it selects short segments of the query sequence (known as seeds) to align to the database sequences to determine whether there is at least a threshold level of matching. If these seeds do not align well, then there is no need to continue that alignment and the next sequence in the database is queried. If the seeds do have a strong alignment, then the alignment is continued for the full sequences. This speed-up prevents the alignment program from spending time trying to align pairs of sequences that will most likely not end up with a high alignment score. Second, BLAST uses heuristics for the alignment procedure rather than using the full Smith–Waterman algorithm. These heuristics speed up the alignment by not filling the full Smith–Waterman matrix for both the initial seed alignments and full alignments. A heuristic approach has two major characteristics. The first characteristic is that it is a problem-solving approach based on practical, well-accepted ways of doing things. The second characteristic is that a heuristic approach might not necessarily result in the exact solution. While it might result

Figure 5.7 NCBI BLAST home page.
(Courtesy of the National Institutes of Health.)

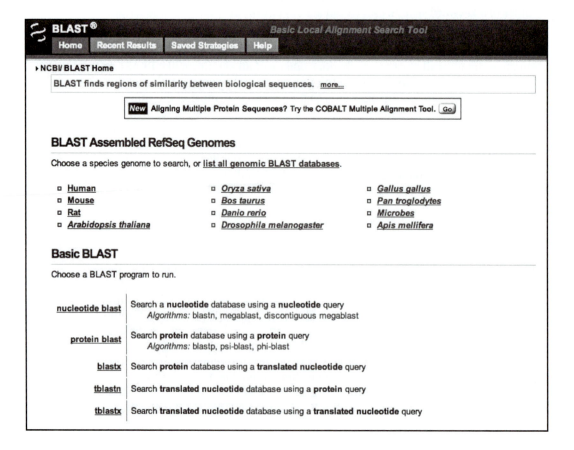

in suboptimal solutions, it is highly operational and applicable to certain problems. We will discuss other heuristic approaches when we detail phylogenetic tree searches in Chapter 8.

The quality of the alignment is represented by an *E*-value, which is similar to the statistical concept of a *p*-value with a slight change. A *p*-value is the probability of achieving a result as significant as the result obtained by chance. Thus a *p*-value of 0.01 for a sequence comparison indicates that there is a 1% likelihood of getting a match as significant as this match by chance. In contrast, the *E*-value determines how likely it is to get a match of a certain level of significance given a database of a certain size. Therefore, an *E*-value of 0.01 indicates that there is a 1% chance of randomly finding a match of that significance in the database. This difference between *E*-value and *p*-value is subtle and not completely important for understanding BLAST results. The main point to remember is that a smaller value (closer to 0) is better. It should be kept in mind that a smaller value is represented as a larger negative exponent in the *E*-value score. Another significant point is that the *E*-value is affected by the size of the database. Therefore, if one were to perform a BLAST search today against all of GenBank and then repeat the same search in a year (when the size of GenBank will have more than doubled), then the resulting *E*-value will be more significant.

The length of the aligned sequence has an impact on the overall *E*-value (**Figure 5.8**). The longer the sequence match, the smaller the *E*-value is in proportion. Note that both hits have perfect matches in their alignments. The upper hit has 52 residues in the matched alignment, and the lower hit has 140 residues in the matched alignment. Note that the lower hit has a smaller ($6 \times 10^{67} << 1 \times 10^{19}$), hence better, *E*-value even though the upper hit does not have any positions in which changes have occurred. These results make sense when we think of the *E*-values as the probabilities of finding a hit as good as it would be by chance alone. In other words, it would be more probable to find a hit by chance alone on a shorter sequence than on a longer one; hence, query sequences that hit longer subject sequences have smaller and better *E*-values.

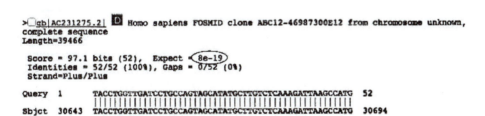

Figure 5.8 Alignments obtained by a single query. Ovals in the figure indicate the E-values for the two "hits." It can be seen that the length of the aligned sequence has an impact on the overall *E*-value. (Courtesy of the National Institutes of Health.)

There are several variations of BLAST for nucleotide and amino acid sequences

Since the nucleic acid sequence database contains accessions that code for genes with protein sequence accessions in the protein database, there are several ways to search the databases. The simplest method involves searching for either a nucleic acid sequence with a nucleotide sequence query or for a protein sequence with an amino acid query. The former is called a "blastn" search and the latter is a "blastp" search. The blastp search is the most straightforward because it compares the string of amino acids in the query with each and every protein in the protein database targeted. The blastn search is slightly more complex because, in addition to searching the database for the exact string of nucleic acids entered, it also searches for the query on the opposite strand of the DNA sequences in the database, a result of the reverse and complementary nature of the double helix of DNA. In essence, when a nucleic acid query sequence (in its 5′ to 3′ direction) is entered into the blastn search, two sequences are actually entered. For instance, if you enter the sequence 5′-GCGCATACGAGCTACGATAGCATCAG-3′ into the query, both this sequence and its reverse complement, 5′-CTGATGCTATCG-TAGCTCGTATGCGC-3′, will be searched.

The second type of search involves querying a protein sequence against a nucleic acid database. This search is called a "tblastn" search, for translated blastn, and it translates the entire nucleotide database in all potential protein translations that are possible. Since there are two strands of DNA and there are three reading frames for each strand (Chapter 2), there are $2 \times 3 = 6$ different potential protein translations of each nucleotide sequence that need to be queried. Because of the added complexity of the six different possible translations, a tblastn search takes significantly longer than a traditional blastn or blastp search.

The next type of search is a blastx search that searches a nucleotide sequence against a protein database. In this case, the query sequence is converted into its six potential translations, and all of them are queried against the database. After the translations, the search proceeds as a normal blastp search.

Finally, the most computationally complex search is called the "tblastx" search: it is used to query a nucleotide sequence against a nucleotide database, but with the comparisons being done on a protein translation of both the query and the database. Thus, the query sequence and the database sequences are first translated into their six potential protein translations. Next, all of these translated protein sequences are compared against each other. While this type of search can be useful for finding distant relationships between sequences (because protein sequence is more conserved than nucleotide sequence), it requires 36 times the computation of a traditional blastn search. Thus, there are five ways of BLASTing your way through the NCBI database (**Table 5.1**).

Performing a BLAST search is straightforward

After all of this talk concerning BLAST, it is time to actually perform a BLAST search. Return to the NCBI Blast home page (http://blast.ncbi.nlm.nih.gov/Blast.cgi). Three BLAST options can be selected for a nucleic acid query: blastn, blastx, and tblastx. To run the blastn application, click on the "nucleotide blast" button on the BLAST home page to bring up the search page (**Figure 5.9**). It contains a box in which the query sequence is pasted (see "Enter accession number, gi, or FASTA sequence"). Alternatively, the button below the query box can be used to upload a sequence from a file. A "job title" can be entered to keep track of the searches, but this is optional. Before the search is run, the database to be searched must be designated. The database can be selected under "Choose Search Set" via the pull-down menu next to the database flag (large oval in Figure 5.9). In order to locate a specific sequence from one of the databases in the pull-down menu, click on the desired database. If the search is meant to capture as many sequences as

Table 5.1 Five ways to perform BLAST searches.

Program	Description
blastp	Compares an amino acid query sequence against a protein sequence database.
blastn	Compares a nucleotide query sequence against a nucleotide sequence database.
blastx	Compares a nucleotide query sequence translated in all reading frames against a protein sequence database. Use this option to find potential translation products of an unknown nucleotide sequence.
tblastn	Compares a protein query sequence against a nucleotide sequence database dynamically translated in all reading frames.
tblastx	Compares the six-frame translations of a nucleotide query sequence against the six-frame translations of a nucleotide sequence database. Note that the tblastx program cannot be used with the nr database on the BLAST Web page because it is computationally intensive.

Source: National Institutes of Health

possible, click on the "Nucleoide collection nr/nt" button. If the search is limited to a sequence from completed genomes only, click on the "Reference genomic sequences (refseq_genomic)." The kind of blastn search then needs to be designated. The "Highly similar sequences (megablast)" search is the most efficient, and this box is checked (square in Figure 5.9). In comparison to a regular BLAST

Figure 5.9 BLAST page. Oval indicates the database being searched. Square indicates the optimization setting. (Courtesy of the National Institutes of Health.)

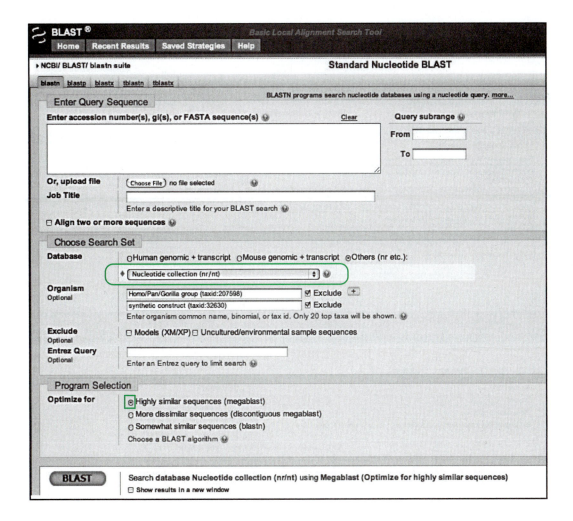

search, a megablast search will run faster, but it will only find closely related sequences and ignore more distant sequences. If the search is conducted with the above information entered, the default parameters are enabled. However, these can be altered by clicking on the "algorithm parameters" button at the bottom of the Web page.

Next we will take a DNA sequence of a gene and BLAST the sequence through the refseq genomic data base. The gene we will use is the human insulin-like growth factor 2 (*IGF2*) gene. It is considered to be one of the shortest genes in the human genome (342 nucleotides translates into 114 amino acids). First, the human *IGF2* nucleotide sequence is copied and pasted into the query box (top panel in **Figure 5.10**). For this BLAST search, we will use the default blastn parameters. To run the search, press the BLAST button at the bottom of the page. It usually takes only a few seconds to 1 min to search for a gene sequence like *IGF2*. Sometimes, though, when doing a tblastx or tblastn search, the search times can be rather long. Remember that these two searches search multiple reading frames. If a search is taking more than 5 min to complete, then most likely there is a problem and you should simply abort the search and start over. The search screen (bottom panel in Figure 5.10) will continuously autorefresh, updating you on the progress of the search.

When the search is completed, a Web page with the search information is generated. Either significant hits will be found or a "No Significant Hits" message will be displayed (in the latter case, the input sequence should be checked as it may not have been entered correctly). The references to the BLAST search engine and other background information are provided at the top of the results page. The actual search report begins with a line giving the length of the query sequence.

Figure 5.10 BLAST page with the IGF2 sequence pasted into the BLAST query box. (Courtesy of the National Institutes of Health.)

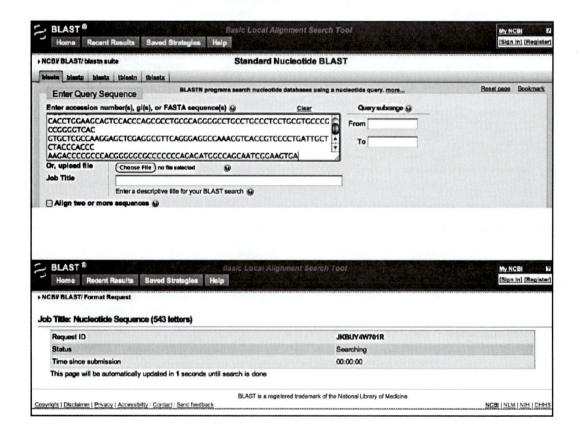

The first result is the "Distribution of BLAST Hits" (**Figure 5.11A**). In this case, we searched only completed genomes, so the hit list is rather small. The Web page shows the *E*-value range of the hit and the relative length of the hit. Colors signify the range of *E*-value. Each of the bars on the Web page representing a sequence is hyperlinked and can be clicked to obtain the actual alignment of the query sequence with the hit. Farther down the page is a button that allows one to view a tree of the sequences. Below that button is a table that summarizes accession numbers, descriptions of the hits, and further metrics (Figure 5.11B).

The actual alignments of the hits, ordered from best to worst, are summarized under the table. If a sequence from the database was used as a query, then that sequence will likely be the first one in the list. As an example, the hit for a gibbon (*Nomascus leucogenys*) sequence (marked with an oval in Figure 5.11B) is shown in **Figure 5.12**.

The data for the horse sequence in FASTA format can be obtained by clicking on the accession number. If all of the sequences are needed from the search, the buttons at the bottom of the search report page can be used (**Figure 5.13**). Click on the "Select All" button and each of the accessions is checked automatically. Next, press the "Get selected sequences" button and a page with a list of all of the sequences from the BLAST search will be generated. Alternatively, to search for specific sequences, place a check mark next to the selected sequences in the list. Once the list of hit sequences is collected, the FASTA files from all of the accessions can be downloaded in the same manner as described earlier in this chapter.

Other BLAST searches with amino acid sequence queries proceed in much the same way as the nucleotide search. A complete tutorial on BLASTing is available on the BLAST home page (Figure 5.7) under the "Help" tab at the top of the page. The first hot button under "Getting Started" is the "Blast short course" button, which leads to a tutorial on how to use BLAST and some of the more advanced aspects of BLAST. Under the heading "About BLAST" is a link to the BLAST glossary, a tool that is useful in communicating the results. We provide Web Feature 5 to familiarize the student with advanced BLAST techniques.

A

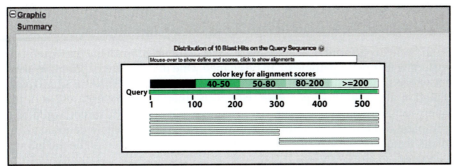

Figure 5.11 Results of BLASTing the *IGF2* sequence. A: distribution of hits. B: list of significant hits.

B

AK232525.1	Sus scrofa mRNA, clone:LVR010062F04, expressed in liver	678	678	100%	0.0	89%
AK240457.1	Sus scrofa mRNA, clone:UTR010075C08, expressed in uter	678	678	100%	0.0	89%
NM_213883.2	Sus scrofa insulin-like growth factor 2 (somatomedin A) (I	678	678	100%	0.0	89%
NM_001114539.1	Equus caballus insulin-like growth factor 2 (somatomedin A	673	673	94%	0.0	90%
AK397635.1	Sus scrofa mRNA, clone: SPL010008B10, expressed in sple	667	667	100%	0.0	89%
AK395456.1	Sus scrofa mRNA, clone: OVRT10009F08, expressed in ova	667	667	100%	0.0	89%

Figure 5.12 Alignment of the *IGF2* gene (labeled Query) and the gibbon "hit" (labeled Sbjct). (Courtesy of the National Institutes of Health.)

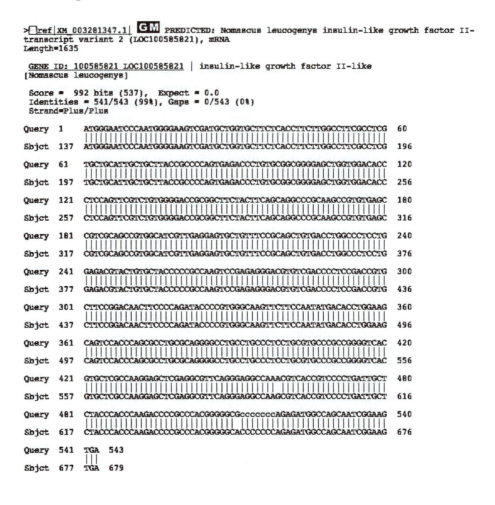

Figure 5.13 Obtaining sequences after a BLAST run. This strip appears above the BLAST-generated alignments after a search. One can select all of the sequences obtained by the search by checking the "Select All" box and pressing the "Get selected sequences" hot button. A new page with a list of all of the sequences with accession information will appear. (Courtesy of the National Institutes of Health.)

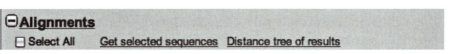

Whole genome alignments can also be performed

In addition to performing alignments of short sequences, much longer sequences, including complete genomes, can be aligned. One of the primary programs used for whole genome alignment is blastz, which is a modified form of BLAST well suited to aligning whole genomes with each other. In a typical BLAST run, the query sequence would be a gene (for example, IGF2), and the target would be a complete genome (for example, the mouse genome). In a whole genome alignment, the query could be the mouse genome and the target could be the human genome.

The goal of a whole genome alignment is to determine the extent of similarity between two genomes and to identify homologous locations between them. The regions of the genome that are found to be homologous are evidence of a lack of divergence between the genomes, and the comparison of several complete genomes leads to a picture of evolutionary conservation across the genome. For example, researchers at the University of California at Santa Cruz (UCSC) have performed a whole genome alignment of 28 vertebrate genomes. By scanning along this alignment, one can determine regions of strong conservation (shared by all 28) or great divergence (shared by only a few genomes). Many researchers have utilized this information for determining gene locations and annotating noncoding regulatory regions.

Summary

- Homology is the cornerstone of phylogenomics and can be established at two levels.
- The first level of homology is the level of genes in genomes, where the goal is to determine which genes or genomic regions are homologous.
- The second level of homology is the level of sequences in genes, where the goal is to determine which residue positions in genes, proteins, and noncoding regions are homologous (orthologous).
- Homology is determined by pairwise alignment of residues a sequence.

- Dynamic programming employs the steps of initialization, filling of the matrix, and traceback to automate the process of alignment.
- BLAST is a widely used tool to identify matches that are inexact but have high similarity. It is optimized for searching a query sequence of nucleotides or amino acids against a large database of sequences.
- Whole genomes can also be compared to determine their level of similarity.

Discussion Questions

1. In Michael Crichton's *The Lost World*, the DNA sequence of a dinosaur gene is displayed. We have translated the sequence into a protein and show it below.

```
MEFVALGGPDAGSPTPFPDEAGAFLGLGGGERTEAGGLLASY
PPSGRVSLVPWADTGTLGTPQWVPPATQMEPPHYLELLQP
PRGSPPHPSSGPLLPLSSGPPPCEARECVMARKNCGATAT
PLWRRDGTGHYLCNWASACGLYHRLNGQNRPLIRPKKRLL
VSKRAGTVCSHERENCQTSTTTLWRRSPMGDPVCNNIHAC
GLYYKLHQVNRPLTMRKDGIQTRNRKVSSKGKKRRPPGGG
NPSATAGGGAPMGGGGDPSMPPPPPPPAAAPPQSDALYAL
GPVVLSGHFLPFGNSGGFFGGGAGGYTAPPGLSPQI
```

Crichton claimed that this sequence was a dinosaur erythroid transcription factor, so we have downloaded the erythroid transcription factor protein from a chicken (like all birds, a chicken is, simply put, a dinosaur).

```
MEFVALGGPDAGSPTPFPDEAGAFLGLGGGERTEAGGLLASY
PPSGRVSLVPWADTGTLGTPQWVPPATQMEPPHYLELLQPPR
GSPPHPSSGPLLPLSSGPPPCEARECVNCGATATPLWRRDGT
GHYLCNACGLYHRLNGQNRPLIRPKKRLLVSKRAGTVCSNCQ
TSTTTLWRRSPMGDPVCNACGLYYKLHQVNRPLTMRKDGIQT
RNRKVSSKGKKRRPGGGNPSATAGGGAPMGGGGDPSMPPPP
PPPAAAPPQSDALYALGPVVLSGHFLPFGNSGGFFGGGAGGY
TAPPGLSPQI
```

Align the two sequences. You can do this by writing or typing the sequences on eight strips of paper (one strip for each line of the sequence) and arranging them on a benchtop so that the amino acids match or do not match. You can use a change to gap cost of 2 to do this. Is there anything strange with the best alignment you can get?

2. Go to BLAST and type in the dinosaur sequence from question 1. Do a BLAST search (blastp) to the protein database. Report your results.

3. Go to BLAST and the blastp (Protein Blast) option and type a short four amino acid sequence into the box of a musician or a musical group like NEIL (for Neil Young, Sedaka, Diamond or any number of musicians with the name) or FLEA. Make sure that the letters you use are in the amino acid alphabet (see Chapter 2). Do a Blast search for the word. Report how many hits you obtain and the quality of the hits. Now type in a five-letter stretch of amino acids like ELVIS or HAYDN or SLASH. Again do a Blast search and report the number of hits and E values. Now do a six-letter name like EMINEM or HANNAH (as in Montana) and a seven-letter name like NIRVANA or an eight-letter name like NICKCAVE. Report your results. Do you have any hypothesis for explaining the distribution of your results?

4. In *Jurassic Park*, the first book in his series, Michael Crichton took a simple sequence from the Internet and tried to pass it off as a dinosaur sequence. A reproduction of the sequence from the book is shown on the next page.

   What organism does this sequence belong to? This problem can be solved by conducting a blastn search with all or part of the sequence.

5. A scientist at the National Institutes of Health, Mark Boguski, wrote a paper on this sequence. Go to Entrez and do a PubMed search for "Boguski." How many publications are in the PubMed database for him? Are all the hits his? Now go to Google Scholar and search for Boguski. How many hits are there? Now restrict the Google Scholar search to Boguski Jurassic. How many hits now? Do any of these hits help you unravel the mystery in Question 1?

```
1      GCGTTGCTGGCGTTTTTCCATAGGCTCCGCCCCCCTGACGAGCATCACAAAAATCGACGC
61     GGTGGCGAAACCCGACAGGACTATAAAGATACCAGGCGTTTCCCCCTGGAAGCTCCCTCG
121    TGTTCCGACCCTGCCGCTTACCGGATACCTGTCCGCCTTTCTCCCTTCGGGAAGCGTGGC
181    TGCTCACGCTGTAGGTATCTCAGTTCGGTGTAGGTCGTTCGCTCCAAGCTGGGCTGTGTG
241    CCGTTCAGCCCGACCGCTGCGCCTTATCCGGTAACTATCGTCTTGAGTCCAACCCGGTAA
301    AGTAGGACAGGTGCCGGCAGCGCTCTGGGTCATTTTCGGCGAGGACCGCTTTCGCTGGAG
361    ATCGGCCTGTCGCTTGCGGTATTCGGAATCTTGCACGCCCTCGCTCAAGCCTTCGTCACT
421    CCAAACGTTTCGGCGAGAAGCAGGCCATTATCGCCGGCATGGCGGCCGACGCGCTGGGCT
481    GGCGTTCGCGACGCGAGGCTGGATGGCCTTCCCCATTATGATTCTTCTCGCTTCCGGCGG
541    CCCGCGTTGCAGGCCATGCTGTCCAGGCAGGTAGATGACGACCATCAGGGACAGCTTCAA
601    CGGCTCTTACCAGCCTAACTTCGATCACTGGACCGCTGATCGTCACGGCGATTTATGCCG
661    CACATGGACGCGTTGCTGGCGTTTTTCCATAGGCTCCGCCCCCCTGACGAGCATCACAAA
721    CAAGTCAGAGGTGGCGAAACCCGACAGGACTATAAAGATACCAGGCGTTTCCCCCTGGAA
781    GCGCTCTCCTGTTCCGACCCTGCCGCTTACCGGATACCTGTCCGCCTTTCTCCCTTCGGG
841    CTTTCTCAATGCTCACGCTGTAGGTATCTCAGTTCGGTGTAGGTCGTTCGCTCCAAGCTG
901    ACGAACCCCCCGTTCAGCCCGACCGCTGCGCCTTATCCGGTAACTATCGTCTTGAGTCCA
961    ACACGACTTAACGGGTTGGCATGGATTGTAGGCGCCGCCCTATACCTTGTCTGCCTCCCC
1021   GCGGTGCATGGAGCCGGGCCACCTCGACCTGAATGGAAGCCGGCGGCACCTCGCTAACGG
1081   CCAAGAATTGGAGCCAATCAATTCTTGCGGAGAACTGTGAATGCGCAAACCAACCCTTGG
1141   CCATCGCGTCCGCCATCTCCAGCAGCCGCACGCGGCGCATCTCGGGCAGCGTTGGGTCCT
1201   GCGCATGATCGTGCTAGCCTGTCGTTGAGGACCCGGCTAGGCTGGCGGGGTTGCCTTACT
1281   ATGAATCACCGATACGCGAGCGAACGTGAAGCGACTGCTGCTGCAAAACGTCTGCGACCT
1341   ATGAATGGTCTTCGGTTTCCGTGTTTCGTAAAGTCTGGAAACGCGGAAGTCAGCGCCCTG
```

Further Reading

Boguski MS (1992) A molecular biologist visits Jurassic Park. *Bio-Techniques* 12, 668–669.

Dynamic Programming Tutorial.
http://www.avatar.se/molbioinfo2001/dynprog/dynamic.html

Needleman–Wunsch slide show:
http://www.maths.tcd.ie/~lily/pres2/sld003.htm

Needleman S & Wunsch C (1970) A general method applicable to the search for similarities in the amino acid sequence of two proteins. *J. Mol. Biol.* 48, 443–453.

Smith–Waterman tutorial:
http://www.clcbio.com/index.php?id=1046

Smith TF & Waterman MS (1981) Identification of common molecular subsequences. *J. Mol. Biol.* 147, 195–197.

Multiple Alignments and Construction of Phylogenomic Matrices

One of the goals of alignment in phylogenomics is to produce a matrix of sequence information that can then be used in a population biology context or in phylogenetic tree-building. The information can then be used in evolutionary comparisons. This chapter describes how we get from raw sequences and genomes downloaded from the database to the phylogenomic matrix. We start out by discussing the differences between pairwise alignment and the ultimate goal of alignment for phylogenomics: multiple sequence alignments. There are several variables to be considered in the process of alignment for phylogenomic analysis. First, there are different alignment procedures and subsequently many different programs that can be used. In addition, there are different parameters that can be inserted into the various alignment programs, and changing these parameters can change the phylogenetic hypothesis that emerges from the alignment. The alignment space then becomes an important aspect of understanding the impact of alignment parameters on phylogenomics. We also examine studies that compare the various alignment programs that exist in the literature. Phylogenetic analysis can also be accomplished through a procedure called optimization alignment. This approach results in a phylogenetic tree without an alignment.

In order to perform phylogenomic analysis, we need a matrix of taxa and their respective sequences. Some phylogenomic studies now utilize hundreds to thousands of taxa and thousands of genes. In order to obtain such a matrix, the sequences from the different species need to all be aligned together through a multiple-sequence alignment. While we have already learned to obtain sequence alignments for two taxa or for two sequences, the task of multiple sequence alignment is more complex since it involves at least three and potentially many more sequences.

Multiple Sequence Alignment

The standard dynamic programming algorithm for aligning two sequences, discussed in Chapter 5, utilizes a two-dimensional matrix for the alignment. Following the same rules for an alignment of five sequences would require a five-dimensional matrix and the filling in of all boxes within every dimension of the matrix. Such high-dimensional matrices grow very large and unwieldy, and they quickly became too large for the limited memory of computers in the 1980s and 1990s, when alignment algorithms were developed. Even with current computers that can handle such matrices, complete filling and traceback of such a matrix would require an extreme amount of computation. For example, a matrix for aligning five 1000-bp sequences would contain 1000^5 entries. As the number of sequences increases, this number quickly becomes too great to even comprehend. Therefore, heuristic algorithms have been developed to perform multiple sequence alignment in a reasonable amount of time. Even so, some computer science advances have allowed for the use of approaches in sequence alignment where the full matrix is actually constructed (Sidebar 6.1). The average biologist, however, usually sticks to standard alignment approaches.

Sidebar 6.1. Smith–Waterman optimization and multiple alignments.

As detailed in Chapter 5, the Smith–Waterman dynamic programming algorithm is fast when two sequences are being aligned; however, it is too slow for the similar tasks of multiple sequence alignment and database searching. To enable a search to be completed in a reasonable amount of time, speed-increasing improvements or short cuts were made to the search algorithms. While these improvements greatly increase the speed of the basic local alignment sequence tool (BLAST), they do come with a cost. The Smith–Waterman algorithm is guaranteed to produce the optimal alignment between two sequences (or a multiple optimal alignment if there are multiple alignments having the same best score). In contrast, BLAST does not necessarily find the optimal alignment, and it might miss a good alignment because of one of the BLAST speed-increasing techniques. Even though this is a concern, it is generally dismissed because most researchers are happy to accept the small decrease in accuracy in exchange for the rapid increase in speed.

However, because the Smith–Waterman algorithm is deterministic, some researchers are actively trying to find methods to speed up this algorithm without sacrificing its accuracy. For example, a field-programmable gate array (FPGA) has been produced to run the Smith–Waterman algorithm. A FPGA is a computer chip that can be modified to run a specific algorithm in its hardware. By having the algorithm integrated into the hardware of a computer, the algorithm can be made to run efficiently and rapidly. In addition, researchers have developed implementations of the Smith–Waterman algorithm that run on graphics cards. The graphics card of a computer is designed to run the same calculation on multiple pieces of data at the same time. For example, a graphics card can rapidly take all of the images on a screen and magnify them. Since Smith–Waterman is a dynamic programming algorithm that can be easily split up into many small pieces where the same code is run on each piece, it is easily adaptable to a graphics card. Some of the graphics card implementations of Smith–Waterman have achieved very strong speed increases. In spite of these improvements, both the FPGA and the graphics card versions of this program remain niche tools that are mostly used by computer scientists.

One of the original tools for multiple sequence alignment that addressed the multidimensionality problem was CLUSTAL. This program was initially developed in 1988, and newer versions of it are still in wide use. At over 20 years old, this is one of the oldest bioinformatics programs still in use. While the progressive alignment approach inherent in CLUSTAL is rapidly being superseded by other programs using iterative approaches, we feel it appropriate to examine this approach briefly. An outline of the procedure for performing an alignment in CLUSTAL is given in Sidebar 6.2. In CLUSTAL, a fast alignment score is calculated between each pair of sequences to identify the relationships among them. These scores are then used to build a guide tree. Finally, the sequences are aligned in pairs iteratively to obtain the multiple alignments. Because of these optimizations, a CLUSTAL alignment is orders of magnitude faster than a straight Smith–Waterman multiple alignment. In addition, this iterative alignment procedure is much less memory-intensive, as there is no multidimensional matrix to store.

In addition to CLUSTAL, many other multiple-sequence alignment programs have been developed (Table 6.1). As with any algorithm, the various approaches make tradeoffs between speed and accuracy. The various alignment programs have been classified according to the characteristic approach each takes to the problem. The first major approach is characteristic of CLUSTAL, where the multiple alignments are built from pairwise alignments. These approaches are called **progressive**. The major problem with this approach is that a gap introduced early in the progressive alignment that is perhaps misplaced cannot be removed later in the process, and this can cause propagation of errors throughout the alignment process. To correct for this problem, **iterative** approaches have been developed that allow for the reassessment of gap placement throughout the alignment procedure. In iterative approaches, the multiple alignments are realigned to eliminate the propagation of errors. Probabilistic approaches have also been developed; this is a subset of progressive approaches, and some also have iterative properties.

for protein coding sequences when the sequences are highly divergent is that the reading frame is easily disrupted by the insertion of gaps that are not in multiples of three. Any inserted gapping not in multiples of three will throw off the reading frame of the nucleic acid sequences and would therefore be biologically meaningless. This problem can be alleviated by first aligning the amino acid sequences of protein coding genes and then using the amino acid alignment as a guide for the nucleic acid sequences. Several programs exist that accomplish the steps to enhance nucleic acid alignments in protein coding regions and result in files formatted for phylogenomic analysis.

Changing Alignment Parameters

In Chapter 5 we discussed alignment costs and "trivial" alignments. We showed alignments of two random strings of Gs, As, Ts, and Cs for a situation where the cost of accepting a change in base identity is so expensive that no gaps are inserted. We also discussed the "trivial" situation where the cost of inserting a gap is so inexpensive that only gaps are inserted to accommodate the alignment of the two sequences. This example demonstrates two concepts. First, any two sequences can be aligned, because of the nature of the alignment procedure. Second, there are a range of parameters that can be used to examine alignment for any set of sequences. This range goes from gaps with infinite cost to changes with infinite cost. We will call the range of costs that can be assigned for an alignment the "alignment space."

So which gap costs does a phylogenomicist choose? One strategy to deal with this problem of being able to change gap costs is to simply be consistent and use a predetermined gap and change cost for all of your alignments. The various programs that are used to accomplish alignment for either protein or DNA sequences have all been set up to have very "reasonable" default conditions (see Table 6.1). But simply taking the default costs and doing an analysis may not be the best approach. Another strategy is to explore the alignment space by conducting alignments at many different costs and comparing these alignments or allowing these multiple alignments to influence each other during phylogenetic analysis. These approaches are discussed below.

Multiple optimal alignments may exist

Even when one cost matrix is used, with a fixed ratio between the mismatch and indel costs, there is a possibility of achieving multiple optimal alignments. This occurs when there is more than one path through the alignment matrix that has the minimal score. Even in fairly simple alignment tasks, there are often going to be a multitude of paths through the alignment matrix that have equal costs. This phenomenon is important because different trees will be generated from the different equally optimal alignments in most cases. In many cases, the choice of an alignment is dictated by a fixed rule, such as always favoring gaps in one of the sequences when backtracking through an alignment matrix to produce only one top alignment. This is done so that the program can easily produce an optimal alignment. In other cases, the choice between which sequence should be given a gap is performed randomly.

Many optimization procedures have more than one maximal solution. This is an important problem in the building of phylogenetic trees, where there are often multiple trees having the same optimal score. Such a situation also occurs when a BLAST search is performed and returns multiple hits with the same E-value. In most cases a researcher will simply choose one of the results, due to the lack of a criterion to distinguish between the results. But it is more correct and more scientific to use empirical rules to choose among the alignments. Two methods that have been used to help deal with this problem are the culling and elision techniques.

Culling and elision are ways to explore the alignment space

Wheeler and colleagues recognized the problem of multiple optimal alignments and offered two solutions to the occurrence of different equally optimal alignments and optimal alignments from different gap and change cost inputs. We will discuss these procedures in the context of variable gap and change costs, but they are equally applicable to processing multiple optimal alignments with the same scores. Both solutions involve exploring the alignment space by doing alignments with different gap costs, saving the alignments, and manipulating the saved alignments, either by culling away columns of the alignment that vary from alignment to alignment or by creating a concatenated matrix of all of the alignments through elision.

Culling simply removes positions that differ from alignment to alignment. These are called "alignment-ambiguous" positions. Once alignment-ambiguous positions are determined, culling removes such columns to produce a reduced matrix without positions that are ambiguous (**Figure 6.2**). It has been shown that the culling approach improves phylogenetic estimation when applied to various computer-generated alignments, but it also reduces resolution because it reduces the amount of phylogenetic information from a matrix as a result of the culling.

Elision is also applied to multiple equally optimal alignments (Figure 6.2). Essentially, the sites in the alignments that are not ambiguous appear in the elided matrix more frequently than other columns, and these alignment-unambiguous columns have greater weight than ambiguous columns. We will discuss weights and character weighting in Chapter 8, but the procedure results in some positions having a larger impact on the phylogeny than others. The alignment-ambiguous columns are weighted by the number of times they appear in the alignments generated to explore the alignment space. For instance, consider a hypothetical pair of sequences, where 10 different gap costs are used to explore the alignment space (**Figure 6.3**). The second and fourth columns are the same in all alignments, so they are weighted by 10. In the first column, two arrangements of the information are found. In alignments 1 and 4–10, the first column is A/A, and in alignments 2 and 3, it is A/−. For this column, the A/A pattern is weighted 8 because it occurs eight times in the 10 elided alignments. The A/− pattern is weighted 2 because it occurs twice among the 10 elided alignments. The third column has the pattern A/− in alignments 1 and 4–10 and the pattern A/A in alignments 2 and 3. For this third column, the A/− pattern would be weighted 8 and the A/A pattern would be weighted 2 in the elision of the 10 alignments.

Figure 6.2 Culling and elision procedures for exploring the alignment space. Upper panel: two different alignments of sequences 1, 2, and 3. Alignment-ambiguous positions are highlighted in green. Note that columns with a white background do not vary between the two alignments. Middle panel: culling of the alignment in which alignment-ambiguous positions are removed). Lower panel: elision procedure is demonstrated. The two alignments are simply concatenated to give a single larger matrix with the number of columns equal to the sum of the number of columns of the two initial alignments. A space is inserted to indicate where the two alignments were concatenated.

```
sequence 1    GGGAATGCGCTAGC-AT-CG--A
sequence 2    GGGACTGCCCGATTGCTACGGGA
sequence 3    GG-ACTG--CGA-TGCTACG-GA

sequence 1    GGGAATGCGCTAGC-AT-CG--A
sequence 2    GGGACTGCCCGATTGCTACGGGA
sequence 3    GG-ACTGC--GAT-GCTAC-GGA

sequence 1    GGGAATGGTA-AT-C-A
sequence 2    GGGACTGCGAGCTACGA
sequence 3    GG-ACTG-GAGCTACGA

sequence 1    GGGAATGCGCTAGC-AT-CG--A GGGAATGCGCTAGC-AT-CG--A
sequence 2    GGGACTGCCCGATTGCTACGGGA GGGACTGCCCGATTGCTACGGGA
sequence 3    GG-ACTG--CGA-TGCTACG-GA GG-ACTGC--GAT-GCTAC-GGA
```

Another method for examining the alignment space was developed by Wheeler. In this approach, called sensitivity analysis, a data set for a group of species with a well-known phylogeny is generated and aligned at several gap and change costs. The alignment space for a data set in this way is explored with a wide range of alignment costs. The congruence of the trees generated from the alignments is then assessed via the known relationships among the phylogeny. The alignments with the highest congruence are judged to be optimal. This approach allows the researcher to determine which alignment parameters produce trees that are congruent with the well-known phylogeny and leads to gap scores that are reasonable for generating matrices for phylogenomic analysis. One drawback of this method is that it will not work unless there is a well-established phylogeny for the species under study.

Choosing an Alignment Program

The choice of which alignment approach to use is not simple. There is no widespread consensus on which of the myriad available alignment programs is "best." Different programs may work better for different data sets or for different purposes. It should also be pointed out that the results of alignment sometimes do not seem reasonable or optimal to the researcher's eye and adjustments to the computer generated alignment are made *a posteriori*. This latter approach simply requires that the alignment program chosen gets the alignment in reasonable shape for the adjustments made by the researcher.

Automated alignment results are frequently adjusted "by eye"

A popular, if not rigorously scientific, strategy is to align your sequences with an alignment program and then adjust the sequence alignments "by eye" to conform to what the researcher thinks is a more reasonable arrangement of the gaps and changes. In this strategy the alignment program is irrelevant because the researcher uses it to get in the ballpark and will simply change the arrangement of the nucleotides according to what looks good to the eye. For instance, the argument for adjusting alignments by eye is based entirely on the poor performance of alignment programs. The alignment step in phylogenetics has prompted the following statement from Morrison:

> "On the one hand, manual alignments are certainly tedious, and they may also be unrepeatable unless there is a clearly stated objective and a protocol. On the other hand, although automatic alignments are speedy and convenient, they are unlikely to fully reflect homology at least under the realistic conditions of <90% sequence similarity. This is an invidious position for a professional to be in, which is why sequence alignment could be considered more as a sport or an art than a science."

Morrison asks the interesting question "Why would phylogeneticists ignore computerized sequence alignment?" The answer to the question is simple: sometimes the alignments obtained simply do not look right. Landan and Graur have characterized the various kinds of problems automated alignment can have. They have also suggested a way to make what they call a reliability check (also called a "reality check") on sequence alignment. They call this the "heads or tails" (HoT) approach, in which the sequences being aligned are reversed to produce a new alignment. The original alignment and the reversed alignment can then be compared. If alignment is not a factor in the generation of information for phylogenetic analysis, then both the original and the reversed data sets should give the same overall phylogenetic hypothesis. Martin et al. show an example using the ankyrin gene for a phylogenetic problem, where the two trees generated from the original data and the HoT-generated alignments are quite different, yet the original information is the same (only reversed). This change is a result of some of the fixed rules used in sequence programs and the randomness of choosing between

alignment cost

1	ATAG A--G
2	ATAG --AG
3	ATAG --AG
4	ATAG A--G
5	ATAG A--G
6	ATAG A--G
7	ATAG A--G
8	ATAG A--G
9	ATAG A--G
10	ATAG A--G

Figure 6.3 Ten alignments of two very short sequences, ATAG and AG. Green boxes indicate the two alignments that have the "--AG" alignment in sequence 2. Note that the other alignments are all the same; it is assumed that these alignments are being tested with different gap and sequence change costs.

optimal alignments. The swapping of the two sequences leads the alignment program to make different choices among the many optimal possibilities. This result suggests that the phylogenetic information inferred from the original data set might be suspect. However, more recent work has challenged whether the HoT method is a viable way of determining the information content of an alignment.

Some researchers have taken the result that alignment programs often result in disparate results to mean that alignment "by eye" is an alternative. Morrison has examined the literature to determine how prevalent this approach is and whether or not specific criteria are employed when alignment by eye is employed. He showed that nearly three-quarters of biologists who use sequence information actually intervene by manually altering the alignments they obtain from computer programs. Unfortunately, his survey showed that alignment by eye is not done with specific criteria in mind in the majority of cases. The alignment by eye approach is also not tenable when genome-level information is considered. Even in small bacterial genomes, over 2 million base pair positions need to be eyeballed if a "by eye" alignment is desired. This state of affairs leaves us with relying on computerized alignments. Morrison points out that biologists need to make clearer the ideas of homology, so that computer scientists can design programs to produce more biologically reasonable alignments.

Alignment programs can be compared by use of benchmark data sets

Several authors have compared alignment programs used in bioinformatics. To do this, they have used what are called "benchmark" data sets. These data sets have been simulated or compiled where the expected results of alignment are known, or at least easily predicted. The idea with these benchmark approaches is that the known alignment can be used as a baseline for assessing the utility of the various alignment programs. The first of these studies, in 1999 by Thompson et al., suggested that the global iterative approaches were superior to other approaches for most kinds of data sets. However, after 10 years of testing alignment programs on benchmark data sets, one group concluded that "There is no single alignment program currently available that consistently outperforms the rest. Ranking is clearly dependent on the test dataset." For instance, as another group showed, global approaches outperform the local approaches in most scenarios. But when isolated motifs or full-length sequences are used, local approaches work better. Overall, two of the alignment programs did perform best on the largest number of test data sets (ProbCons on 18 out of 43 different test data sets and MAFFT on 9 out of 43 test data sets).

Another alignment question, not realized until recently because of the availability of large amounts of sequence, is how the different programs perform on data sets with large numbers of taxa. Liu et al. suggest that "as the number of sequences increases, the number of alignment methods that can analyze the datasets decreases. Furthermore, the most accurate alignment methods are unable to analyze the very largest datasets we studied, so that only moderately accurate alignment methods can be used on the largest datasets." Using several benchmark ribosomal RNA data sets, Liu et al. found that SATe and MAFFT perform the best on these large benchmark data sets, ranging from 110 to 78,000 taxa. Both SATe and MAFFT, however, require longer run times. Web Feature 6 gives detailed instructions on how to use CLUSTAL and MAFFT.

Dynamic Versus Static Alignment

Other methods that do not involve multiple sequence alignment may be used to generate phylogenies. **Optimization alignment** (OA) is a dynamic procedure, where alignments are not generated; instead, the approach gives "dynamic" homology statements that are relevant only at the nodes of a tree. The procedure is dynamic because the homology statements change from one tree topology to another, and hence the alignments are dynamic from node to node in the tree.

Fixed-state optimization (FSO) takes contiguous stretches of nucleotides and treats them like states of a larger character. For instance, it is possible with FSO that the entire 18S rDNA gene is a single character, or the first exon of a protein coding gene might be a single character. The actual sequences of the individuals would then be the character states. With different character states for all taxa in the analysis, this might seem a silly approach, as all taxa would have different character states and there would be no phylogenetic informativeness to the data. But if different costs are imposed for different transformations from one sequence to another, then there is information in the approach. Optimizing the fixed-state approach is done through a Sankoff optimization approach, and hence the optimal tree can be assessed and chosen. In this approach, alignment is needed in order to assess the cost of transforming from one state to another.

These two approaches and multiple sequence alignment (MSA) can be placed into context when alignment as coupled to phylogenetic analysis is represented in two dimensions (**Figure 6.4**). The first dimension is whether the homology statements are for single residues or for stretches of residues (such as genes or exons). The second dimension is whether the homology statements are static (as in MSA and FSO) or dynamic (as in OA). We will return to this two-dimensional representation when we discuss phylogenetic analysis of genome-level sequences. While FSO is an appealing choice for sequence alignment, it has the drawback of requiring vast amounts of computational power. These resources are often out of the reach of a typical researcher who performs alignments on a desktop computer.

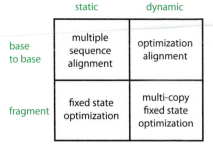

	static	dynamic
base to base	multiple sequence alignment	optimization alignment
fragment	fixed state optimization	multi-copy fixed state optimization

Figure 6.4 Modes of sequence homology. In this representation, alignment as coupled to phylogenetic analysis is represented in two dimensions. The first dimension is whether the homology statements are for single residues (base to base) or for fragments or stretches of residues (such as genes or exons). The second dimension is whether the homology statements are static (as in multiple sequence and fixed state optimization) or dynamic (as in optimization alignment). (Adapted from W.C. Wheeler, *Cladistics*, 17: S3–S11, 2001. Courtesy of John Wiley & Sons.)

Summary

- Multiple sequence alignment is more complex and requires different tools than pairwise sequence alignment.

- Selection and implementation of costs to produce an alignment are somewhat arbitrary. Most alignment programs allow for insertion of user-defined alignment parameters. Alignments should use multiple optimality criteria to explore the alignment space.

- Choosing an alignment program is an important first step. While we have described a progressive-only algorithm in this chapter (CLUSTAL), we suggest that researchers explore other algorithms and approaches to alignment. The recent development of iterative approaches has rendered the progressive alignments nearly obsolete.

- Changing alignments can change the phylogenetic hypothesis obtained, and methods are needed to accommodate this phenomenon. Cull and elision are techniques that can be used to deal with variation of alignments.

- Alignment by eye, while appealing to our visual nature as humans, is not an acceptable approach unless every step of the ocular alignment is recorded and reported. Alignments by eye can rarely be replicated from one researcher to the next, and repeatability is a hallmark of good science.

- Other methods that do not involve multiple sequence alignment, such as optimization alignment and fixed-state optimization, may be used to generate phylogenies.

Discussion Questions

1. Describe how the culling and elision procedures would be applied to a case where multiple optimal alignments were obtained by use of the same scoring function.

2. Take the following sequence alignments for three different alignment costs.

   (a) Cull them and characterize how many positions were culled as a result of being alignment-ambiguous.

   (b) Elide them and characterize the weights of each column in the two initial alignments.

```
alignment 1
sequence 1    CGATCGATCA
sequence 2    CGA-CG-TCA
sequence 3    CGACCGA-CA
sequence 4    CGAT-GACCA


alignment 2
sequence 1    CGATCGATCA
sequence 2    CGAC-G-TCA
sequence 3    CGACCGA-CA
sequence 4    CGAT-GACCA


alignment 3
sequence 1    CGATCGATCA
sequence 2    CGA-CG-TCA
sequence 3    CGACCGA-CA
sequence 4    CGATGAC-CA
```

3. Using Table 6.1, group the various alignment programs by their default parameters.

4. Discuss the differences and similarities of alignment procedures as depicted in Figure phy-6.01/6.1. Are there other ways to classify the alignment approaches implied by the figure?

Further Reading

Chenna R, Sugawara H, Koike T et al. (2003) Multiple sequence alignment with the Clustal series of programs. *Nucleic Acids Res.* 31, 3497–3500.

Gatesy J, DeSalle R & Wheeler W (1993) Alignment-ambiguous nucleotide sites and the exclusion of systematic data. *Mol. Phylogenet. Evol.* 2, 152–157.

Larkin MA, Blackshields G, Brown NP et al. (2007) Clustal W and Clustal X version 2.0. *Bioinformatics* 23, 2947–2948.

Linder CR, Suri R, Liu L & Warnow T (2010) Benchmark datasets and software for developing and testing methods for large-scale multiple sequence alignment and phylogenetic inference. *PLoS Curr.* 2 (November 18), RRN1195 (DOI 10.1371/currents.RRN1195).

Liu K, Linder CR & Warnow T (2010) Multiple sequence alignment: a major challenge to large-scale phylogenetics. *PLoS Curr.* 2 (November 18), RRN1198 (DOI 10.1371/currents.RRN1198).

Morrison DA (2009) Why would phylogeneticists ignore computerized sequence alignment? *Syst. Biol.* 58, 50–158.

Phillips A, Janies D & Wheeler WC (2000) Multiple sequence alignment in phylogenetic analysis. *Mol. Phylogenet. Evol.* 16, 317–330.

Sankoff D (1972) Matching sequences under deletion/insertion constraints. *Proc. Natl. Acad. Sci. U.S.A.* 69, 4–6.

Thompson JD, Higgins DG & Gibson TJ (1994) CLUSTAL W: improving the sensitivity of progressive multiple sequence alignment through sequence weighting, position-specific gap penalties and weight matrix choice. *Nucleic Acids Res.* 22, 4673–4680.

Wernersson R & Pedersen AG (2003) RevTrans: multiple alignment of coding DNA from aligned amino acid sequences. *Nucleic Acids Res.* 31, 3537–3539.

Wheeler WC, Gatesy J & DeSalle R (1995) Elision: a method for accommodating multiple molecular sequence alignments with alignment-ambiguous sites. *Mol. Phylogenet. Evol.* 4, 1–9.

Wise MJ (2010) Not so HoT - heads or tails is not able to reliably compare multiple sequence alignments. *Cladistics* 26, 438–443.

Genome Sequencing and Annotation

Since current sequencing machines cannot handle full chromosome-size pieces of DNA, the sequencing is performed upon small segments of the genome, which are then assembled. This assembly takes large computational resources and, despite the continuing advances in computer power, remains one of the most time-consuming tasks of whole genome sequencing. After a genome is sequenced, it is annotated to identify its functional regions by use of a variety of computational tools. In this chapter we will explain the approach to genome sequencing that was used for the Human Genome Project as well as the approaches that are currently utilized for next generation sequencing (NGS).

Genome Sequencing

At present, despite ongoing advances in computing power, sequencing machines cannot handle full chromosome-size pieces of DNA. Thus sequencing is performed upon smaller segments of the genome, which are then assembled. Various techniques have been developed to accomplish whole genome sequencing.

Small viral genomes were the first to be sequenced

The general technique for DNA sequencing that was first used on phage and bacterial genomes is the shotgun cloning method. Raw genomic DNA is broken down into small pieces and cloned into plasmids, which were then fed into the sequencing machine separately. Traditional Sanger sequencing machines (named after Frederick Sanger, who co-invented DNA sequencing) can produce reads of 500–800 base pairs (bp), so the fragments to be sequenced were this size. In order to increase fidelity, researchers would often use a larger piece of DNA and sequence both ends of the fragment so that they would be able to connect the two sequences. This technique is known as paired-end sequencing, and it is extremely powerful. For example, if 500 bp of sequence were produced from each end of a fragment of DNA of approximately 2000 bp (within a range of 1800–2200 bp), then it would be known that those two sequences were approximately 1000 bp apart from each other.

The small pieces are then reassembled to form the genomic sequence, via assembly techniques that take this 1000 bp gap into consideration. For eukaryotic sequences this technique was used, and indeed continues to be used, when small portions of a genome, such as a specific gene, are being investigated. Indeed the paired end approach, described in detail below, is being used extensively with next-generation sequencing approaches.

Bacterial artificial chromosome-based sequencing employs a "divide and conquer" strategy for larger genomes

The shotgun technique was thought to be sufficient for viral genomes, but was also deemed to be too difficult for a large eukaryotic genome like the human genome. The human genome is 3 billion bp in size, so to completely cover it with

500 bp reads, 6 million sequencing reads would be required. Thus the assembly of the genome would be the equivalent of a 6 million piece jigsaw puzzle. (In actuality there was a need for many times this number of reads, for statistical reasons that will be explained later in this chapter.) Even for a computer, a 10 million piece jigsaw puzzle where the junctions between the pieces are not clearly defined is very difficult.

To accommodate this concern, the initial strategy for the Human Genome Project consisted of breaking the big problem into smaller ones. That is, researchers would first map the genome into smaller and smaller units. For humans, and other eukaryotes, the genome is naturally organized into chromosomes, and these subdivisions were an effective way in which to break up the genome so it would be easier to handle with respect to generating information. For example, the genome could be broken into segments averaging between 0.2 billion and 0.05 billion bp; these are the approximate sizes of the largest (chromosome 1) and the smallest (chromosome 22) chromosomes in the human genome.

After the genome was separated by chromosome, an additional strategy was utilized. This approach involved dividing the chromosomes into smaller pieces that could be readily copied by bacteria. These pieces are known as bacterial artificial chromosomes (BACs), and as the name implies, they were analogous to bacterial chromosomes except that they contained a portion of a human chromosome. Each of the BACs was inserted into the bacterium *Escherichia coli*, which assumed the BAC was one of its own chromosomes and made numerous copies of it. Basically, the bacteria were co-opted as a biological photocopier for DNA. After the bacteria had made copies of the BAC, they were ground up and the BACs were extracted.

Portions of each BAC were then sequenced and restriction-mapped to provide a BAC map through a chromosome (a cartoon of a BAC map is shown in **Figure 7.1**). After these BACs had been mapped to the chromosome, each BAC was then fragmented into small pieces that were randomly sequenced and assembled as we have explained above. The use of BACs greatly reduced the size of the problem that needed to be tackled computationally. A BAC is usually several hundred thousand bp in size, so if one were using 500 bp reads, it would produce a few hundred or a few thousand sequencing reads that needed to be assembled. While it is not a trivial task to assemble these reads, it is much easier than trying to assemble tens of millions of reads. In essence, BAC-based sequencing is based upon the philosophy of "divide and conquer."

This BAC-based technique was successfully utilized for genomes such as *Caenorhabditis elegans* and *Drosophila*, which are each several hundred megabases in size. The drawback of this technique is the time and effort required to produce all of the BACs for the genome. For a human-sized genome, there is a need for thousands of BACs, all of which needed to be created, cloned, and sequenced individually.

The Human Genome Project

The BAC-based approach was the initial strategy adopted for sequencing the human genome. After the project was underway, J. Craig Venter decided that he would try to sequence the genome straight from the raw sequence without using BACs, in a technique known as whole-genome shotgun sequencing. Instead of keeping track of the fragments before sequencing, this method consisted of

Figure 7.1 Example of bacterial artificial chromosome map. Depiction of the chromosome (top); a physical map (middle); and the set of bacterial artificial chromosome (BAC) clones that "covers" the chromosome (bottom). The cross-hatching on the chromosome represents the centromere of the map. The arrows represent the BAC clones and indicate the direction in which the chromosomal sequence was inserted into the BAC cloning vector.

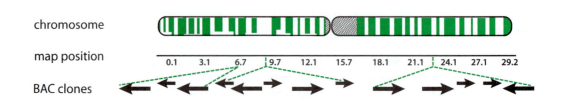

randomly shearing the genome into pieces of manageable size for sequencing and then reassembling the genome after the sequencing was completed. In fact, when Venter first proposed this approach for bacteria, his requests for funding were turned down because review panels felt that it would never work. However, while Venter's proposal to sequence the genome of *Haemophilus* with this method was under review, his laboratory completed the process remarkably quickly. This genome was relatively small at 2 MB in size. Dr. Venter next sequenced the larger *Drosophila* fruit fly genome using the shotgun technique. The *Drosophila* genome is 120 MB in size, which, while larger than the *Haemophilus* sequence, is still much smaller than the human genome. Therefore, people were skeptical that computers were up to the task of dealing with the much larger human genome.

This disagreement led Dr. Venter to break off from the public Human Genome Project and instead to found a company named Celera to privately sequence the human genome using the whole-genome shotgun technique. He intended to sequence the genome faster and then to patent it and make profit by charging people for access to the sequence. At the same time, the public genome project continued to use the BAC-based method. Eventually, in 2000 the two genome sequences were completed around the same time, and in a White House ceremony, President Clinton declared that there was a tie between the two sequencing projects. Incidentally, the Celera business plan failed, and as of now it is a small drug-development company.

There are, however, two important legacies of the Celera genome sequencing effort. First, the computational needs of shotgun sequencing, while substantial, were not impossible. Computer programmers were able to develop innovative algorithms that were able to take all of the sequencing reads and assemble them into a genome. At the same time, increases in computational speed following Moore's law (which states that the average computing power will double every 18 months) made the problem more tractable. The second legacy of the Celera experiment was that all succeeding genome projects for various species with large genomes have utilized a whole-genome shotgun strategy.

Statistical Aspects of DNA Sequencing

When sequencing is performed, the segment that is processed by the sequencing machine is basically chosen randomly from the library of available sequences. This is because the genomic DNA is shredded into small pieces and then the pieces are individually sequenced. In order to make sure that the entire genome is sequenced, one needs to make sure that each position in the genome is contained in at least one of these randomly selected segments. A good example for this problem is to take an ordinary die and determine how many rolls it will take to get each of the numbers between 1 and 6 at least once. Assuming the die is fair, each number has the same probability of being produced. But, since each role is random, it is likely that it will take more than 6 rolls to get the complete set of each of the 6 numbers. Try this for yourself and determine how many rolls it takes you to get all of the numbers.

Lander–Waterman statistics tell how many sequencing reads are needed to cover a genome

Two scientists, Eric Lander and Michael Waterman, noticed this statistical problem and wanted to determine how many sequencing reads would need to be randomly chosen to cover a genome. They produced the Lander–Waterman theory that answers this question in terms of the size of the genome and the sequencing reads. Interestingly, the theory was actually developed to analyze restriction maps of DNA that were the main technology for understanding DNA sequence before sequencing became readily available. Even so, the theory is readily applicable to sequencing.

Figure 7.2 Lander–Waterman calculation. On the *y*-axis, the number of contigs refers to the number of pieces that the assembled genome will theoretically contain. The *x*-axis describes the level of coverage of the genome sequence. (Reprinted from E.S. Lander, M.S. Waterman, *Genomics*, 2(3):231–239, Copyright 1988, with permission from Elsevier.)

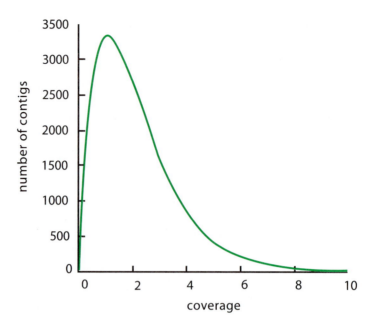

A plot resulting from the Lander–Waterman calculation is shown in **Figure 7.2**. On the *y*-axis, the number of **contigs**, or contiguous fragments, refers to the number of pieces that the assembled genome will theoretically contain. The *x*-axis states what level of coverage of the genome sequence will be produced. Of course these numbers are theoretical statistics, so one might actually get the entire genome to assemble into 1 contig at 2× coverage, but this is highly unlikely. Based upon this calculation, the sequencing of the human genome for the initial Human Genome Project produced sequence equivalent to 7.5 times the complete genome size.

In addition to the issue of random selection of sequencing reads, there is also the loss of overlap sequence that needs to be accounted for. If the sequencing reads are 500 bp in length, and they are expected to overlap by 20 bp for the assembly, then in reality, two assembled reads would only produce 980 bp of sequence rather than 1000 bp. While this may appear to be a small amount of sequence, it is not insignificant when dealing with a large genome.

Sequence Assembly

As we have mentioned, the assembly of sequencing reads is one of the main computational bottlenecks in genome sequencing. The basic technique of assembly is to take the sequence reads and attempt to identify overlaps between them. This overlapping of reads is due to the multifold coverage of the genome that is generally produced due to the Lander–Waterman calculations. As an example, take these three hypothetical sequencing reads.

```
1. AGCTGCAGCGCGAGCG
2. CGAGCGTGATGCATGC
3. CATGCAGCGATTAGAT
```

These three reads can be lined up in the following fashion

```
1. AGCTGCAGCG CGAGCG
2.          CGAGCG TGATGC CATGC
3.                        CATGC AGCGATTAGAT
```

and would therefore be assembled into the following contig:

```
AGCTGCAGCG CGAGCG TGATG CATGC AGCGATTAGAT
```

Obviously this is a simplification, since it involves only three sequences rather than the millions that are normally utilized in a eukaryotic assembly. With such a large number of reads, heuristics and indexing strategies are utilized to simplify the number of comparisons that need to be done between the reads. A straight comparison of each of a million reads with each of the other reads would require 1 million factorial comparisons, which is clearly unrealistic.

One key characteristic of an assembly algorithm is the amount of overlap required between sequence reads to make an assembly. In our example, we had only 5 or 6 bp of sequence to overlap, but in reality a much longer overlap is usually required. If the overlap is too short (such as 5 bp), then many incorrect matches will be made because the same 5 bp sequence occurs many different places within the human genome. On the other hand, if too long an overlap is required, then many true matches between sequences will not be assembled since there are no reads exceeding the required threshold. In practice, a researcher will utilize a variety of different overlap thresholds to determine which value is optimal for a particular genome.

In addition to the use of overlapping sequences, paired-end reads are greatly beneficial for assembly. Knowing that two sequences are a certain distance apart because they are from the same read allows for the assembly algorithm to fill in the distance between them and to help provide an anchor for the sequence. In practice, a mixture of different gap sizes is useful for properly assembling a genome to provide different levels of structure.

Paired ends are also important for assembling sequences that are highly repetitive. There are many regions of the human genome that contain heterochromatin or tandem repeats. The repetitive structure of these regions makes them extremely difficult to assemble by an overlap technique. If there are paired reads that span one of these repetitive regions, then even if the intervening sequence cannot be assembled, at least the length of the repetitive region is known and the gap in the assembly can be demarcated.

Next-Generation Sequencing

Recently, new sequencing machines have been introduced that have vastly greater throughput in sequence production as compared to the traditional Sanger sequencers. However, the reads produced by these machines are much shorter. In general, the reads produced by any of these products (Illumina, SOLiD, Ion Torrent) are less than 100 bp in length. These machines have been used to sequence the genomes of many humans as well as a variety of large eukaryotic genomes. Since the sequencing reads are so short, their assembly is complicated. If a 20 bp overlap is required to assemble two reads, then two adjacent assembled reads will only produce a sequence of 180 bp. This represents a 10% (20/200) loss of sequencing data. Fortunately, the computer programmers have come to the rescue again by developing innovative algorithms that can reconstruct a complete human genome from these short reads. One of the main tools to facilitate this assembly is the use of paired-end reads. In this technique, the two ends of a DNA fragment whose approximate size is known are simultaneously sequenced. Thus, pairs of sequence that are each separated from each other by approximately 2000 bp can be produced. This end-pairing was used in the initial human genome sequencing, but it has become much more important for the short reads produced by next generation sequencing. Even so, the complete resequencing and assembly of a human genome is very computationally intense, and speeding up this procedure is an active research area.

Since the speed and price of DNA sequencing are rapidly decreasing, it has been promised that, in the coming years, it will be possible to sequence a human genome for $1000 in a few hours. One key to this process will be the ability to assemble and annotate this genome rapidly to allow for its analysis. A list of genome assembly programs can be found in **Table 7.1**. The list is separated into assemblers that deal with longer input sequences, such as from Sanger sequencing, and two kinds of "next-generation" sequence assemblers. The first kind, called *de novo*, refers to assemblers that do not need an existing scaffold to accomplish the assembly. The second group requires an existing scaffold of sequences, either from other longer

Table 7.1 Genome assembly programs.

Program	Source
Sanger Reads	
TIGR Assembler	Institute for Genomic Research (TIGR)
phrap	University of Washington
Celera Assembler	Celera Genomics
Arachne	Broad Institute
Phusion	Sanger Center
Atlas	Baylor College of Medicine
Next-Generation Shorter Reads – *de novo* (no scaffold)	
ABySS	Canada's Michael Smith Genome Sciences Centre, C++ as source
ALLPATHS	Broad Institute
Edena	Illumina Genome Analyzer
EULER-SR	University of California, San Diego
MIRA2	MIRA (Mimicking Intelligent Read Assembly)
SEQAN	Tobias Rausch
SHARCGS	Max Planck Institute for Molecular Genetics
SSAKE	Canada's Michael Smith Genome Sciences Centre
Velvet	European Bioinformatics Institute (EMBL-EBI)
Next-Generation Shorter Reads Using Existing Scaffold	
BFAST	Nils Homer, Stanley F. Nelson, and Barry Merriman (UCLA)
Bowtie	Ben Langmead and Cole Trapnell
BWA	Heng Lee
ELAND	Illumina author Anthony J. Cox
Exonerate	Guy St C Slater and Ewan Birney (EMBL)
GenomeMapp	1001 Genomes project
GMAP	Thomas Wu and Colin Watanabe (Genentech)
gnumap	Brigham Young University
MAQ	Heng Li (Sanger Centre)
MOSAIK	Michael Strömberg (Boston College)
MrFAST	University of Washington
MUMmer	Stefan Kurtz et al. (TIGR)
RMAP	Andrew D. Smith and Zhenyu Xuan (CSHL)
SeqMap	Hui Jiang (Stanford)
SHRiMP	Stephen Rumble (University of Toronto)
SWIFT	Kim Rasmussen ands Wolfgang Gerlach

reads or from an existing full-genome assembly from a species closely related to the one being sequenced.

It is important to point out that the assembly algorithms for classical longer read sequences used in the sequencing of the human genome differ from methods now used to assemble short reads generated from next-generation technology. Specifically, the former uses an approach called "overlap consensus." This approach is described above and has difficulties with sequence repeats and redundancy. It takes the sequences "as is" and attempts to find the best match with the whole read. The latter approach is better for short reads and is called the de Bruijn graph approach. This approach uses what are called k-mers. A k-mer is a subset of the sequences being assembled. For example, the DNA sequence GATTTCAC has only one 8-mer in it. The sequence also has two 7-mers (GATTTCA and ATTTCAC), three 6-mers (GATTTC, ATTTCA, and TTTCAC), and so on. A set of subsequences is then created for all of the short reads in the data set. Once these subsequences have been generated at a specific k-mer value, the shortest path is computed through all of the sequences. This approach is computationally intense compared to the consensus overlap approach. The most important parameter in the de Bruijn approach is the k-mer value, as different k-mer values will give different assemblies.

The de Bruijn graph approach circumvents the problems of overlap consensus assembly. Rather than using the reads "as is" and trying to link them, the k-mers (all subsequences of length k within the reads) are computed and the reads are represented as a path through the k-mers. Such a paradigm handles redundancy better than the overlap consensus approach and makes the computation of paths more tractable.

In consensus overlap assembly, overlaps between reads are used to create a consensus sequence. In a de Bruijn graph, all reads are broken into k-mers and a path between the k-mers is calculated. Assemblers that use this approach have been released recently, including a modification of the long-read assembler Euler, ALLPATHS, and Edina. It is likely that the de Bruijn graph assemblers will improve vastly. Furthermore, the use of paired-read technologies will be better handled by de Bruijn graph assemblers, thereby creating potential for vast improvement in assemblies and alignments. Reports from primary papers that describe these methods support this view and indicate that the de Bruijn graph assemblers assemble larger contigs and can handle sequencing errors and complex genomes better than their counterparts.

Gene Finding and Annotation

Once a genome has been assembled, the next step is to identify and annotate the genes. This is a very difficult task for a eukaryotic genome, such as the human genome, in which genes constitute only a small portion of the genome. A further complication is the fact that eukaryotic genes contain exons and introns. The exons code for protein sequence, while the introns are excised from the mRNA sequence before it is translated into protein.

Gene finding can be accomplished via extrinsic, *ab initio*, and comparative approaches

There are three main approaches that are used for gene finding in a eukaryotic genome: extrinsic, *ab initio*, and comparative. The **extrinsic** approach uses known mRNA sequences or protein sequences as a guide to annotation. The assembled genome is searched against all known mRNAs and proteins in a database for that species via BLAST (see Chapter 5). All of the hits are collected and mapped. This approach requires extensive amounts of preliminary information derived from mRNA (usually from expressed sequence tags or ESTs; see Chapter 2) or from protein sequences.

The **ab initio** approach to gene identification involves certain "telltale" signs of genes in genomic sequence such as transcription-factor binding sites and initiation sequences. For example, certain features, such as a TATA box and a CpG island, are found upstream of coding genes. In addition, all proteins begin with a methionine having the ATG codon and end with one of the three stop codons within the same reading frame. A basic program to identify coding genes could therefore be constructed to look for a TATA box and a CpG island followed very closely by an ATG. This gene would then be predicted to continue until the first in-frame stop codon. In reality, *ab initio* gene-finding programs are much more sophisticated than this example, but they rely on the same principles.

The **comparative genomics** approach, which has become more popular with the availability of many whole genome sequences, takes advantage of the evolutionary conservation of genes in genomes. Since genes have many more constraints on their sequence variability than the noncoding portions of the genome, regions of different organisms that have highly similar sequences are likely to be coding sequences. Because natural selection will conserve certain regions of genes by purifying selection (see Chapter 13 for a discussion of selection at the genomic level), it stands to reason that these regions will be conserved in related organisms. The comparative genomics approach works by aligning the sequences of two or more genomes and identifying the conserved portions.

None of these approaches are perfect, and they are often used together to annotate a newly sequenced genome. The main choice of a gene-finding method concerns the amount of information that is available for the genome of interest. If there is extensive EST data for the organism, then the extrinsic approach will be very helpful. However, this is not the case for many organisms since large numbers of ESTs are needed to properly annotate a genome, and such libraries are not available except for well-studied genomes. If the genomes of many related species have been sequenced, then the comparative approach will be very fruitful. For example, 12 *Drosophila* fruit fly genomes have been sequenced and annotated. If one were to sequence a thirteenth *Drosophila* genome, it would be straightforward to annotate by use of the existing sequences. Even for genomes that are more distant, the comparative approach is fruitful. Finally, if no EST data are available and there are no annotated genomes of similar species, an *ab initio* approach can be used. Even though the extrinsic and comparative methods are preferred when the needed data are available, there is still a strong value to the *ab initio* approach. First, there will always be gaps in an EST library. For an EST to be collected, a researcher has to have examined a specimen expressing that mRNA. A gene that is, for example, only expressed in human fetus toes at embryonic day 55 will not be found in an EST library unless a researcher has examined mRNA from the toes of day 55 human fetuses. Secondly, all species are distinct from each other genetically. While it is often the case that the genes of two species are very similar and the differences between the species are caused by small variations in gene sequence or expression, this is not always the case. There can be genes that are unique to a particular species and do not have similar sequences in any other species. A good example of this is the large number of olfactory receptor genes (genes used in the process of smelling) that are present in rodents but are missing in other mammals. One final challenge for gene-finding programs is to identify exon–intron junctions. This is a very difficult task since the boundaries of exons are not very well defined. We demonstrate the utility of gene finding programs in Web Feature 7.

Gene functional annotation helps to determine gene function

After the genes in a genome have been identified, the task still remains to determine their function. This is usually performed via a comparative approach. Each gene that is identified (through any of the methods outlined above) is queried by use of a BLAST search against the GenBank database (see Chapter 5), which contains all of the known annotated DNA sequences. A gene that matches strongly to an annotated gene of a different species is usually given the name of the annotated gene.

Another method to determine the function of a gene is by analyzing the motifs that it contains. A protein motif is a small portion of a protein that performs a specific function and is generally well conserved. As a specific example, one of the most common DNA binding domains in a protein is the zinc-finger domain. While two proteins might be from species that are very distant from each other evolutionarily, their zinc finger domains would be very similar to each other. Outside of these domains, there is much less conservation of the protein sequence. In order to annotate proteins based upon the presence of conserved domains, there are Websites such as Pfam. This site contains multiple sequence alignments (see below) along with hidden Markov models that are used to classify protein domains. As of summer 2012, Pfam contained references for almost 15,000 protein domains.

Gene ontology facilitates the comparison of genes

With over 25,000 genes in the typical vertebrate genome and a tenth that in most bacterial genomes, how do we make sense of the functions of this many genes on a whole genome scale? Another problem that arises is that the gene names for one organism might differ greatly from the orthologous genes in another genome. How do we reconcile these naming differences, or do we simply put up with this tower of Babel?

One way to solve this problem is to create a new language or "ontology" for how we look at genes and their functions on a genomic scale. While this might seem a daunting task, it is no different than the task facing biologists in the naming of species about fifty years ago. Over the past several decades, taxonomic commissions have been established to regulate the naming of organisms, such as the International Commission on Zoological Nomenclature, the International Code of Nomenclature for Algae, Fungi, and Plants, and the International Committee on Taxonomy of Viruses. Without these commissions, each laboratory or scientist working on the common house mouse (*Mus musculus*) could have given this organism a different name. Imagine a scientific community where there are 100 different names for the same organism!

An important development over the last decade with respect to genome-level sequences, therefore, has been the development of a database that attempts to clarify the names and functions of genes. Ashburner once suggested that biologists would rather "share a toothbrush than a gene name." Unfortunately, this attitude has generated confusion in genetic studies due to the assignment of different names to the same gene, which significantly diminishes the ability to cross-reference studies. In response to this problem, Ashburner and his colleagues founded a consortium known as Gene Ontology or GO, established in 1999. The mission of the GO is "...to address the need for consistent descriptions of gene products in different databases." The GO Consortium developed a Website to consolidate the information on genes from a wide variety of model organisms, which is examined in detail in the Web Feature for this chapter. This Website offers many tools to access and process data, and although it requires some effort to master its navigation, there are several operations that can be viewed in the context of comparative genomics. There are several ways to browse the GO database, but the official browser is called AmiGO (Ami Gene Ontology). We will look at two specific functions of the GO database in the Web Feature for this chapter.

The database can be searched for the name of a gene or a gene product. For the example in the Web Feature exercise, we use the keyword "uroplakin" as a query. There are five types of uroplakin proteins. UPK1A and UPK1B are tetraspanins (they span the membrane four times in eukaryotes) and have related sequences; therefore, they come from the same gene family. The other three uroplakins (UPK2, UPK3, and UPK3A) are related to each other and come from a different gene family. In the membrane, UPK1A interacts with UPK2, while UPK1B interacts with UPK3 and UPK3A. We will use the *UPK2* gene as an example for this chapter. The GO search returns many genes from eukaryotes, because if the gene is part of a gene family, then members from the entire family are returned.

The most important aspect of the GO Website is its ability to classify a gene based on its function by use of the gene ontology framework. This system offers a **controlled vocabulary**, with a set of standard terms and definitions (that is, words and phrases) used for retrieving, storing, and indexing information. GO also defines the terms and delineates the **relationships** between the terms, which is characteristic of a **structured vocabulary**. This type of vocabulary denotes that any single gene has a hierarchical classification of names that is very similar to the way taxonomists and systematists classify organisms. The most specific terms for a gene are called "children," and the related term above it is called a parent. There are three superparents within the ontology classification scheme: biological process (BP), cellular component (CC), and molecular function (MF) (Sidebar 7.1). All genes in GO can be traced back to one or more of these three superparents. The GO system enables the classification of genes on the basis of function. The structured vocabulary for the *UPK2* gene can be accessed and the GO number for this gene is GO:0030855, which also represents a group of genes that have the function of epithelial cell differentiation.

Two hierarchies for *UPK2* are obtained from the GO Website (**Figure 7.3**). In the case of *UPK2*, the superparent is Biological Process (BP). Note that only two hierarchies of the more than 10 returned for *UPK2* are depicted. The children of the superparent BP are "cellular process" and "developmental process." These two terms are then parent to "cellular developmental process" and "anatomical structure development," respectively. While the GOs for a particular gene (in this case, *UPK2*) resemble a taxonomic hierarchy, any GO term can have more than one parent term or more than one hierarchy.

The GO terms can be used to explore lists of genes for function; for example, in conducting a microarray analysis of all of the genes that show up-regulation, the

Sidebar 7.1. Parent gene ontology categories.

A **cellular component** is just that, a component of a cell, but with the proviso that it is part of some larger object; this may be an anatomical structure (for example, rough endoplasmic reticulum or nucleus) or a gene product group (for example, ribosome, proteasome, or a protein dimer).

A **biological process** is a series of events accomplished by one or more ordered assemblies of molecular functions. Examples of broad biological process terms are cellular physiological process or signal transduction. Examples of more specific terms are pyrimidine metabolic process or α-glucoside transport. It can be difficult to distinguish between a biological process and a molecular function, but the general rule is that a process must have more than one distinct step. (A biological process is not equivalent to a pathway; at present, GO does not try to represent the dynamics or dependencies that would be required to fully describe a pathway.)

Molecular function describes activities, such as catalytic or binding activities, that occur at the molecular level. GO molecular function terms represent activities rather than the entities (molecules or complexes) that perform the actions and do not specify where or when, or in what context, the action takes place. Molecular functions generally correspond to activities that can be performed by individual gene products, but some activities are performed by assembled complexes of gene products. Examples of broad functional terms are catalytic activity, transporter activity, and binding; examples of narrower functional terms are adenylate cyclase activity and Toll receptor binding.

(Quoted from the Gene Ontology Website: http://www.geneontology.org)

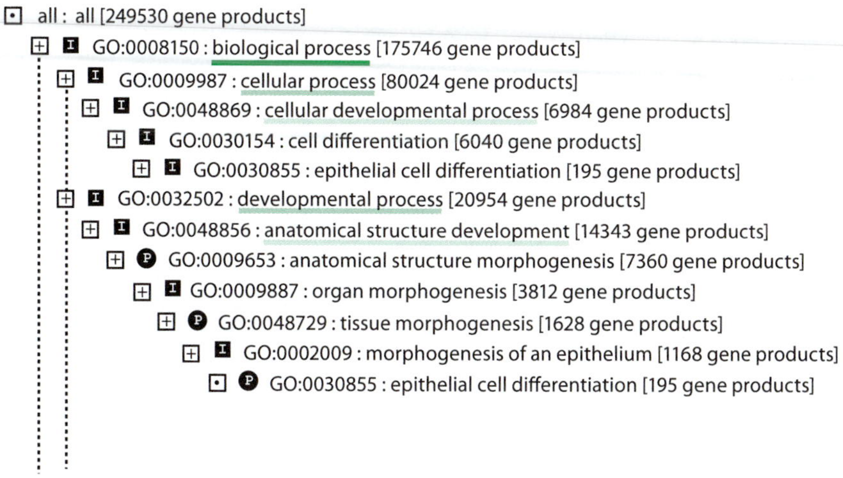

⊡ all : all [249530 gene products]
  ⊞ Ⓘ GO:0008150 : biological process [175746 gene products]
    ⊞ Ⓘ GO:0009987 : cellular process [80024 gene products]
      ⊞ Ⓘ GO:0048869 : cellular developmental process [6984 gene products]
        ⊞ Ⓘ GO:0030154 : cell differentiation [6040 gene products]
          ⊞ Ⓘ GO:0030855 : epithelial cell differentiation [195 gene products]
    ⊞ Ⓘ GO:0032502 : developmental process [20954 gene products]
      ⊞ Ⓘ GO:0048856 : anatomical structure development [14343 gene products]
        ⊞ Ⓟ GO:0009653 : anatomical structure morphogenesis [7360 gene products]
          ⊞ Ⓘ GO:0009887 : organ morphogenesis [3812 gene products]
            ⊞ Ⓟ GO:0048729 : tissue morphogenesis [1628 gene products]
              ⊞ Ⓘ GO:0002009 : morphogenesis of an epithelium [1168 gene products]
                ⊡ Ⓟ GO:0030855 : epithelial cell differentiation [195 gene products]

Figure 7.3 Two GO ontologies for UPK2. The dark green indicates the root parent "biological process." The medium green indicates the "children" for each of the two ontologies displayed: "cellular process" and "developmental process." These children both have children of their own, as indicated by the light green. (Adapted from AmiGO Hub, Web Presence Working Group. S. Carbon, A. Ireland, C.J. Mungall, S. Shu, B. Marshall, S. Lewis, *Bioinformatics,* 25(2):288–289, 2009. Courtesy of *Bioinformatics.*)

GO terms for the annotated genes can be obtained. These lists can then be used to see if a particular category or categories of genes are overrepresented in a micro-array experiment. We explore this approach further in Chapter 18.

A Phylo-View of Genomes Sequenced to Date

The combination of next-generation approaches with the Sanger sequencing during the last 5 years has led to huge advances in the number and kinds of genomes sequenced. In addition, the expressed sequence tag (EST) approach has been adapted to the next-generation technology, and a large number of transcriptomes for organisms exist. This section details the various genome sequencing projects that have been accomplished to date. We will leave metagenomics (sequencing ecological microbial communities) and metabiomes (sequencing organismal microbial communities) to Chapter 17.

In Chapter 2, we discussed the technologies involved in DNA sequencing. The advances in and costs of sequencing technology are important aspects of comparative genomics. Generating the first draft of the human genome required the expenditure of 3 billion dollars. However, with the development of new sequencing techniques, the prospect of producing less expensive whole genome sequences is a reality. Some liken the expansion of DNA sequencing technology to the growth of the Internet. George Church suggested that, just as the Internet had a tipping point in its expansion in 1993 because of cost reduction, so will DNA sequencing.

Currently, in order to compare different organisms, proxies for whole genomes are often used due to the current cost of sequencing These genome proxies consist of shallow genome sequencing (sequencing over 10 different genes), whole plastid genome sequences (mostly sequences from chloroplasts), microsatellites, or expressed sequence tag sequencing.

Perhaps the best way to demonstrate the extent of genomes sequenced is to use a phylogenetic tree or "tree of life." A schematic tree of life with the number of species sequenced for the major groups of taxa in the tree as of March 2012 is shown in **Figure 7.4**. Because taxonomic level varies from major group to group (that is, a family of plants might mean something completely different when compared to a family of animals) we have tried to focus on major organismal groups for this figure and have followed the NCBI major groups for the most part. The green oval gives the number of full plastid or mitochondrial genomes and the white circles give the number of full nuclear or bacterial chromosomal genomes. These numbers will no doubt have increased since our construction of this tree.

Figure 7.4 Phylogenetic tree. A wide variety of organisms and the number of finished genome projects for each group of taxa are shown. White circles and ovals designate chromosomal sequencing projects, while green circles and ovals designate organelle genomes (mitochondrial and chloroplast). lopho, Lophotrochozoa; ecdys, Ecdysozoa.

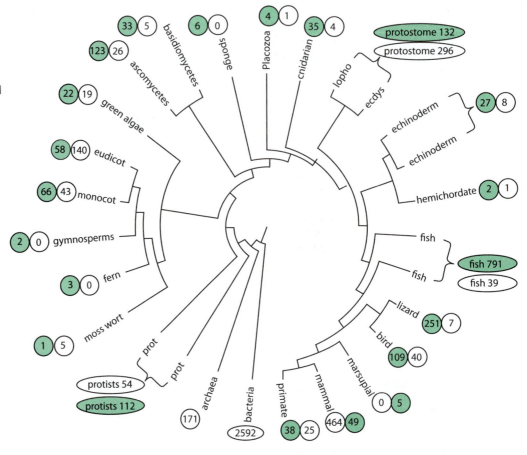

Summary

- Genome sequencing cannot at present be accomplished on a single large piece of DNA such as a whole chromosome, so the task is subdivided into smaller parts. New technologies are being developed to sequence longer and longer templates.

- Shotgun and bacterial artificial chromosome-based methods have been used to sequence larger genomes, as in the Human Genome Project.

- Lander–Waterman statistics estimate how many sequencing reads will be needed for an entire genome.

- Assembly of sequence fragments remains a computationally intensive step of the sequencing process.

- Genes can be located in the genome by extrinsic, *ab initio*, and comparative approachess. Gene functional annotation helps to determine gene function, and gene ontology facilitates the comparison of genes between organisms.

- Next-generation DNA sequencing will continue to get better and cheaper and will expand further the infomatics aspect of genomics.

- Genomes sequenced thus far can be viewed as part of a "tree of life."

Discussion Questions

1. Assemble the following fragments.

   GCATCGATGCATGCACATC

   ACATCGATGCATG

   ACGACGACTACGATGACAC

   TAGCTAGCATCG

   GCATGCTAGCTAGCATCGATCG

   ATCGACGACGACTAC

   TCGACGACGAC

2. Use a six-sided die to reconstruct a Lander–Waterman graph. This should be a class exercise. Work in groups of two. For one group do a Lander–Waterman experiment set, by rolling the die 6 times for 10 sets. Keep track of the number of times each value on the die comes up, and treat the die rolls as fragments of DNA. When contiguous numbers are next to each other, then this is a contig. For instance, a result might be the following:

   A. 1, 2, 3, 4, 5, 6

   B. 1, 1, 2, 3, 4, 5

   C. 1, 1, 2, 3, 5, 5

   D. 1, 1, 1, 3, 3, 5

   E. 2, 2, 2, 4, 4, 4

   F. 1, 1, 1, 1, 1, 1

   In the context of genome sequencing we could interpret these results as

   A. 1 contig (1–6)

   B. 1 contigs (1–5)

   C. 2 contigs (1–3, 5)

   D. 3 contigs (1, 3, and 5)

   E. 2 contigs (2 and 4)

   F. 1 contig (1 only)

The groups in your class should then do the following sets of rolls:

group	no. of rolls	sets of rolls
1	2	10
2	3	10
3	4	10
4	5	10
5	6	10
6	9	10
7	12	10
8	15	10
9	21	10
10	27	10

Now take all of the results from the class and graph the number of contigs versus the number of rolls from each of the groups in the class. The graph should have 6, 9, 12, etc., on the x-axis and there should be 1–3 contigs on the y-axis. Does this resemble the Lander–Waterman graph in Figure phy-7.02/7.2? If so, why? If not, why not?

3. Refer to Figure 7.4. The numbers for plastid, mitochondrial, and nuclear genomes were collected in March, 2012. Divide the taxa among the class and revise these numbers in class, using the searches of the NCBI Genome database.

Further Reading

Adams MD et al. (2000) The genome sequence of *Drosophila melanogaster*. *Science* 287, 2185–2195.

Ashburner et al. (2000) Gene ontology: tool for the unification of biology. *Nat. Genet.* 25, 25–29

Davidsen T, Beck E, Ganapathy A et al. (2010) The comprehensive microbial resource. *Nucleic Acids Res.* 38, D340–345.

Drosophila 12 Genomes Consortium (2007) Evolution of genes and genomes on the *Drosophila* phylogeny. *Nature* 450, 203–218.

Fleischmann R, Adams M, White O et al. (1995) Whole-genome random sequencing and assembly of *Haemophilus influenzae* Rd. *Science* 269, 496–512.

The International Human Genome Sequencing Consortium (2001) Initial sequencing and analysis of the human genome. *Nature* 409, 860–921.

Lander ES & Waterman MS (1988) Genomic mapping by fingerprinting random clones: a mathematical analysis. *Genomics* 2, 231–239.

Pearson H (2001) Biology's name game. *Nature* 411, 631–632.

Pop M (2004) Shotgun sequence assembly. *Adv. Comput.* 60, 193–248.

Pop M, Salzberg SL & Shumway M (2002) Genome sequence assembly: algorithms and issues. *IEEE Comput.* 35, 47–54.

Pop M, Phillippy A, Delcher AL & Salzberg SL (2004) Comparative genome assembly. *Briefings Bioinf.* 5, 237–248.

Roberts M, Hunt BR, Yorke JA et al. (2004) A preprocessor for shotgun assembly of large genomes. *J. Comput. Biol.* 11, 734–752.

The 1000 Genomes Project Consortium (2010) A map of human genome variation from population-scale sequencing. *Nature* 467, 1061–1073.

Venter JC et al. (2001) The sequence of the human genome. *Science* 291, 1304–1351.

Tree Building

As we discussed in Chapter 3, trees have become an important aspect of most of modern biology. This chapter examines the methodology used in applying the information obtained about organisms to the creation of branching diagrams. Tree building involves implementing step-by-step procedures, or algorithms, that receive input data, pass the input data through the finite steps of the procedure, and provide a numerical or graphical solution at the end. Tree building can be thought of in two dimensions. The first dimension concerns whether characters or similarity are used to construct the tree. The second concerns how the tree is built. One possible way to build a tree is by use of a deterministic algorithm that has a single cycle of computation which will yield a single solution. Alternatively, an iterative way to build a tree uses an approach with sub-algorithms that produce many solutions that are then sorted using an optimality criterion (or criteria). The three major approaches to phylogenetics that we will discuss—distance, parsimony, and likelihood analysis—fit into this two-dimensional scheme. Distance analysis compresses the original information from an organism, whether it is anatomical, behavioral, or molecular, into an overall similarity measure that is then processed by algorithms like UPGMA or neighbor joining. The algorithms used for distance (sometimes called phenetic) analysis are a series of steps that attempt to place species that are more closely related to each other into clusters. Parsimony and maximum likelihood analysis use characters like the four bases of DNA sequences or the 20 amino acids found in proteins. These two approaches, along with the distance-based approach called minimum evolution, then look at all the possible trees (or use programs that do this heuristically) for the taxa being examined. For parsimony analysis, the procedure attempts to map the characters onto the various trees and then chooses the tree or trees that take the fewest steps to accomplish the mapping. For likelihood analysis, the procedure uses a model of sequence change to estimate the likelihood, given the various trees and the model that is used. Minimum evolution assesses the total distance that would be needed for a particular tree topology and chooses the tree where the amount of distance needed is minimal. Before going into the mechanics of each of these methods we discuss the nature of the characters involved in phylogenomics (nucleotides and amino acids). Because nucleotides and amino acids have been studied in an evolutionary context for half a century, and because the biology of how these characters change are thought to be well-known, models have been developed that describe how sequences change. These models have been incorporated into phylogenetic analysis, not only for estimating the distances between species for distance analysis and likelihood analysis but also into parsimony analysis as schemes for how characters are transformed during evolution. The difference between models and character transformation is subtle, and both have become integral parts of phylogenetic analysis.

Distances, Characters, Algorithms, and Optimization

Tree building is the process of organizing entities into a hierarchical branching structure (see Chapter 3). The best way we have found to describe the somewhat

Figure 8.1 Schematic drawing of the structure of analysis space in phylogenetics. The analysis space takes the form of a 2 × 2 matrix, where distances and characters are two classes of data source, or not, and optimality criterion are approaches to phylogenetic analysis.

	distances	characters
no optimality criterion	neighbor joining	
optimality criterion	minimum evolution	parsimony likelihood Bayes' rule

bewildering complexity of data analysis in phylogenomics is to look at the problem from two perspectives or "dimensions" (**Figure 8.1**). The first dimension is whether characters or distances are used, and the second takes into account whether an optimality criterion is used. The main distinction between using similarity versus characters is that distance methods examine all of the pieces of data collectively, while character methods examine each piece or column of data individually. These differences between distance- and character-based approaches result in the structuring of how data are compiled and input into the different analyses. Both approaches use a set of aligned sequences as input. Distance methods compress all of the individual character information into a single metric for each pair of species, resulting in an asymmetric matrix of pairwise distances of each taxon in the analysis to all others. Character-based approaches use the raw sequence data as input into the analysis (see Figure 8.1).

In the second dimension, phylogenetic methods can be classified on the basis of whether an optimality criterion is applied. The difference between these two kinds of approaches is that the non-optimality approach simply follows a set of rules and goes through only one set of instructions before stopping. Therefore, this kind of approach is very fast because it simply proceeds to the final answer through the steps inherent in the deterministic algorithm. Several authors have pointed out that the result is not really a phylogenetic tree but rather a phenogram that summarizes the pairwise distances in the initial similarity matrix. Others argue that if the similarity measures used to generate the phenogram are reasonable indicators of relatedness, then the resulting phenogram could also be a reasonable and reliable representation of phylogeny. The optimality criterion approach, on the other hand, compares all possible solutions (that is, all possible trees for a given number of taxa) with an objective optimal solution in mind. Optimality criterion approaches are greedy with respect to analysis time because they compute costs for a large number of different solutions and then impose a comparison step, where the optimality criterion is compared for all of the trees assessed and used to pick the most reasonable solution.

The number of trees grows with each additional taxon

When using an optimality criterion, the cost of every possible tree for the entities involved needs to be assessed. This aspect of optimality approaches makes these methods a computational nightmare for larger numbers of taxa. The generalized equation for determining the number of bifurcating trees (where each branch point has exactly two new branches) is shown below:

$$(2n-3)!! = \frac{(2n-3)!}{2^{n-2}(n-2)!}, \text{ for } n \geq 2$$

where n is the number of entities in the tree. As you can see here, if $n = 3$, then the number of trees equals $(6-3)!/2(1!) = (3 \times 2 \times 1)/2(1) = 6/2 = 3$. For four taxa, the number of trees would be $(8-3)!/2^2(2!) = 5!/4(2) = 120/8 = 15$, and so on.

While three taxa can only produce three possible rooted trees (see below), there are over 2 million possible rooted trees that can be produced from 10 taxa. These numbers rapidly increase, and with only 20 taxa, the number of possible taxa exceeds the number of seconds in a billion years. This means that if a researcher were looking at trees of 20 taxa and could look at 1 tree per second, it would take him or her over a billion years to examine each tree individually. Fortunately, there are fast algorithms that have been developed to construct trees with large numbers of organisms in a reasonable amount of time, and we will discuss some of these algorithms in this chapter.

Trees can be rooted by several methods

In the above calculations, we have used the term "rooted." What does rooting mean and how do we do it? When we build trees, *a priori* we select the subjects or taxa we want to analyze. These taxa are called "ingroups". For instance, if we are interested in three taxa, a, b, and c, we call these three taxa the **ingroup**. Without further information or input into the analysis, we can infer only that these three entities are equally related to each other in a single unrooted tree (**Figure 8.2**). Rooting simply means taking an unrooted network or tree and selecting a place where the ancestor of all the ingroups connects to the tree. We can do the rooting arbitrarily (although this would circumvent some of the purpose of doing phylogenetic analysis), or we can use more analytical approaches to root the tree. The best way to root this tree is to use an **outgroup**, a fourth taxon that is known to be "outside" or less related to the three ingroup taxa than any of the ingroup taxa are to each other. In this way, a tree is constructed from both ingroup and outgroup taxa and the root is then placed between the ingroup and the outgroup.

Outgroups are always outside the group of organisms for which relationships are examined (the ingroup). Outgroups allow polarization of the changes that have occurred in the character state data because the outgroup, by definition, determines which characters are more like the outgroup and identifies which taxa have character states that imply more recent common ancestry with the outgroup.

Another way to root a tree is called "midpoint rooting." This approach estimates the distances between all pairwise comparisons in the tree. The longest pairwise distance is then halved and the point that is midway between the two taxa in the comparison is chosen as the root for the tree. This midpoint approach is illustrated in **Figure 8.3**.

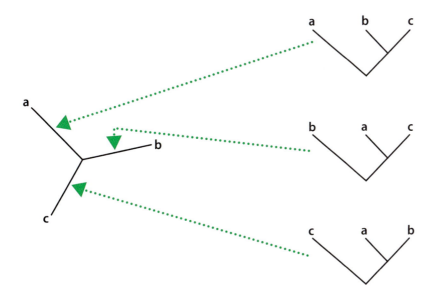

Figure 8.2 An unrooted tree for three taxa (a, b, and c). Arrowheads point to three possible rooting points. The roots can be imposed by adding an outgroup or by arbitrary rooting at the arrowheads. Dotted lines from each arrowhead lead to the topology of the three taxa if a root is placed in these locations.

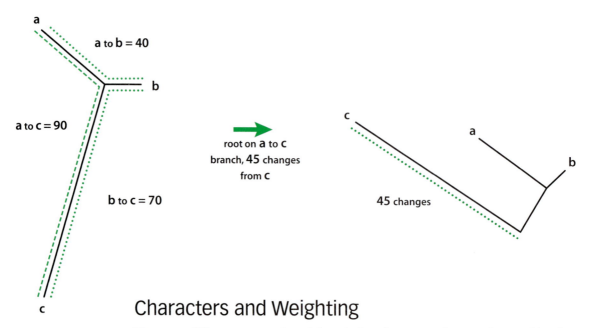

Figure 8.3 An example of midpoint rooting. The tree to be rooted (left) has three taxa (a, b, and c), and the length of the branch indicates the amount of change that has occurred on the branch. Dotted lines indicate the two smaller paths of the tree. The path from a to b is 40 changes, and the path from b to c is 70 changes. The dashed line indicates the longest path (a to c), which is 90 changes. The arrow points to the tree (right) that has been rooted at the midpoint (45 changes on the path from a to c). The dotted line on the tree is 45 changes long.

Characters and Weighting

There are different categories of descriptive characters that can be used in phylogenetic analysis. Characters can be either **discrete** or **continuous**. Discrete characters can be **multistate** or **binary** (two states). A binary character would simply be scored as present or absent for each taxon (usually 0 for absent and 1 for present). For example, a binary character for vertebrate animals would be the presence of wings. This character would be present for birds (1) and absent for most mammals (0). A discrete character can be a numerical or descriptive variable, "10 teeth" versus "20 teeth" or "long" versus "short", but these need to have a certain defined number of states for the tree-building algorithms to utilize them. A multistate character has more than two possible states. For instance, 0 legs versus 2 legs versus 4 legs would be a multistate character with three states (0, 2, and 4).

In contrast to the discrete characters, a numerical or continuous character is not limited to a defined set of positions. An example of a continuous character would be weight. Organisms on Earth have a very great range in weight, from almost immeasurably light bacteria to extremely heavy blue whales. Thus, a character for weight would have a wide range and generally not be discrete. Even so, researchers will often create bins to artificially discretize their numerical data to simplify it and allow for easier analysis. This discretization would involve marking each organism for having a weight in a particular range: between 1 and 100 grams, between 100 and 1000 grams, etc., all the way up to including the weight of a blue whale. After discretization of the data, there might be only 10 possible values for the character rather than the almost infinite true number of values. This approach is called gap coding.

Character states in molecular data may include the presence of genes and the sequence of nucleotides or amino acids

For molecular data, the presence or absence of a particular gene in a genome would be considered a binary character, since an organism either possesses or lacks the gene. Another kind of discrete character in molecular biology would be the nucleotide or amino acid present at a particular location in a gene sequence. There are four nucleotides in DNA, so there would be four character states. Some molecular systematics argue that a fifth character state exists for DNA, corresponding to the "gap" introduced into DNA sequence alignments to maintain primary character homology. For amino acids there are 20 amino acids that can be found in any position in a protein sequence, and hence there are 20 character states (or 21, if gaps are counted) in amino acid data.

An example of a numerical character in genomics would be the expression level of a gene. This expression can vary greatly (see the discussion concerning microarrays) and is generally expressed as a real number. But, as with anatomical numerical variables, gene expression is often reduced to a simple discrete variable by calculating the ratio between the observed expression and a defined reference. In this case, the expression would be counted as 2× or 3× the reference level, for example. In some cases, gene expression data are simplified even more to become a binary variable where a gene is counted as either expressed or not expressed (that is, turned on or turned off).

Some discrete and numerical character states are ordered

Discrete and numerical character states can be **ordered**. Certain character state transitions can be thought of as being more likely to occur than others. For instance, for insects, gaining wings is thought to be much more difficult evolutionarily than losing them. A good simple example of thinking about ordered characters in this way is called Dollo parsimony. There are two kinds of Dollo parsimony, "up" and "down." Up Dollo parsimony assumes that regaining a complex attribute state ($1 \rightarrow 0 \rightarrow 1$) is not allowed, and only single $0 \rightarrow 1$ and multiple $1 \rightarrow 0$ state changes are allowed. In essence, this method gives greater weight to single $0 \rightarrow 1$ changes in a lineage and to any $1 \rightarrow 0$ changes than the multiple $1 \rightarrow 0 \rightarrow 1$ kinds of changes. Down Dollo parsimony is simply the reverse reasoning.

When attributes exist as multistate or continuous, the character states can also be ordered. For instance, for a morphological attribute with three states—absent, small, and large (A, S, and L)—this character can be coded so that the progression from $A \rightarrow S \rightarrow L$ is the best transition based on, say, knowledge of developmental biology. This ordering amounts to giving $A \rightarrow S$ and $S \rightarrow L$ changes greater weight than $A \rightarrow L$ changes. Continuous character states, for example, values of 0.23, 0.45, 0.56, and 0.67 mm for the length of the wings of an insect, can also be arranged as ordered character states, $0.23 \rightarrow 0.45 \rightarrow 0.56 \rightarrow 0.67$. If the researcher knows about the developmental biology of the organism and can justify the assumption that size follows an incremental trajectory, then this ordered arrangement of the continuous character state data can be imposed.

This approach of ordering characters is controversial since it presupposes that it is more likely for morphological characters to first appear small (S) and then to grow larger (L). It might be possible that this character appeared first in a very large state and then decreased in size over evolutionary time. But this assumption is generally acceptable for traits where there is very strong external evidence that they were incorporated in an ordered manner.

Characters can be weighted relative to one another

Another distinct kind of weighting involves choosing certain kinds of characters (but not character states) that appear to be more informative than others, based on experience or from an understanding of the biology or chemistry of certain changes and giving those characters more weight. An example of this approach can be found in molecular data sets where a systematist might want to weight stem regions of structural RNAs as more important than loop regions. This is because the stem regions of the RNA are thought to be much more important for binding and reactivity than the loop regions, and hence would be more stable. A molecular systematist might decide to count the stem positions in a dataset as being, say, 5 or 10 times more important than loop positions.

The relative weight of certain classes of characters (sometimes called **process partitions**) has been a subject of intense debate. The most pointed discussion centers on organismal systematics, where authors argue for the weighting of morphological or anatomical attributes over molecular attributes. This argument is based on the idea that anatomical attributes are more reliable than molecular data because they tend to converge less frequently than molecular ones do. The logic for this claim is that anatomical characters converge more rarely than molecular ones, or that anatomical characters are more carefully chosen than molecular characters so that they are less prone to convergence. DNA sequences can back-mutate rather easily and converge in character state, whereas an anatomical character like the presence or absence of wings is much more stable over longer evolutionary periods. However, as we will see in subsequent chapters, anatomical characters can and will mislead phylogenetic inference.

Which characters should be used?

Put simply, there are two ways to think about the character state information. Either the characters are equally weighted, or some form of departure from equal weighting is imposed. Since living organisms have a limited but enormous number of attributes that can be exploited for phylogenetic analysis, we also need to discuss the quality of these characters we use in phylogenetic and phylogenomic analysis. Because different regions of the genome evolve at different rates, some phylogeneticists suggest that we cannot make universal use of all regions of the genome for all problems. In other words, some regions of the genome might be more appropriate for very closely related organisms, and other regions might be more appropriate for distantly related organisms.

In addition, knowledge of the molecular biology of sequences that we use in phylogenomics can be applied to assess the utility of certain sequences for specific phylogenomic problems. The knowledge we have about molecular processes can therefore be used to model the evolutionary process and hence use the model to more precisely utilize the molecular information for reconstructing phylogeny. This process involves understanding how we might prefer some characters over others or, in other words, how we might weight characters or model processes that affect the quality of our data.

Before the advent of molecular biology and the availability of tools to investigate genetic characteristics of organisms, researchers were restricted to the use of anatomical characters for studies. These characteristics formed the basis of nearly all phylogenetic trees up until the early 1980s, when systematists really started to use DNA sequence information in their trees. It might at first seem that anatomical characters that can be measured and quantified are somewhat limited in number, and indeed, the average numbers of anatomical characters in such early studies are usually double-digit and occasionally triple-digit. Even so, some very robust phylogenies were constructed, for example, for the major relationships of plants, mammals, and many other organismal groups. Aspects of these morphologically based phylogenies have held up and are still recognized as valid even after the introduction of molecular techniques.

The choice of which characters to use in a phylogenetic analysis is usually based on how variable the characters are. The amount of sequence divergence relative to divergence time for a rapidly evolving gene and a slowly evolving gene in a group of closely related organisms is shown in **Figure 8.4**. Note that the slowly evolving gene is linear all the way out to 1 billion years (1000 million years ago [MYA] or 10^9 years), while the rapidly evolving gene "plateaus" at about 50 MYA (or 5 $\times 10^7$ years ago). The plateau represents the fact that the DNA sequence is saturated or that multiple hits are occurring in the sequences at the same positions. The slowly evolving sequence does not plateau because multiple hits at the same position in the slowly evolving molecule are not accruing. This figure suggests that when using DNA sequence information for phylogenetics, an examination of the relationships of major animal groups (say sponges, cnidarians, insects, and vertebrates) would not focus on rapidly evolving molecules such as the D-loop of the mitochondrial genome. Such a study would probably rely on a slowly evolving source of characters like the 18S ribosomal RNA gene, which evolves extremely slowly. On the other hand, a study of sibling species of *Drosophila* would avoid information that was slow to change over evolutionary time, such as the nuclear

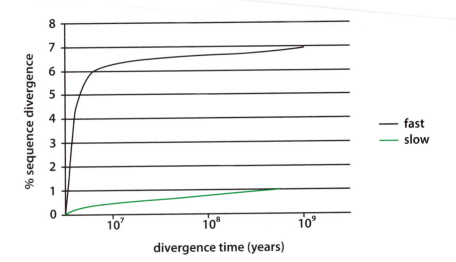

Figure 8.4 Graph showing the difference between slowly and rapidly evolving genes.

18S ribosomal RNA genes or actin genes, where very little if any variation occurs at this taxonomic level. Rather, an analysis of these sibling species might include the rapidly evolving D-loop of the mitochondrial genome.

A matrix of DNA sequences is used to illustrate different tree-building approaches

In the matrix provided below, the relationships of three mammalian species—cat, dog, and human—are analyzed from the same data set. These data are derived from a 5-amino-acid stretch in a gene found in all vertebrates called *CD9*. This gene encodes a protein of about 250 amino acids and is important in cell surface reception, but we are going to use only the data in the matrix below as an example. Before both distance- and character-based analyses are applied to this matrix, one additional species needs to be added. For the matrix we will use in the following examples, the same short sequence for opossum, a marsupial, is included as an outgroup. These DNA sequences are used to construct a tree that best represents the data and hence reveals the relationships of the organisms. Here is the short 15-base-pair sequence for the four species. The numbers above the sequences refer to numbers of bases from the beginning of the short sequence. The positions shown in boldface type are sites in the DNA sequences that vary, and the positions in lightface type are invariant sites with respect to the four taxa.

```
                      1     1
                5     0     5
cat_CD9     ACTCAACCTTCCAGC
dog_CD9     ACCCAGCCTTCTAGC
human_CD9   AATAATAATTCCAGC
opossum_CD9 AACAACAATTCTAGT
```

In the three-taxon example (dog, cat, and human) with an outgroup (marsupial), the possible number of rooted trees is only three. In this example, the three trees would include the dog with the human, the cat with the human, and the dog with the cat (see **Figure 8.5**). One way to represent these is in a branching diagrams, as shown. Or you can simply represent these trees with parenthetic notation or what is called the Newick tree format. So the tree that represents the dog with the human is represented as (cat, (dog, human)), the cat with the human is represented as (dog, (cat, human)), and the dog with the cat is represented as (human, (dog, cat)). With the above matrix, we will see how distance- and character-based approaches behave when these data are analyzed in terms of the relationships

Figure 8.5 Three possible rooted groupings of three species. Cat + human (left); dog + human (center); dog + cat (right).

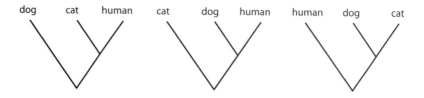

among cats, dogs, and humans. We will start with parsimony (one of the character-based approaches), then examine distances, and conclude with likelihood.

Basics of Maximum Parsimony Analysis

Parsimony and likelihood approaches share a great deal of methodological similarity. However, the differences between these approaches are critical to discerning the relative merits of each. The differences reside in the optimality criterion of each approach, and this aspect has an impact on the philosophical and methodological approaches of both. As discussed earlier, the optimality criterion of likelihood approaches is to find the tree that is most likely to produce the observed data depending on the model. In likelihood approaches, probabilistic models of sequence change are imposed on the system to compute the likelihoods and implement the optimality criterion. Parsimony takes a completely different approach because it utilizes a simpler optimality criterion.

Parsimony simply attempts to minimize the number of evolutionary character-state changes that are inferred from the data on a particular tree topology. It is based on the well-known principle of Ockham's razor, established by William of Ockham in the fourteenth century, which states that explanations that require the fewest assumptions are better than those that require many assumptions. For building trees, this means that the tree topology with the shortest number of evolutionary changes on it is the most parsimonious and satisfies Ockham's razor. Walter Fitch, a computational biologist at the University of California, Riverside, developed an algorithm (Fitch's algorithm) to parallel the earlier theory, which is described in more detail below.

The first step in building a parsimony tree involves examining the characters for patterns. This can be done without even looking at a tree diagram. In a parsimony context, characters are defined as either informative or uninformative. An informative character is a character whose state varies across taxa and is **derived** in a subset of the taxa. Derived, in the context of phylogenetic analysis, simply means that it is different from the ancestral state. For instance, hair is a derived character for mammals, in comparison with all vertebrates. Such a character therefore provides information concerning the placement of those taxa with the character in the tree. An uninformative character is one that does not change between taxa, or is derived in only a single taxa, or is derived in all ingroup taxa, and is therefore not useful for parsimony based tree building.

These patterns are easy to "see" when three ingroups and a single outgroup are studied. There is actually no need to designate that the trees we are evaluating are rooted, but for the purposes of clarity, we make the distinction between rooted and unrooted trees. If more than 20 taxa are included in an analysis, recognizing what a character is doing relative in the taxa can be quite difficult. In the *CD9* example, several of the characters behave the same way, as seen below.

			1		1
		5	0		5
cat_*CD9*		ACTCAACCTTCCAGC			
dog_*CD9*		ACCCAGCCTTCTAGC			
human_*CD9*		AATAATAATTCCAGC			
opossum_*CD9*		AACAACAATTCTAGT			

Fitch's algorithm uses set theory

Throughout the rest of this discussion, we are simply counting steps to determine the best fit of characters on trees. This approach uses Fitch's algorithm mentioned above. It is a very straightforward, yet incredibly useful algorithm that generates counts of the number of steps a character has on a tree. The algorithm uses set theory to implement the evaluation of intersection and union of the bases on the tips of the tree as a means to count the number of changes on a tree. Intersection refers to the state in which the two bases are not the same while union refers to the state in which the two bases are the same. Let's start with a simple matrix of one character:

species	A	B	C	O
character	T	G	T	G

Here are the steps involved in evaluating the number of steps for a single character on a single tree. First, the tips of the tree are labeled with the character states in the matrix. In this case, the species at the tips would be A, B, C, and O, with the bases T, G, T, and G labeled respectively. Let's say we are examining the tree where A and B are sister taxa. Second, the algorithm looks at the node that represents the common ancestor of A and B. Species A has a T and species B has a G. Therefore, the possible states at the first node placing A with B are G and T. According to set theory, this is not an intersection, but rather a union. Fitch's algorithm assigns a cost of one every time a union is required and zero every time an intersection is found. The next step is to evaluate the union or intersection of the common ancestor of C and O. Again in this case, the ancestor is the union of G and T, so we count another step for a total of two steps. This procedure is repeated until the root is met at the final count. To obtain this final count, we assess the two ancestors we have described above. For A and B, the algorithm states are G and T and for species C with O, the states are G and T. Here we have an intersection and the zero count is added to the total count, which yields a total of two steps for this character. A few rules have been added to Fitch's algorithm to accelerate the process. One, if all taxa have the same state, then the score is zero and there is no need to apply the algorithm to these characters. Two, if only one of the taxa is different from the rest, then there is no need to go through the algorithm for that character, and the score is one. Finally, if one can recognize patterns that are similar in the distribution of characters, then the number of steps can be inferred for those characters with similar patterns. For instance, if the pattern for species A, B, C, and O is T, G, T, and G, then any character with that pattern will have two steps on the tree that places species A and B together. Patterns like G, T, G, T, or A, C, A, C, or C, T, C, T will have two steps on the tree with species A and B together, because they have the same distribution pattern as T, G, T, G.

Let's apply Fitch's algorithm to the *CD9* example. It is important to know that the algorithm also helps one visualize the hypothesized changes that occur along the branches of the tree. This approach of inferring (or reconstructing) character states at nodes in a tree is discussed below. For each of the possible three trees, we use the algorithm to determine how many steps a particular character adds to sum of steps and then explain how the characters might have evolved on the three trees.

There are three trees that we need to evaluate for number of steps (see **Figure 8.6**). Recall that when a pattern in the characters is the same, the score for those characters will be the same for all of the trees being assessed. We first notice that characters 1, 5, 9, 10, 11, 13, and 14 are identical in all four taxa. Using Fitch's algorithm, we know that each of these characters will impart a score of 0 to the total. Next, look at character 15, which has a T in the opossum and a C in all of the ingroups: cat, dog, and human. One of the addenda to Fitch's algorithm is that for any character where only one taxon is different from the rest, the algorithm will always show a score of 1, regardless of the topology of the tree. Character 15 falls

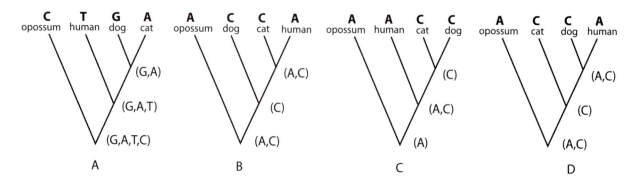

Figure 8.6 Fitch algorithm applied to characters 6, 2, 4, 7, and 8.
A: Character distribution of character 6 on the phylogenetic hypotheses dog + cat. B, C, and D: Character distributions for characters 2, 4, 7, and 8 on the three possible trees for these three taxa. The Fitch sets are given at the nodes in the trees in parentheses.

into this category. Character 6 is the next character examined, and it is shown mapped on the tips of the tree (Figure 8.6A). The Fitch sets are shown next to the nodes where they are inferred. Moving from one node to the next creates unions, which means that this character will add three steps to the total. When the algorithm is applied to the other two possible trees, the same applies.

Next we can look at characters 2, 4, 7, and 8, where the cat and the dog have the same base (C) and the human and opossum have a second and same base (A) in the position. This means that each of these four characters will have the same impact on sums. Figures 8.6B through D show the Fitch sets for each node for the three possible trees for characters 2, 4, 7, and 8. The trees with dog + human and cat + human show two unions and one intersection; hence, two steps are imparted by these characters for these two trees. However, Figure 8.6C shows that the tree with dog + cat shows a single union and two intersections, thus imparting a single step for this topology.

The only characters left are 3 (where the cat and human have the same base (T) and dog and opossum have the same base (C)) and 12 (where the cat and human have the same base (C) and dog and opossum have the same base (T)). While the patterns for characters 3 and 12 do not coincide with respect to nucleotide identity, they do have the same pattern of the cat and human (same base) as the dog and opossum. **Figure 8.7** shows the Fitch sets for the three possible topologies for character 3 (character 12 would have the same patterns except that the Cs and Ts would be reversed). Note that the tree placing cat with human (Figure 8.7A) has one union and two intersections; therefore, this tree will receive a score of one added to the total for both character 3 and 12. Both trees in Figures 8.7B and 8.7C have two unions and intersections; thus, a score of two is added to the sum for these trees for characters 3 and 12.

The scores for all of the characters for the three possible trees are shown in **Figure 8.8**. Therefore, using the Fitch algorithm, we see that the dog + cat topology is the most parsimonious (12 steps), followed by the cat + human tree (14 steps), with the dog + human tree (16 steps) being the least parsimonious.

A second way to look at the parsimony approach is to examine how the characters have changed in an evolutionary context on the tree; in other words, how ancestral

Figure 8.7 Fitch sets for three topologies for character 3. Character distributions for characters 2, 4, 7, and 8 on the three possible trees for these taxa. The Fitch sets are given at the nodes in the trees in parentheses.

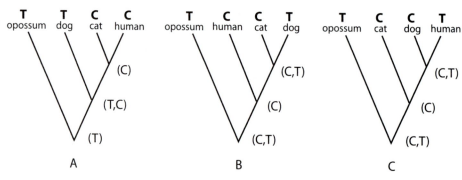

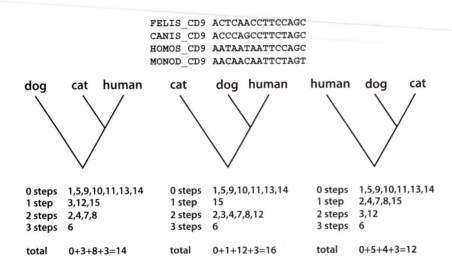

```
FELIS_CD9  ACTCAACCTTCCAGC
CANIS_CD9  ACCCAGCCTTCTAGC
HOMOS_CD9  AATAATAATTCCAGC
MONOD_CD9  AACAACAATTCTAGT
```

Figure 8.8 Summary of the maximum parsimony operations performed for Figures 8.6 and 8.7.

dog	cat	human	cat	dog	human	human	dog	cat

0 steps	1,5,9,10,11,13,14	0 steps	1,5,9,10,11,13,14	0 steps	1,5,9,10,11,13,14
1 step	3,12,15	1 step	15	1 step	2,4,7,8,15
2 steps	2,4,7,8	2 steps	2,3,4,7,8,12	2 steps	3,12
3 steps	6	3 steps	6	3 steps	6
total	0+3+8+3=14	total	0+1+12+3=16	total	0+5+4+3=12

character states are reconstructed. Since we determined that the tree with dog + cat is the most parsimonious, we will refer to that tree for the next few steps. First look at positions 1, 5, 9, 10, 11, 13, and 14. As discussed above, the character states (that is, whether the base is G, A, T, or C) for each of these positions are identical in all four taxa, which are the three ingroups (A, B, C) and the single outgroup (O). These characters are called invariant and are phylogenetically uninformative since they do not have any variation between the taxa. Now look at position 15. It is a T in the opossum and a C in all of our ingroups: cat, dog, and human. This character is also phylogenetically uninformative because no change occurs within the ingroups. Position 15 tells us nothing about how dog, cat, and human are related to each other in a parsimony context.

In positions 2, 4, 7, and 8, the cat and the dog have the same base (C) and the human and opossum have a second and same base (A) in the position. Positions 3 and 12 reveal that cat and human have the same base (a T for character 3 and a C for character 12) and dog and opossum have the same base (a C for character 3 and a T for character 12). All six of these sites are considered phylogenetically informative because they will tell us (as we shall see below) something about the relationships of the ingroups.

The next site to examine is position 6. Note that this position has a different base for all of the taxa in our matrix: cat, dog, human, and opossum have A, G, T, and C, respectively. Phylogenetically, this position is uninformative, because the changes from the ancestral state all can best be explained as occurring on the branches leading to the individual species in the tree.

When we map characters onto a tree, the base state in the outgroup is taken as a starting point, and the changes that explain the end states in the three ingroup taxa are mapped on the tree in question (**Figure 8.9**). For each example, boxes are used to represent the common ancestral base and ovals are used to represent a change. In these examples, the number of steps is calculated by counting the number of ovals that are mapped onto the tree. For character 15, this position is a T in the opossum outgroup and a C in all of the ingroups. The best way to explain this character on the dog + cat tree is to hypothesize a single mutation (from a T to a C) in the common ancestor of all the ingroups (Figure 8.9A).

Explaining base positions 1, 5, 9, 10, 11, 13, and 14 (which are invariant in all four taxa) in the dog + cat tree is trivial since they do not result in mapping steps (in fact, no steps need to be added to the cat + human or dog + human trees either). For position 6 (which is different in all four taxa), if the characters are not weighted or rescored, the information in this position is uninformative in a parsimony context, it adds three steps to the overall length of the tree (Figure 8.9B). If the outgroup is a C, the rest of the character states can best be explained as occurring

Figure 8.9 Parsimony 3. Boxes indicate character states in ancestors, and ovals represent changes in the DNA sequence. A: Most parsimonious explanation for character 15 on the dog + cat tree. B: Most parsimonious way to represent character 6 changes on the dog + cat tree. C: Most parsimonious way to represent character change on the dog + cat tree for character characters 2, 4, 7, and 8. D and E: Most parsimonious ways to represent character change on the dog + cat tree for character 12.

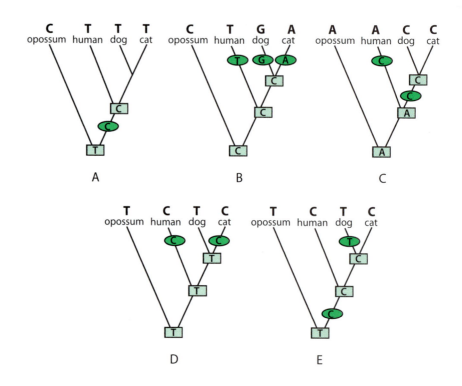

on the lineages leading to each of the ingroups after divergence from any of the common ancestors. This explanation would mean that there are three steps (a C to T change in the human lineage, a C to G change in the dog lineage, and a C to A change in the cat lineage; Figure 8.9B) that occur after the divergence of these taxa from the common ancestor of placental mammals. These three steps are represented by the ovals.

In positions 2, 4, 7, and 8, the opossum and human have the same base (A) and the cat and dog have the same base (C). Since all four positions (2, 4, 7 and 8) have the same pattern of distribution of bases (i.e., cat and dog share a base while opossum and human share a different base), all four of these characters give the same inference (Figure 8.9C). Note the single oval on the branch leading to the common ancestor of cat and dog. This mapping pattern indicates that, for base positions 2, 4, 7, and 8, a single step on the dog + cat tree is required to explain the bases in the ingroups for these positions in the sequence.

For the last two positions (3 and 12), opossum and dog have the same base, and cat and human have the same base (but it is different from dog and opossum). The distribution of the bases in position 12 for the ingroup taxa on the dog + cat tree are shown in Figures 8.9D and 8.9E (the distribution for character 3 can be obtained simply by substituting C for T and vice versa). Note that there are two ways to represent the character reconstruction for this character (and character 3 as well). Each of these two positions, regardless of how the character is reconstructed, requires two steps to explain the dog + cat tree.

In summary, the results for the dog + cat tree are shown below.

										1	1	1	1	1	1	TOT
position	1	2	3	4	5	6	7	8	9	0	1	2	3	4	5	.
number of steps	0	1	2	1	0	3	1	1	0	0	0	2	0	0	1	12

When the *CD9* data are applied to the dog + human tree, the number of steps on this tree for base positions 1, 5, 6, 9, 10, 11, 13 , 14, and 15 is the same as that with the cat with dog tree (see the positions in boldface type above). For base positions

2, 4, 7, and 8, the mapping of these base changes on the dog + human (not shown) shows that two steps are required to explain the base distributions in the ingroups. Once again, only positions 3 and 12 remain, and these two positions require two steps each to explain the base distributions in the ingroups, illustrated below.

										1	1	1	1	1	1	TOT
position	1	2	3	4	5	6	7	8	9	0	1	2	3	4	5	.
number of steps	0	2	2	2	0	3	2	2	0	0	0	2	0	0	1	16

For the final possible topology, cat + human, the same mapping procedure can be applied. The number of steps obtained for this topology is shown below.

										1	1	1	1	1	1	TOT
position	1	2	3	4	5	6	7	8	9	0	1	2	3	4	5	.
number of steps	0	2	1	2	0	3	2	2	0	0	0	1	0	0	1	14

Rescoring is a simple way to weight characters in parsimony

The following example is a good one to demonstrate how weighting behaves in parsimony. Recall character 6 in the *CD9* sequences above. We designated this position as uninformative because each taxon had a different base. However, this site can be informative if a weighting scheme is imposed on the characters or if the characters are rescored relative to some molecular criterion. For instance, bases C and T share a single-ring pyrimidine structure, while bases A and G have a two-ring purine structure. If the characters in this position are rescored as being either pyrimidines (C = T = Y) or purines (A = G = R), then the position would look like this:

```
cat_CD9        a = R
dog_CD9        g = R
human_CD9      t = Y
opossum_CD9    c = Y
```

How would weighting affect the results? If all the Cs and Ts in the matrix are rescored as pyrimidines (Y) and all the As and Gs are rescored as purines (R), this would weight transitions (where a pyrimidine is changed to another pyrimidine or a purine is changed to another purine, causing a minimal change in structure) over transversions (where a pyrimidine is changed to a purine or vice versa, causing a significant structural change). The matrix would then appear as

```
                  1       1
              5   0       5
cat_CD9      RYYYRRYYYYYYRRY
dog_CD9      RYYYRRYYYYYYRRY
human_CD9    RRYRRYRRYYYYYRRY
opossum_CD9  RRYRRYRRYYYYYRRY
```

Instead of eight sites with variable sites, there are now only five. All five of these positions (positions 2, 4, 6, 7, and 8) have a pattern in which the cat and dog share the same type of base (purine or pyrimidine), while the opossum and human both have the opposite type of base.

The schematic below summarizes the number of steps required for each of the three trees, given the data in the revised purine/pyrimidine matrix. Clearly, the dog + cat topology is the most parsimonious under this weighting scheme, but in this case the other two trees are equally less parsimonious.

position	1	2	3	4	5	6	7	8	9	1 0	1 1	1 2	1 3	1 4	1 5	TOT .
dog + cat	0	1	0	1	0	1	1	1	0	0	0	0	0	0	0	5
cat + human	0	2	0	2	0	2	2	2	0	0	0	0	0	0	0	10
dog + human	0	2	0	2	0	2	2	2	0	0	0	0	0	0	0	10

Distance Methods

Distance approaches are also called phenetic approaches. Such methods use distances to construct a tree and can start with either raw distance or similarity measures (as from DNA–DNA hybridization or immunoprecipitation) of species or with discretely distributed attribute information (such as molecular sequence information) that is then transformed into distance or similarity matrices. A more modern version of phenetics is the widely used neighbor joining (NJ) method (discussed in detail below), which can routinely take discrete DNA or protein sequence data and transform them into a matrix of distances that can then be used in the NJ algorithm.

The first step in distance analysis is to compute the pairwise distances for each taxon in the matrix to all of the other taxa. This step in essence compresses the data matrix from 15 character positions for each taxon to six pairwise distance measures (one for each pair of taxa). There are many ways to compute distances between taxa by use of DNA or amino acid sequences; for instance, we can simply compute the raw or absolute distances for pairwise comparisons. Here is what the distance matrix would look like if the absolute number of changes between each pair of taxa is computed:

		1 0	1 5
	5	0	5

cat_*CD9*	ACTCAACCTTCCAGC
dog_*CD9*	ACCCAGCCTTCTAGC
human_*CD9*	AATAATAATTCCAGC
opossum_*CD9*	AACAACAATTCTAGT

	cat	dog	human	opossum
cat	–			
dog	3	–		
human	5	7	–	
opossum	8	6	4	–

These pairwise distances, known as raw unscaled distances, are computed by counting the differences between two sequences in the matrix. As an example, when the dog and cat sequences are compared, positions 3, 6, and 12 differ between cat and dog, so 3 is entered in the matrix for the distance of cat to dog. Within this matrix, the diagonal is left blank because it would measure the differences between a sequence and itself, which is implicitly zero. Additionally, the upper triangle of the matrix is left blank since it would simply be a mirror of the lower half of the matrix.

We can also use a method of computing distances that involves application of a sequence evolution model, which addresses three aspects of sequence change discussed above. The first aspect takes into account the possibility that the different kinds of base changes (such as transitions versus transversions, discussed earlier) have different evolutionary dynamics and hence the different changes can be placed into different classes. The second aspect of sequence evolution concerns base composition of the sequences being examined. Some models assume equal base frequencies in sequences ($f(A) = 0.25, f(G) = 0.25, f(C) = 0.25$, and $f(T) = 0.25$), while other models take base composition bias into consideration. The third aspect involves the site-specific heterogeneity of sequence change, which will be explored in the discussion about likelihood.

To demonstrate how more complex distance measures are made from sequences, we will focus on two of these parameters: transformation probability and base frequency. The first step is to calculate the frequencies of specific changes by dividing the number of changes in each cell of the matrix by the total number of nucleotides in the aligned sequences. For the example of cat and dog

cat_CD9   ACTCAACCTTCCAGC

dog_CD9   ACCCAGCCTTCTAGC

there are three base changes (underscored above). One is a T to C, the second is an A to G, and the third is a C to T. Each change occurs once out of 15 bases for a frequency of 1/15 or 0.067. The frequency matrix then appears as follows:

	G	A	T	C
G	–	0	0	0
A	0.067	–	0	0
T	0	0	–	0.067
C	0	0	0.067	–

The entries along the diagonal have been removed as these frequencies are not involved in the calculation of distance. The raw or uncorrected distance is the sum of all entries in the above matrix ($0 + 0 + 0 + 0.067 + 0 + 0 + 0 + 0 + 0.067 + 0 + 0 + 0.067 = 0.2$). By repeating this procedure for each pair of organisms, we are able to obtain a complete scaled raw distance matrix:

	cat_CD9	dog_CD9	human_CD9	opossum_ CD9
cat_CD9	---			
dog_CD9	0.20	---		
human_CD9	0.33	0.47	---	
opossum_CD9	0.53	0.40	0.27	---

Either this matrix or the matrix calculated above for raw uncorrected distances can be used to construct a distance tree. While they would present different lengths for the tree, they would yield directly proportional branches.

Corrections for multiple hits may be introduced

One problem involving the use of uncorrected raw matrices is that they do not take into account multiple changes that could occur at the same position in the sequence, and hence they may underestimate the amount of evolution that has occurred between two sequences. Perhaps the forerunner of all models for DNA sequence change is the Jukes–Cantor model (JC). This model accounts for multiple hits to positions in the sequences under examination. In this model of DNA sequence evolution, the rate of nucleotide substitution is the same for all of the 12 possible transformations of nucleotides. A correction for multiple hits in this model is needed because, even though we might observe the following case in a direct comparison

species 1   G ———→A
species 2   G ———→A

the actual nature of the changes might be

species 1   G —→ C —→ T ———→A
species 2   G ———————————→A

The "hidden" changes in species 1 need to be accommodated by the Jukes–Cantor model. If two species have sequences that are effectively randomized, they would be expected to match at about 0.25 of the sites in the sequence by chance alone. This would be a distance of 0.75. Using this expectation, this JC correction is given as

$$d_{xy} = -(^3/_4) \ln (1 - ^4/_3 D) \tag{8.1}$$

where d_{xy} is the JC-corrected distance we want and D is the simple raw difference between two sequences, which maximizes at 0.75. Hence the maximal value that d_{xy} can take on is when $D = 0.75$. When $D = 1.0$ is substituted into the equation, this gives us $d_{xy} = -(^3/_4) \ln (1 - ^4/_3) = -(^3/_4) \ln (-^1/_3)$, which is an undefined quantity.

To avoid an undefined logarithm of 0 or of a negative number, D must be less than $^3/_4$ or 0.75. This is another way of saying any two random sequences will have a distance of 0.75 and a similarity of 0.25, given that all changes among the four bases are equally probable.

Corrections can be made by using evolutionary models

Other models of evolutionary change can be incorporated into the analysis. One of the simplest models accommodates the transition/transversion problem and is demonstrated by the Kimura two-parameter (K2P) model. The K2P model assumes that transitions are more frequent than transversions. Sidebar 8.1 shows how the distance measured for cat and dog is calculated with the K2P model. The entire distance matrix for this approach is shown below.

	cat	dog	human	opossum
cat	---			
dog	0.255	---		
human	0.477	0.733	---	
opossum	0.936	0.589	0.381	---

Equations for the computation of distances for these models and other more complicated models can be estimated for the four sequences in the example to yield a plethora of distance matrices. This poses a problem in that the researcher should then ask which matrix (which distance measure) is the best. When we discuss likelihood methods, we will introduce a method for determining which model is the best for a particular analysis.

Sidebar 8.1. Using K2P to calculate a distance.

The relationship of the transition (P) and transversion (Q) frequencies and how they affect genetic distances is shown in the following equation:

$$D(K2P) = \tfrac{1}{2} \ln [1/(1 - 2P - Q)] + \tfrac{1}{4} \ln [1/(1 - 2Q)]$$

$$(8.2)$$

where $P = e + b + l + o$ (purine to purine or pyrimidine to pyrimidine changes) and $Q = c + d + g + h + i + j + m + n$ (purine to pyrimidine or pyrimidine to purine changes). The lowercase letters refer to frequencies defined in the matrix below.

	G	A	T	C
G	a	b	c	d
A	e	f	g	h
T	i	j	k	l
C	m	n	o	p

We will calculate the K2P distance for cat to dog. In this comparison, there are three transitions (l, o, and e) and no transversions. If there are three transitions, $P = l + o + e = 3/15 = 0.20$. If $Q = 0$, the second term in the K2P equation is 0. The K2P distance then is

$$D(K2P) = \tfrac{1}{2} \ln [1/(1 - 2(l + o + e))]$$

$$(8.3)$$

or $\tfrac{1}{2} \ln [1/1 - 2(0.2)] = \tfrac{1}{2} \ln [1/(1 - 0.4)] = \tfrac{1}{2} \ln [1/0.6] = \tfrac{1}{2} \ln [1.667] = \tfrac{1}{2}[0.511] = 0.255$. Each cell of the pairwise distance matrix for the dog to cat, human, and opossum can be similarly calculated to give the matrix below:

	cat	dog	human	opossum
cat	---			
dog	0.255	---		
human	0.477	0.733	---	
opossum	0.936	0.589	0.381	---

In the K2P model, we assumed that all base frequencies are equal ($= 0.25$). Another model that takes into consideration base composition bias is called the Felsenstein 81 (F81) model. The matrix generated by use of F81 is shown below and calculated in Sidebar 8.2.

	cat	dog	human	opossum
cat	---			
dog	0.23700	---		
human	0.45382	0.75750	---	
opossum	1.01255	0.58685	0.34055	---

Sidebar 8.2. Using the F81 model to estimate distances.

The F81 model ignores transition versus transversion bias and computes sequence distances by using the following equation:

$$F81 = -B \ln [1 - (ham/B)] \qquad (8.4)$$

where $B = 1 - [f(A)^2 + f(C)^2 + f(G)^2 + f(T)^2]$, $f(N)^2$ is the frequency of each base N squared, and ham is the "hamming distance" between the two sequences. Hamming distance is simply the percentage of sites that vary from one sequence to the other in a pairwise comparison.

For the cat to dog comparison, $f(A) = 0.23$, $f(C) = 0.47$, $f(G) = 0.10$, $f(T) = 0.20$, and ham $= 3/15 = 0.2$, so we compute B as $1 - (0.053 + 0.221 + 0.010 + 0.040) = 1 - 0.324$

$= 0.676$. Inserting this result into the F81 equation yields $F81 = -0.676 \ln [1 - (0.2/0.676)] = -0.676 \ln (1 - 0.296) = -0.676 \ln (0.704) = -0.676 (-0.351) = 0.237$.

The rest of the F81 matrix is shown in the text. If we had assumed that all base frequencies were equal [that is, $f(A) = f(C) = f(G) = f(T) = 0.25$], then B is equal to $1 - (0.0625 + 0.0625 + 0.0625 + 0.0625) = 1 - 0.25 = 0.75$ or $3/4$. The F81 equation then reduces to

$$F81 = -\tfrac{3}{4} \ln (1 - [ham/(\tfrac{3}{4})])$$
$$= -\tfrac{3}{4} \ln (1 - [\tfrac{4}{3}(ham)]) = JC \qquad (8.5)$$

This equation is simply the Jukes–Cantor equation that corrects for multiple hits with equal base frequencies.

Neighbor joining is a stepwise based approach to tree-building

We will explore the construction of a tree using the most popular distance tree-building approach, the NJ method. This is a stepwise deterministic method that uses a distances, and therefore can be computed rapidly for even a large number of taxa. This method includes several steps that are reiterated throughout the construction of the tree. The algorithm, which starts with the unscaled raw distance matrix, is outlined below; details are given in Sidebar 8.3.

- Starting from the initial distance matrix, calculate Q values for each distance in the matrix to create a new matrix called a "Q matrix" (see Sidebar 8.3, step 1, for calculations).

- Find the pair of taxa in the new Q matrix with the smallest distance. In the example, there are two Q distances that are equally small: cat to dog (-26) and human to opossum (-26). When this occurs, choose one randomly.

- Start building the tree by making a node that joins the two taxa that are nearest neighbors from the previous step (join the nearest neighbors). For this example, we are using cat to dog as the first nearest neighbor. Specifically, we join cat and dog in the first node. This node is now a combination of cat and dog and will be referred to as the cat/dog node. The combination node will be used to continue building the tree (Sidebar 8.3, step 2).

- Determine the score for the new node. Calculate the distances from cat and dog to the new cat/dog node (see Sidebar 8.3, step 3).

- Cycle the algorithm, considering the pair of joined neighbors as a single taxon and using the distances calculated in the previous step (Sidebar 8.3, step 4).

- Start the cycle over. Calculate a new Q matrix as in Sidebar 8.3, step 5.

- Join neighbors.

The tree constructed from the raw uncorrected distances by the NJ approach is shown in **Figure 8.11**. Similarly, NJ trees for the K2P matrix and the HKY matrix are also shown. Note that these trees should be considered unrooted; that is, we have not designated a root in any of them. If we did designate a root, it would be the opossum, and the topology of the networks shown in Figure 8.11 would change.

Minimum evolution uses minimal distance as a criterion to choose the best solution among multiple trees

Note that NJ and its relatives like UPGMA are deterministic algorithms. Such algorithms will give a single tree when applied to a distance matrix. However an approach that uses minimal distance as a criterion to choose the best tree can be applied, and this is called the minimal evolution criterion. Minimum evolution (ME) requires the production of multiple trees and their evaluation according to the optimality criterion that the tree has the minimal amount of evolution on its branches. If the number of taxa in an analysis is less than 11 or so, ME requires that the sum (S) of all branch lengths for every tree be computed and compared. A branch length in distance analysis refers to the amount of change that is apportioned to a particular branch in a tree. The tree with the smallest S is then chosen as the best solution or the tree with the minimal amount of "evolution" on its branches. Distances or branch lengths are usually computed with the NJ algorithm. Many ME searches start with the NJ tree and implement branch swapping on the initial NJ tree. Alternatively, a starting tree that the researcher thinks is the most appropriate can be used, or a random tree can be used as a starting tree. If the number of taxa is greater than 11, then heuristic searches are necessary because of the large number of possible trees that would need to be considered. We will discuss heuristic searches in Chapter 9, where we will also discuss the

Sidebar 8.3. Neighbor joining, step by step.

Step 1. Calculate Q values for each distance in the matrix to create a new matrix called a "Q matrix." The new Q values are calculated by the following general equation:

$$Q_{ij} = (r-2)d_{ij} - \Sigma d_{ik} - \Sigma n d_{jk} \tag{8.6}$$

where r is the number of taxa in the matrix and d represents distances. The subscript letters refer to the various pairwise comparisons that can be made: i and j are the pair being considered, and k refers to taxa other than i or other than j. From the raw unscaled distances (number of nucleotide changes) in the following matrix, these values are calculated in detail.

	cat	dog	human	opossum
cat	---			
dog	3	---		
human	5	7	---	
opossum	8	6	4	---

In our specific case, the Q value for the cat to dog distance is calculated as follows:

$$Q(\text{cat,dog}) = (r-2)[d(\text{cat,dog}) - \Sigma d(\text{cat, other taxa}) - \Sigma d(\text{dog, other taxa}) \tag{8.7}$$

with the values $r = 4$, $d(\text{cat, dog}) = 3$, $\Sigma d(\text{cat, other taxa}) = 3 + 5 + 8 = 16$ (sum of distances from cat to dog, human, and opossum), and $\Sigma d(\text{dog, other taxa}) = 3 + 7 + 6 = 16$ (sum of distances from dog to cat, human, and opossum) as follows: $Q(\text{cat, dog}) = (4-2)3 - 16 - 16 = -26$. The new Q matrix, with each Q value calculated in this manner, is as follows:

	cat	dog	human	opossum
cat	---			
dog	-26	---		
human	-22	-18	---	
opossum	-18	-22	-26	---

Step 2. Determine the score for the new cat/dog node. The distances from cat and dog to this new cat/dog node are calculated from the following general equations, where f and g are the paired taxa, k represents taxa outside the pair, and u is the newly generated node. Note that here $r = 3$, as cat, dog, and cat/dog are the taxa we need to consider.

$$d_{fu} = \tfrac{1}{2}(d_{fg}) + \tfrac{1}{2(r-2)}[\Sigma d_{fk} - \Sigma d_{gk}] \tag{8.8}$$

$$d_{gu} = \tfrac{1}{2}(d_{fg}) + \tfrac{1}{2(r-2)}[\Sigma d_{gk} - \Sigma d_{fk}] \tag{8.9}$$

In our specific case, the distance from cat to the new cat/dog node is

$$d(\text{cat, cat/dog}) = \tfrac{1}{2}d(\text{cat, dog}) + \tfrac{1}{2(r-2)}[\Sigma d(\text{cat, other taxa}) - \Sigma d(\text{dog, other taxa})] \tag{8.10}$$

So $d(\text{cat, cat/dog}) = \tfrac{1}{2}(3) + \tfrac{1}{2(3-2)}[16 - 16] = 1.5$ = distance from cat to new node.

In our specific case, the distance from dog to the new cat/dog node is

$$d(\text{dog, cat/dog}) = \tfrac{1}{2}(d(\text{cat, dog}) + \tfrac{1}{2(r-2)}[\Sigma d(\text{dog, other taxa}) - \Sigma d(\text{cat, other taxa})] \tag{8.11}$$

So $d(\text{dog, cat/dog}) = \tfrac{1}{2}(3) + \tfrac{1}{2(3-2)}[16 - 16] = 1.5$ = distance from dog to new node.

Step 3. Calculate the distance of all taxa outside this pair to the new node (cat/dog node). Use the following general equation, where f and g are the paired taxa, k represents taxa outside the pair, and u is the newly generated node:

$$d_{uk} = \tfrac{1}{2}[d_{fk} - d_{fu}] + \tfrac{1}{2}[d_{gk} - d_{gu}] \tag{8.12}$$

In our specific case, the distance from the new cat/dog node to the outside taxon human will be

$$d(\text{cat/dog, human}) = \tfrac{1}{2}[d(\text{cat, human}) - d(\text{cat, cat/dog})] + \tfrac{1}{2}[d(\text{dog, human}) - d(\text{dog, cat/dog})] \tag{8.13}$$

So $d(\text{cat/dog, human}) = \tfrac{1}{2}[5 - 1.5] + \tfrac{1}{2}[7 - 1.5] = \tfrac{1}{2}[3.5] + \tfrac{1}{2}[5.5] = 1.75 + 2.75 = 4.5$ = distance from cat/dog to human.

In our specific case, the distance from the new cat/dog node to the outside taxon opossum will be

$$d(\text{cat/dog, opossum}) = \tfrac{1}{2}[d(\text{cat, opossum}) - d(\text{cat, cat/dog})] + \tfrac{1}{2}[d(\text{dog, opossum}) - d(\text{dog, cat/dog})] \tag{8.14}$$

So $d(\text{cat/dog, opossum}) = \tfrac{1}{2}[8 - 1.5] + \tfrac{1}{2}[6 - 1.5] = \tfrac{1}{2}[6.5] + \tfrac{1}{2}[4.5] = 3.25 + 2.25 = 5.5$ = distance from cat/dog to human. The distance from human to opossum remains 4. The revised matrix is shown below.

	cat/dog	human	opossum
cat/dog	---		
human	4.5	---	
opossum	5.5	4	---

The connection between dog and cat is illustrated in the top panel of Figure 8.10.

Step 4. Calculate new Q matrix in the second cycle of NJ. The Q value for the cat/dog to human distance is calculated from Equation 8.6 in step A, with the values

Sidebar 8.3. Neighbor joining, step by step (continued).

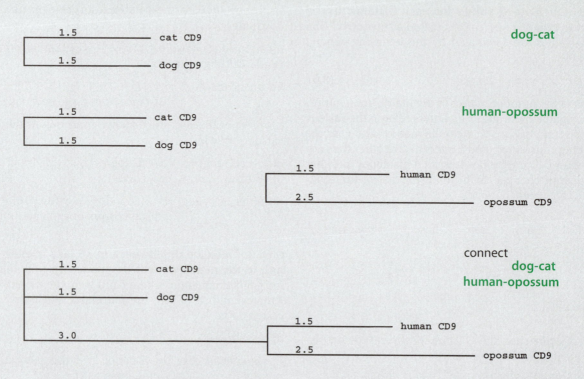

Figure 8.10 Neighbor joining (NJ) connections. Top: Absolute distances calculated in Sidebar 8.3 allow for the connection of cat with dog in a NJ tree. Middle: Absolute distances calculated in Sidebar 8.3 allow for the connection of human with opossum in a NJ tree. Bottom: The cat/dog and human/opossum nodes are joined by the NJ approach.

$r = 3$(cat/dog, human, and opossum), d(cat/dog, human) = 4.5, Σd(cat/dog, other taxa) = 4.5 + 5.5 = 10, and Σd(human, other taxa) = 4.5 + 4 = 8.5, as follows:

Q(cat/dog, human) = $(r-2)[d$(cat/dog, human)$]$
$- \Sigma d$(cat/dog, other taxa) $- \Sigma d$(human, other taxa)

(8.15)]

So Q(cat/dog, human) = $(3-2)[4.5] - 10 - 8.5 = -9.5$.

The new Q matrix, with each Q value calculated in this manner, is as follows:

	cat/dog	human	opossum
cat/dog	---		
human	-14	---	
opossum	-14	-14	---

Step 5. Joining and recalculating distances in third cycle. Since the entries are equal in the new Q matrix, human and opossum are arbitrarily chosen as the next nearest neighbors. When all Q entries are equal in an NJ computer run, the computer will randomly choose one of the pairs as the next neighbors to join.

a. Join human and opossum at a second node.

b. Calculate the distance of each taxon in the pair to this new node, using Equations 8.8 and 8.9 introduced in step B. Note that here $r = 3$, as human, opossum, and human/opossum are the taxa we need to consider.

In our specific case, the distance from human to the new human/opossum node would be

d(human, human/opossum) =
$^1/_2[d$(human, opossum)$] + ^1/_{2(r-2)}[\Sigma d$(human, other taxa) $- \Sigma d$(opossum, other taxa)   (8.16)

Thus d(human, human/opossum) = $^1/_2[4] + ^1/_{2(3-2)}$ $[(4.5+4) - (5.5+4)] = 2 + ^1/_2[-1] = 1.5 =$ distance from human to the new node.

In our specific case, the distance from opossum to the new human/opossum node would be

d(opossum, human/opossum) =
$^1/_2[d$(human, opossum)$] + ^1/_{2(r-2)}[\Sigma d$(opossum, other taxa) $- \Sigma d$(human, other taxa)   (8.17)

Thus d(opossum, human/opossum) = $^1/_2[4] + ^1/_{2(3-2)}[(5.5+4) - (4.5+4)] = 2 + ^1/_2[1] = 2.5 =$ distance from opossum to new node.

Sidebar 8.3. Neighbor joining, step by step (continued).

The connection between human and opossum is illustrated in the middle panel of Figure 8.10.

c. Calculate the distance of all taxa outside this pair to the new node, using Equation 8.12 introduced in step 3. In our specific case, the distance from the single outside taxon, cat/dog, to the new human/opossum node is

d(human/opossum, cat/dog) = $^1/_2$[d(human, cat/dog) − d(human, human/opossum)] + $^1/_2$[d(opossum, cat/dog) − d(opossum, human/opossum)]   (8.18)

Thus the distance from cat/dog to the new human/opossum node will be d = $^1/_2$[4.5 − 1.5] + $^1/_2$[5.5 − 2.5] = $^1/_2$[3.0] + $^1/_2$[3.0] = 1.5 + 1.5 = 3.0 = distance from cat/dog to the new node.

d. The resulting matrix is trivial, as shown below. The cat/dog node is joined to the human/opossum node at a distance of 3.0.

	cat/dog	human/opossum
cat/dog	---	
human/opossum	3.0	---

The connection between cat/dog and human/opossum nodes is shown in the bottom panel of Figure 8.10.

problem of tree islands and inadequate search procedures. These problems can be overcome for ME the same way they can for other methods like likelihood and parsimony. The ME tree is an unrooted tree, but in most cases it will be presented as a rooted tree for convenience.

Figure 8.11 Three NJ trees. Absolute distances (top), Kimura two-parameter (K2P) distances model (middle), and Hasegawa, Kishino, and Yano (HKY) distances model (bottom) were used to construct the trees.

absolute distance

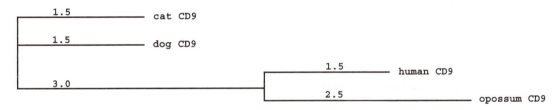

K2P

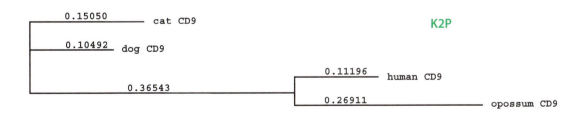

HKY

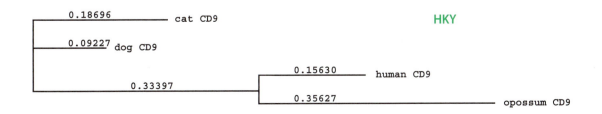

Basics of the Maximum Likelihood Approach

Likelihood approaches are extremely computationally intense, as they require a character-by-character assessment of probabilities of character-state change. We use the following conditional probability statement to implement likelihood: $L(W|D) = P(D|W)$.

In other words, the likelihood of a set of model parameters (W) having specific values from observation (D) is equal to the probability of those data (D) occurring under the given model parameters (W). In the context of evolutionary trees, this means that maximum likelihood computes the probability that the observed sequences in a matrix were generated by a chosen model. It also means that likelihoods for all possible trees for the taxa have to be evaluated with the optimality criterion, with the tree with the highest likelihood being the preferred tree.

There are several aspects of sequence change that can be modeled in likelihood analyses to estimate likelihood (L) given a specific tree topology: for example, branch length, base composition, transformation probabilities, and site-specific heterogeneity. Specifically, maximum likelihood (ML) methods evaluate branching order and branch lengths of a tree in the context of a probability that a proposed model of evolution and the tree would give rise to the data. ML explicitly estimates a probability from observed data under a specific model. In this sense there is an undeniable linkage among the model, the tree, and the data; therefore, if a poor or inappropriate model is chosen, a poor tree will be generated. This linkage between model and tree is the reason phylogeneticists have developed methods to assess the "appropriateness" or likelihood of models fitting data. In this section, we describe the ML approach and discuss the assessment of model appropriateness in a likelihood context. This description is not exhaustive but should provide the basic framework for the approach.

ML begins with a data matrix of aligned sequences. We focus on the *CD9* matrix given above, but amino acid matrices are equally amenable to likelihood analysis, although the models are much more complicated and computationally intense because of the potential presence of 20 amino acids as compared to four nucleotides. For three ingroup taxa and an outgroup, the evaluation in an ML context involves only three trees: cat + dog, cat + human, and dog + human (**Figure 8.12**). The problem is reduced to the following: if we have a matrix of sequence information, a model for sequence evolution, and a tree, we can ask: "what is the probability that the tree we are examining could have generated the data in the matrix under the model we impose, in contrast to another tree?"

To compute this probability, we need to examine each alignment position or column in the matrix separately to evaluate the likelihood at individual sites that, given the tree and the model, the evolutionary process would have generated the data. To do this, we calculate the likelihood for each site as the sum of the probabilities of every possible reconstruction of ancestral states under the model imposed. Branch lengths, transformation probabilities, base composition, and rate heterogeneity are all factors in calculating these probabilities.

Figure 8.12 Maximum likelihood schemes. The left panel shows how the likelihood of the first position in the sequence matrix in this chapter is analyzed in a general context. The first position in the example has A in all taxa. N1 and N2 represent the possible bases that could exist at the nodes. The middle panel shows a specific situation for computing the likelihood of the first position. The right panel shows the generalized approach to analyzing the second nucleotide position. N1 and N2 stand for the possible bases that can be at the nodes in this network.

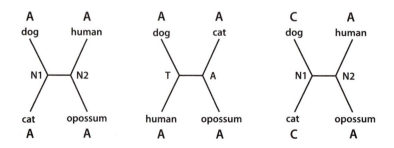

Once the probabilities for each site are computed, the likelihood of the tree is estimated by taking the likelihoods at each site and multiplying them by each other. However, this step is usually accomplished by summing the log likelihoods at each site, which are then reported as the log likelihood of the complete tree. This approximation is valid because the summed probabilities are miniscule and little error is introduced by the approximation. Once these log likelihoods are estimated for each tree, the one with the highest likelihood is chosen as the maximum likelihood tree.

When the ML approach is applied to the *CD9* example, remember that, for three ingroup taxa, there are three trees to evaluate. For this example, we use the K2P model, which assumes transition versus transversion differences. There are no base composition departures [$f(G) = f(A) = f(C) = f(T) = 0.25$] and no site-specific rate heterogeneity. There are two ways to establish the difference in probability of transitions and transversions in the K2P model. The first involves imposing a ratio that the researcher feels is appropriate for the taxa under examination. For instance, one might impose the assumption that the probability of a transition is twice the probability of a transversion from previous knowledge of how a particular gene sequence might evolve. The second and safer approach is to simply compute the probabilities from the data themselves.

In order to compute the probabilities by site, we need to compute the probability for each and every possible base transformation given the topology and base states at the terminals. This calculation is conducted for the first two positions in the *CD9* sequence example:

	1	2
cat_*CD9*	A	**C**
dog_*CD9*	A	**C**
human_*CD9*	A	**A**
opossum_*CD9*	A	**A**

Note that the first column does not vary. This result does not mean that the probability is trivial or simple to estimate. Rather, we need to calculate the probability of every potential base transformation that could have produced As in that position in all four species. The left panel in Figure 8.12 shows the first topology we will evaluate that implies dog with cat. The four species all have an A in the first position. N1 and N2 can be any combination of the four possible bases. There are 16 possible arrangements of G, A, T, and C in the N1 and N2 positions in the figure. Therefore, there are 16 possible likelihoods we need to compute for this single base position.

N1	N2	N1	N2	N1	N2	N1	N2
A	**A**	C	A	G	A	T	A
A	C	**C**	**C**	G	C	T	C
A	G	C	G	**G**	**G**	T	G
A	T	C	T	G	T	**T**	**T**

The middle panel in Figure 8.12 depicts one of these results on the tree, where the ancestral state at the node connecting dog and cat is a T and the ancestral node connecting human and opossum is an A. The likelihood for this reconstruction is the product of the likelihoods of the transformations on the tree. Since there are five branches, five probabilities need to be calculated for the tree for this position: $p(AT|1)$, $p(TA|2)$, $p(TA|3)$, $p(AA|4)$, $p(AA|5)$. These probabilities are multiplied by each other because they are linked probabilities. Likewise, for each of the 16 possible combinations for N1 and N2, there are five branch probabilities to compute. The probability for each of these 16 possible states for N1 and N2 is computed the same way: by multiplying the five individual branch probabilities. After these 16 probabilities are produced, the final probability for the entire site is computed by *adding* the 16 probabilities together because they are **independent** probabilities. When this computation is complete, we have a probability for the first position in the alignment. The scheme one would follow for computing a probability in the second position is shown in the right panel in Figure 8.12.

As discussed above, we need to calculate 16 probabilities for the 16 different combinations of N1 and N2, each of which requires estimating five internal branch probabilities. Calculation of the likelihood probabilities for all positions for the three tree topologies we need to consider for the aligned *CD9* sequences in the cat with dog tree is shown in Sidebar 8.4. In the examples above (Sidebar 8.4), we have also calculated the log of each likelihood (ln *L*) for each position in the sequence. These calculations are useful because they help estimate the overall likelihood of each tree, which is simply the sum of the 15 ln *L* values for each tree. When we sum these calculations for the three trees, the results are

```
K2P     dog + cat     cat + human     dog + human
-------------------------------------------------------
-ln L     57.82079       63.31416        65.66072
```

The tree with the maximum likelihood is actually the tree with the lowest ln *L* value, which in this example is the dog + cat topology (left panel of Figure 8.12). As we did for distances, we can also conduct a ML search for the best tree under the *F81* model. Such a search yields the following ln *L* values for the three possible topologies:

```
F81     dog + cat     cat + human     dog + human
-------------------------------------------------------
-ln L     55.98813       58.39242        60.69938
```

Once again, the dog + cat topology is the preferred ML topology, but notice the gap between the maximum likelihood topology and the other two has been reduced as the result of implementing a different model. We will delve into more complex ways of treating nucleotide and amino acid data in Chapter 9.

Transformation and Probability Matrices

There are many ways to adjust the various character-state changes that may occur, which range from rescoring the characters to using models. One of the best ways is to use a cost matrix that summarizes the weight of all changes that can occur during evolution of the sequences.

Transformation cost matrices address the probability of character state changes in parsimony analysis

In some cases, the probability of character-state change can be assessed from the background biology of the sequences from the taxa under examination. The probabilities can be assigned arbitrarily, drawn from some pre-established knowledge of how the characters change (like "I think losing wings is 100 times easier than gaining them"), or based on empirical observations. The latter, less arbitrary probabilities are gleaned from models. The simplest model would be that all possible changes amongst nucleotides are equally likely. This is called the Jukes–Cantor (JC) model. A slightly more complex model comes from knowing the simple biochemistry of nucleic acids and that transitions (pyrimidine to pyrimidine changes and purine to purine changes) occur more frequently than transversions (in which a pyrimidine changes to a purine or vice versa) (**Figure 8.13**).

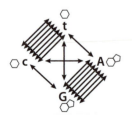

Figure 8.13 Relative frequency of transitions to transversions. The frequency is represented by the number of double-headed arrows. For instance, A to G changes (transitions) will occur more frequently than c to A or t to G changes (transversions).

If we have this information in hand, then there is a range of ways that we can incorporate the knowledge into how we analyze the data. At one extreme we can simply say that transitions are not reliable because they change faster and hence more convergence is possible. We would then be able to justify eliminating transitions from the phylogenetic matrix. In other words, we give a weight of 1 to any characters that are transversions versus a weight of 0 to transitions. This kind of analysis in parsimony is called "transversion parsimony." The simplest way to implement this weighting scheme for DNA sequences would be to rescore the characters as we did above into Ys (for pyrimidines) and Rs (for purines). At the other extreme, we can ignore the information on transitions and transversions and simply weight

Sidebar 8.4. Calculating likelihoods in sequences, position by position.

These calculations are for the tree defined by dog + cat:

site	1	2	3	4	5
L	0.092689	0.012235	0.002855	0.012235	0.092689
ln L	-2.378510	-4.403471	-5.858832	-4.403471	-2.378510

site	6	7	8	9	10
L	3.768×10^{-4}	0.012235	0.012235	0.092689	0.092689
ln L	-7.883793	-4.403471	-4.403471	-2.378510	-2.378510

site	11	12	13	14	15
L	0.092689	0.002855	0.092689	0.092689	0.019142
ln L	-2.378510	-5.858832	-2.378510	-2.378510	-3.955876

The whole process then needs to be repeated for the tree that places cat + human:

site	1	2	3	4	5
L	0.080318	0.003066	0.006376	0.003066	0.080318
ln L	-2.521759	-5.787514	-5.055288	-5.787514	-2.521759

site	6	7	8	9	10
L	2.433×10^{-4}	0.003066	0.003066	0.080318	0.080318
ln L	-8.321043	-5.787514	-5.787514	-2.521759	-2.521759

site	11	12	13	14	15
L	0.080318	0.006376	0.080318	0.080318	0.016905
ln L	-2.521759	-5.055288	-2.521759	-2.521759	-4.080171

A final set of likelihood calculations for the tree that places dog + human is depicted:

site	1	2	3	4	5
L	0.061592	0.003438	0.002662	0.003438	0.061592
ln L	-2.787218	-5.672832	-5.928679	-5.672832	-2.787218

site	6	7	8	9	10
L	3.415×10^{-4}	0.003438	0.003438	0.06159 2	0.061592
ln L	-7.982193	-5.672832	-5.672832	-2.787218	-2.787218

site	11	12	13	14	15
L	0.061592	0.002662	0.061592	0.061592	0.026801
ln L	-2.787218	-5.928679	-2.787218	-2.787218	-3.619319

every character in the DNA sequences equally, as in the JC model. In between these two extremes are a broad range of weighting schemes and models that can be imposed to deal with the transition/transversion phenomenon and other nucleotide transformation schemes.

These other schemes and models might involve quantifying the degree to which transversions are favored over transitions. The nucleotide state transformation matrix below is an example of weighting transversions over transitions, where the numbers in the matrix represent the cost in the number of steps such changes would incur in a phylogenetic analysis.

	G	A	T	C
G	0	2	1	1
A	2	0	1	1
T	1	1	0	2
C	1	1	2	0

For instance, in the matrix above, a change from a G to a G would cost 0. A change from a G to an A is a transition (purine to purine) that occurs at higher frequency than the other category of nucleotide change (transversions) and would cost twice as much as a transversion. Transitions are listed in this matrix as having a cost of 2 because such changes might not be simple single events and might result in convergence and will be misleading with respect to the actual phylogeny of the entities in the analysis. On the other hand, a change from a G to a T only costs 1 because these changes are rarer and less likely to be the result of back mutations. Back mutations occur as a result of the mutation process in cells. They refer to changes from one base state to another and back again to the original base. The upper half of the matrix is simply a mirror image of the lower half because directionality is not considered in this kind of weighting matrix.

The exact values in a matrix like this can be very specific to the genes and taxa being examined. As an example, a popular approach in understanding the biology of populations is to reconstruct the relationships of maternal lineages within a species using mitochondrial DNA (mtDNA) D-loop sequences. The mtDNA D-loop is an untranscribed region of the mitochondrial genome that resides at the origin of replication of the mtDNA and evolves very rapidly. The evolution of this region is so rapid that the frequency of transitions relative to transversions is very high. In some cases the ratio of transitions to transversions is between 10:1 and 20:1. This observation would force the systematist to rely heavily on transitions *and* any of the very rare transversions that might occur in reconstructing the maternal genealogy and perhaps lead to the following matrix:

	G	A	T	C
G	0	20	1	1
A	20	0	1	1
T	1	1	0	20
C	1	1	20	0

As we said above, when the divergence of organisms deepens, the transitions are less reliable because some of the nucleotide positions become saturated with change (Figure 8.4) and exhibit convergence; thus, the apparent frequency of transitions to transversions is lower. In this case, weighting transversions *over* transitions in phylogenetic analysis is preferable. These transition to transversion ratios can be estimated from the actual sequence data or can be established *a priori* by calculating the relative times of divergence with a calibrated ratio. This example leaves out a potential fifth character state, a gap introduced by alignment, but this character state is more difficult to incorporate into the matrix based on frequency of its occurrence and on the relatively unknown dynamics of gap to base and base to gap changes. While some researchers are developing models to accommodate gaps, they have been treated as missing data in most ML analyses.

Amino acid transformations in proteins can be similarly weighted by use of a broad array of matrices. The genetic identity matrix discussed in chapter 6 is one of the simplest models and is based on the genetic code. Other types of matrices are based on quantification of the kind and frequency of amino acid transformations in proteins in the database. The first matrix constructed for this purpose was the Dayhoff matrix. Cost matrices such as the PAM and BLOSSUM matrices can also be adapted to weight characters in protein-based phylogenetic studies.

Generalized probability matrices incorporate probabilities rather than integral weighting values in likelihood analysis

The previous section describes character-state transformation matrices for parsimony analysis. Likelihood analyses use similarly constructed transformation probability matrices to implement models of evolution for proteins and nucleic acids. Distance analysis also uses a similarly constructed cost or probability matrix to compute the initial distances from the aligned sequences that are then used in the various distance approaches. However, instead of placing integer or weighting values in the matrix, probabilities or frequencies are used. A generalized transition to transversion two-parameter (α and β) transformation matrix is shown below. In this matrix, α represents the probability of the occurrence of transitions and β represents the probability of the occurrence of transversions.

	G	A	T	C
G	0	α	β	β
A	α	0	β	β
T	β	β	0	α
C	β	β	α	0

Another way to look at the transformation matrices is to treat each cell in the 4×4 matrix as an independent variable. There are four state changes in DNA sequences (five, if gaps in alignments are counted). Therefore, there are 16 potential state changes (shown in the matrix in Sidebar 8.1). The frequency of each change is represented with an italicized lowercase letter. In the matrix, e, b, l, and o are transitions; a, f, k, and p require no change; and the rest (c, d, g, h, i, j, m, and n) are transversions.

	G	A	T	C
G	a	b	c	d
A	e	f	g	h
T	i	j	k	l
C	m	n	o	p

Note that some of the changes involve the same base states but are given a different letter. For instance, o (frequency of C to T changes) is just the opposite of l (frequency of T to C changes). Such generalized matrices can be used to expand the parameters that are used in maximum likelihood and distance analysis and are called generalized time reversal (GTR) models.

Various models that have been used in likelihood and distance analyses and their relationships to one another are shown in **Figure 8.14**. In general, the transformation probabilities in these matrices are calculated from the data, and the

Figure 8.14 Hierarchical relationships of the generalized time reversal family of substitution models. These models are used in maximum likelihood searches and in calculating distances for distance methods. Arrows indicate the class or identity of an assumption of molecular evolution incorporated into a model of nucleotide change that converts the model with the least number of assumptions (JC) to the models with more assumptions. The Jukes and Cantor (JC) model is the least assumptive model in the maximum likelihood hierarchy. The Felsenstein (F81) model assumes equal base frequencies. In the Kimura two-parameter (K2P) model, the assumption of a single substitution type is made. The K2P model can give rise to the Hasegawa, Kishino, and Yano model (HKY (F84)), where the more restrictive assumption of two substitution types is made, and the Kimura three-substitution type model (K3ST), where the K2P model has a further assumption of equal base frequencies applied. The F81 model also gives rise to the HKY (F84) model. The Tamura and Nei (TrN) model is related to the HKY (F84) model with the further assumption of two substitution types. In the SYM model, three substitution types are assumed to be part of the model. Finally, the generalized time reversal (GTR) is achieved.

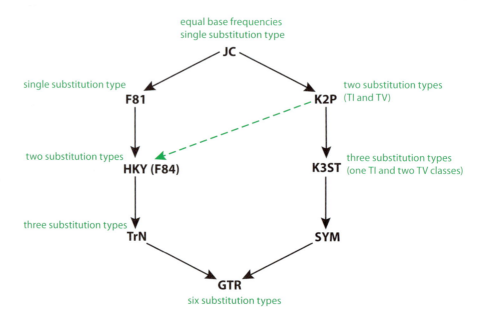

appropriateness of particular models relative to one another is determined by a least-squares test of the log likelihoods of each model. Algorithms have been developed that can be used for the assessment of fit for different models and to determine the most appropriate model to use for phylogenetic analysis in a likelihood context.

The ratio of the four bases is an important aspect of sequence evolution models. It stands to reason that if there is a large AT to GC bias (as there usually is in mitochondrial DNA sequences), certain kinds of base transformations are more likely than others. Many of the models shown in Figure 8.14 take these base composition biases into account, while some assume equal ratios. It also stands to reason that not all positions in a DNA or protein sequence will have equally probable site-specific variation. In these cases, there are functions that can be used to calculate the probability of site-specific heterogeneity.

Summary

- Data analysis in phylogenomics has two dimensions: whether characters or distances are used, and whether an algorithm or an optimality criterion is used.

- Character states in molecular data may include the presence or absence of genes and the sequence states of nucleotides or amino acids, and these can be weighted relative to each other. Characters themselves can be weighted relative to one another.

- Maximum parsimony analysis is a character-based method that attempts to minimize the number of evolutionary character-state changes on a particular tree topology by use of a mathematical optimality criterion. It is based on the well-known principle of Occam's razor.

- Distance analysis compresses the original information from an organism into an overall similarity measure that is then processed by a deterministic algorithm. Corrections for multiple hits and other models

of evolutionary change can be incorporated into distance analysis.

- Neighbor joining is an algorithm-based distance method for tree building that attempt to place species that are more closely related to each other into clusters.

- Minimum evolution assesses the total distance that would be needed for multiple tree topologies and chooses the one where the amount of distance needed is minimal.

- The maximum likelihood approach involves character-by-character assessment of the probabilities of change and is thus computationally intensive. The procedure uses a model of sequence change to estimate the likelihood, given the various trees and the model, that the data would exist.

- Transformation and probability matrices are used to address the probabilities of character-state changes.

Discussion Questions

1. Take the following 7-base sequences and construct NJ, MP, and ML trees for each. Use the JC model to estimate distances for the NJ approach. Use the JC model with no other corrections for the basic ML analysis. The numbers above the bases represent the positions in the small gene.

```
      1 2 3 4 5 6 7
TxX   G A A A A A A
TxY   G A A A A T T
TxZ   C C A A T T A
O     C C A T T A T
```

2. Map the following characters onto the three possible trees for the three ingroup taxa (I1, I2, and I3). You should be able to map it two different ways. Are they different with respect to number of steps implied for the three trees?

```
I1    A
I2    A
I3    T
O     T
```

3. Arrange the different models depicted in Figure 8.14 in an order that reflects the nesting of one model within another. How does this arrangement of models differ from the diagram in Figure 8.14?

4. What other aspects of sequence change can you think of that should be incorporated into a model other than the base transformation models we have talked about in this chapter?

Further Reading

Farris JS (1983) The logical basis of phylogenetic analysis. In Advances in Cladistics II (NI Platnick VA Funk eds) pp 7–36. Columbia University Press.

Felsenstein J (1981) Evolutionary trees from DNA sequences: a maximum likelihood approach. *J. Mol. Evol.* 17, 368–376.

Felsenstein J (1982) Numerical methods for inferring evolutionary trees. *Q. Rev. Biol.* 57, 379–404.

Rzhetsky A & Nei M (1992) A simple method for estimating and testing minimum-evolution trees. *Mol. Biol. Evol.* 9, 945–967

Saitou N & Nei M (1987) The neighbor-joining method: a new method for reconstructing phylogenetic trees. *Mol. Biol. Evol.* 4, 406–425.

Robustness and Rate Heterogeneity in Phylogenomics

So far we have learned to mine the databases and to manipulate accessions from the database for phylogenetic analysis. In the chapters to follow, we will address several matters of phylogenetics that have become issues as a result of the large amounts of information brought on by genomics. In Chapter 8 we discussed the issue that as the number of taxa grows larger, the number of trees that need to be assessed in the context of optimality becomes astronomically large. Consequently, methods for ensuring that tree searches are efficient and robust have been developed. In addition, phylogenomics has created a situation where all the genes of a genome are made available for analysis. However, not all genes in a genome have the same history. In addition, even within a gene, the various base pairs and their encoded amino acids do not evolve uniformly. Consequently, we need to discuss the corrections that are possible for taking evolutionary heterogeneity into consideration. Another issue relevant to phylogenomics, first recognized in systematics in the 1980s, involves assessing the robustness of inferences we make concerning phylogenies and also, to a certain extent, population-level inferences. In phylogenetic analysis there are actually two ways that scientists assess robustness of an inference. Resampling techniques have allowed the assessment of robustness in the context of bootstrapping and jackknifing. However, these approaches do not give the researcher a statistic that can be applied to the probability of relationships shown in a phylogenetic tree. Other approaches have been developed to assess the statistical significance of inferences made in phylogenetics.

So Many Trees, So Little Time

In explaining distance, likelihood, and parsimony approaches, we have focused on a simple, small number of taxa. Modern systematic analysis, using genomics and high-throughput approaches, has greatly expanded the number of taxa and base pairs or amino acid residues to be analyzed. The number of trees increases rapidly as the number of taxa rises, and the production of an optimal tree becomes an nondeterministic polynomial time (NP)-complete problem. NP-complete problems are a large class of computational problems spanning math and science that all evade efficient direct solutions: they cannot be solved exactly in any reasonable amount of time. However, heuristics have been produced that can provide approximate solutions. A few of the more clever ways to increase the efficiency of searches in the face of this computational problem are described below.

Tree space allows trees to be grouped by optimality

The trees that can be generated from a particular number of taxa are distributed in what is called "tree space" (**Figure 9.1**). Tree space is a nebulous concept, but it can represent the distribution of trees in three dimensions. The axes in the horizontal plane are not important other than to state that, by changing certain parameters, the position in the tree space will change on these axes. If two points in the graph are close to each other, then they have very similar characteristics. The important axis, however, is the vertical axis on which tree optimality is graphed. In this way,

Figure 9.1 Schematic diagram of "tree space." The *x*- and *y*-axes represent tree topology coordinates. The highlighting on the graph represents optimal fits, with darkest highlighting showing the shortest tree and lightest highlighting representing extremely suboptimal tree lengths. Two islands of "good" solutions are shown. Island II has the better solutions, as it is a "higher" peak. The various trees present different topologies, and the numbers below the trees represent the number of steps that tree takes. The tree space can be entered at a large number of places with starting trees. Once tree space is entered, the search moves the solutions to the nearest highest peak. For starting tree a, the search ends up on island I. For starting tree b, the search ends up on island II, the peak with the highest optimality (tree with the fewest steps).

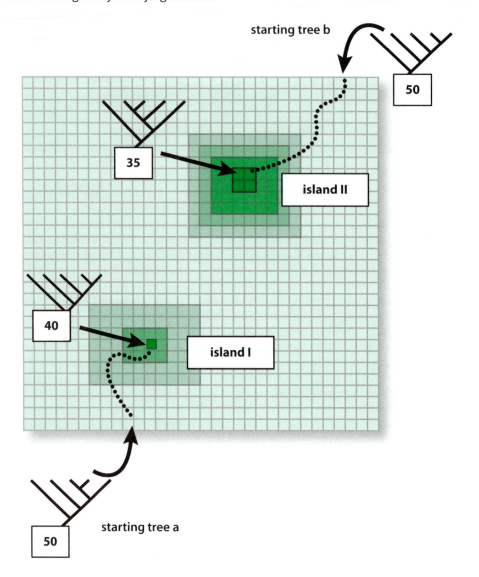

trees can be grouped into "islands" with optimality relationships, which reflect the degree of relatedness of the trees. The surface of this graph resembles a mountain range with valleys and peaks. The peaks represent optimality solutions, and there may be more than a single peak in any given tree space. The peaks also differ in height, meaning that there are different regions of tree space with solutions that are good but not necessarily the best. Another rule about tree space is that from a given location in tree space, movement can occur only to the nearest highest optimality peak. Therefore, unless there is a way to move about the tree space surface when searching for an optimal tree, the same peak will always be reached, and it may not necessarily be the optimal solution. This is a classic mathematical problem of dealing with local and global maximum values.

There are three basic issues with tree space against which systematists contend. The first issue is getting into the tree space in the first place, known as the starting tree problem. The second problem concerns reaching the locally optimal location in tree space. The whole point of the exploration of tree space is to find optimal solutions. In other words, how does one move from a valley to an optimality peak in the tree space? The final problem relates to the question that, upon reaching an optimality peak, is it the best one can do? Is there another peak elsewhere in the tree space that reflects a more parsimonious or more likely solution?

Selection of a starting tree is the first challenge

A large number of taxa in a phylogenetic analysis means that there is an even larger number of starting trees from which to choose. As an example, consider a data matrix with 20 species in it. Unlike the three-taxon cases discussed in Chapter 8, where one can simply visualize the three different trees required to impose an optimality criterion, there are nearly 10^{21} trees to consider. One method for obtaining a starting tree is to take the first taxon in the matrix and add it to the second, then add those two to the third, and then those three to the fourth, and so on, to the 20th taxon in the matrix. Now if we made a new starting tree by adding taxa to the starting tree randomly (for example, take the 19th taxon and add it to the second taxon, add those two to the fifth taxon, etc., until all 20 taxa have been added to the new starting tree), then we would have a new starting point in tree space. We can make as many starting trees as we like so that each different starting tree will land us in a different starting spot in the tree space. Most starting trees may not even be close to the most parsimonious tree or to one of the optimality peaks.

Another approach to obtain a starting tree is to use a computationally rapid search such as neighbor joining (NJ) to provide the initial tree. This initial tree, while conforming to the NJ algorithm, might not necessarily (and indeed, in most cases with large numbers of taxa, will not) be the best solution for parsimony or likelihood. But while the optimal NJ tree will not necessarily be the most parsimonious or likely tree, it should place us within a fairly optimal location in tree space. This is because, even though NJ and parsimony and likelihood are very different methods of determining trees, in many cases the optimal solutions are fairly similar to the NJ solution.

Peaks in tree space can be reached by branch swapping

Here we address the second problem: how is the starting tree altered so as to climb to an optimality peak? There are three ways to move or swap branches around a starting tree: nearest-neighbor interchange (NNI), subtree pruning and regrafting (SPR), and tree bisection and reconnection (TBR). These approaches are described below. To summarize these methods, the order from least efficient to most efficient is NNI → SPR → TBR. Efficiency here refers to more completely examining tree space. This is also the order of computational complexity, so the most efficient approach is also the computationally most difficult. It is hoped that, by traversing tree space using one of these methods, a better solution will be obtained. These techniques follow an iterative approach. First, several simple changes to the starting tree are determined using one of the branch swapping techniques. Next, the best tree produced through the first round of swapping is chosen. This tree then serves as the starting tree for the next round of branch swapping. This process is repeated until an optimal or one of the optimal trees is obtained. The stopping point for the iteration is usually either when a certain threshold is reached or when no better trees are obtained after several iterations. For example, if the tree starting one iteration has an arbitrary score of 15, and all of the possible trees for the swapping of that tree also have a score of 15, then the algorithm might stop since it does not see any opportunities for improvement. At the same time, since all of the trees have the same score of 15, they are all equally optimal and none of them could be termed the "best" tree as compared to the others. Therefore, any of the final trees can be said to represent the relationships between the entities under consideration.

The NNI approach takes a starting tree and recognizes that each interior branch of the starting tree actually defines a local region of four trees that are called "subtrees." NNI then takes one interior branch at a time and exchanges it with the next nearest branch, of which there are always two for an interior branch (**Figure 9.2**).

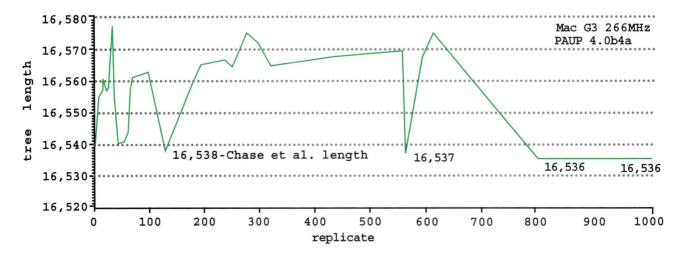

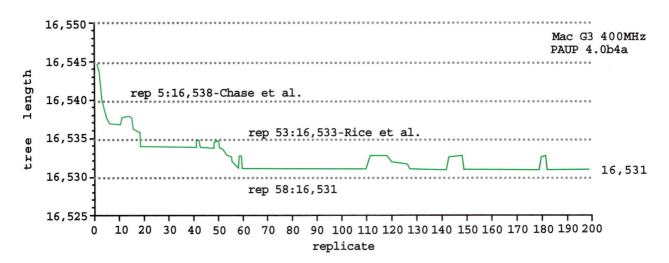

Figure 9.5. Nonratchet versus ratchet analysis of the *rbcL* plant database. Upper panel: a heuristic search with 1000 random additions and a tree bisection and reconnection branch swapping approach. The graph shows that the best island is reached after 800 replicates and the tree length is 16,536 steps. This run took 19 hours to complete. Lower panel: the ratchet analysis, which took 4 hours and resulted in a tree five steps shorter than the heuristic search in the top panel. (Courtesy of Derek Sikes.)

we discussed in previous chapters, the chemical nature of the individual amino acids in a protein dictate the three-dimensional structure of proteins. Proteins contain folds and sheets that are stable, functionally necessary, and more resilient to evolutionary change than the loop regions in proteins. Amino acid sequences also exhibit rate heterogeneity. Given these examples, how is rate heterogeneity incorporated into ML models?

Rate heterogeneity can be included in likelihood models by use of a γ distribution

Fortunately, a computational framework exists that allows for the inclusion of rate heterogeneity into ML models. The approach takes advantage of what is called a γ distribution dictated by a mathematical equation. This distribution forms specific shapes when the amount of change in a sequence is depicted relative to the number of positions changing (**Figure 9.6**). In computing these distributions, there is a factor in the equations that controls the shape of the distributions, known as an α shape parameter. By computing or predesignating the α parameter, one can take into account the shape of the γ distribution for a particular data set.

Table 9.1 Values of ln *L* for three possible trees for two models with γ correction using a preset α = 0.5.

	tree 1, dog + cat	tree 2, cat + human	tree 3, dog + human
K2P + G model			
-ln *L*	55.62047	61.11105	62.08981
F81 + G model			
-ln *L*	54.50018	56.43914	57.67398

Thus, if this general phenomenon is incorporated into the probability statements in the likelihood models, then rate heterogeneity can be accommodated.

There are two ways to accommodate α in ML analyses. First, a predetermined α parameter that approximates reality can be used. Likelihood analyses that adhere to this approach use an α parameter of 0.5, which approximates the curve shown in the middle panel in Figure 9.6, where very few sites show a significant amount of change and most show little change. Second, an α parameter can be calculated from the data using the model and tree shapes to compute a maximum likelihood for the gamma distribution. The α parameter is then obtained from this computation.

With only a few taxa in a data matrix, this calculation is fairly simple. However, in the presence of many taxa, the computation is time-consuming, as a full likelihood analysis is needed. Consider the dog/cat/human/opossum example introduced in Chapter 8. The K2P model and the F81 model can be evaluated with an α parameter of 0.5 and by computing α from the data themselves.

The difference between the first analysis described in Chapter 8 for K2P and the current analysis *with* rate heterogeneity as a γ distribution (where α = 0.5) is that the likelihoods calculated site by site are computed incorporating the γ distribution in the calculation. This analysis yields the ln *L* values for the three possible trees in the example for K2P and F81 shown in **Table 9.1**.

Note that, for both transformation models, the dog + cat tree has the highest likelihood (recall that the tree with the maximum likelihood has the lowest ln *L* value), but for the F81 model and a γ distribution, the gap between the ln *L* values of the three topologies has closed considerably as compared to the analyses that omitted the γ distribution. When we calculated ln *L* for these two models by estimating the α parameter from the data, the results in **Table 9.2** were obtained.

Again, the dog + cat tree is the most likely, but note that the dog + human tree has the second highest likelihood, whereas in the previous comparisons, the cat + human tree had the second highest likelihood. Modern systematic studies usually

Figure 9.6 Three γ distributions for site-specific rate heterogeneity. The upper panel shows a γ distribution where most positions change at more or less moderate but equal rates and very few positions (at the tails of the curve) change very rapidly or very slowly. The middle panel shows a γ distribution that is more common for DNA sequences, where most of the characters change very little, with all of the changes being localized to a small number of sites. The lower graph shows a "flat" γ distribution, where equal numbers of sites fall into rate classes that increase steadily, with no one rate class being overrepresented in the sequence.

Table 9.2 Values of ln *L* for three trees for K2P and F81 models with γ correction using an α calculated from the data.

	tree 1, dog + cat	tree 2, cat + human	tree 3, dog + human
K2P model			
-ln *L*	52.97719	60.93737	59.99660
F81 model			
-ln *L*	53.17687	56.43317	56.20828

involve hundreds of taxa, and these studies pose special problems in terms of the methods used to evaluate optimality criteria tree by tree.

The likelihood ratio test can be used to determine which models are more likely or probable than others

One other problem that needs to be addressed in terms of constructing data matrices is the number of models used and the proportion of likelihoods to be estimated. Given these considerations, what keeps an ML analysis from turning into a Tower of Babel because of all of the different models that can be applied?

Fortunately, for likelihood, we can assess which models are more likely or probable than others. In order to develop an approach to assess the best model, we first need to understand the relevant statistical methods. The first tool is the likelihood ratio test (LRT), which assesses the fitness of likelihood models. The statistic is given as

$$\delta = -2 \log \Lambda \tag{9.1}$$

where

$$\Lambda = \max [L_s \text{ (simpler model)}|\text{data}]/\max[L_c \text{ (alternative model)}|\text{data}]$$
$$= 2 (\ln L_c - \ln L_s) \tag{9.2}$$

where L_s is the likelihood score computed for the simpler (or parameter-poor) of the two models being compared and L_c is the likelihood score computed for the more complex (or parameter-rich) of the two models being compared. The value of this ratio is equal to or greater than 0, as the likelihood under the parameter-rich model is equal to or higher than the likelihood under the simpler model.

This approach requires that the models be related in a certain way so that characteristics such as "parameter poor" and "parameter-rich" can be assessed easily. Specifically the models should be what is called nested. For example, the F81 model can be reduced to the Jukes–Cantor (JC) model when the base compositions are set equal to 0.25 for all frequencies. Therefore, the JC model is nested within the F81 model. In other words, nested models are created when the null hypothesis of one model is a special case of the alternative hypothesis. The δ statistic obtained from the LRT is distributed as a χ^2 distribution (or related to χ^2 distributions). A χ^2 distribution is a kind of binomial distribution where only two variables are allowed (say "success" or "failure," or in the case of systematic approaches that use optimality criteria, "optimal" or "not optimal"), with q degrees of freedom, where q is the number of parameters removed from the parameter-rich hypothesis to derive the simpler model. Figure 8.14 shows how the various models that are used by systematists are related to each other and provides the number of parameter differences between each successive model.

One final point to consider in calculating the LRT is that, since the models need to be nested, the LRT must be accomplished on the same tree topology. However, once the models are compared and one is judged the most appropriate, this model can then be applied to generate the final tree. The LRT is the least complicated test used to determine appropriateness of model. Other, more complex approaches can also be applied.

Several programs can rapidly compare models

Keith Crandall and David Posada have developed rapid model testing approaches and have created programs for such testing called Modeltest (for nucleotides) and Prottest (for amino acids). In the original Modeltest program there were 14 basic models for which four different sets of parameters were imposed to give 56 basic models that Modeltest could compare. Even more models have subsequently been added to this list. As an example of how the LRT works, we discuss below four tests involving the *CD9* example. A significance cutoff of $p < 0.01$ is used to assess the significance of the LRT.

The first test compares models with different base frequency parameters; specifically, the JC model and the F81 model. The F81 model reduces to the JC model when base frequencies are equal, while the F81 model accommodates differing base frequencies. Therefore, JC is the simpler model (the null hypothesis) and F81 is the more parameter-rich model. One model can easily reduce to another and hence the latter model is nested in the former simply by making an assumption about the parameters of the first model, inserting values into the first model that correspond to the assumptions and doing the math. If, after the math is accomplished, the model now looks like another simpler model, then the first model has been reduced to the simpler latter model.

The $-\ln L$ score for the F81 maximum likelihood tree (dog with cat) was 55.9883. The $-\ln L$ score for the JC model is 59.7302. Using the equation for Λ given above, we have $2(-55.9883 - (-59.7302)) = 2[3.7419] = 7.4838$. The degrees of freedom for this comparison is simply the number of parameters in the richer model that need to be removed to obtain the simpler model. The F81 model accommodates four base frequencies; therefore, three parameters need to be removed to reduce this model to the JC model. In this case, the degrees of freedom are 3. Examination of a χ^2 table reveals that, with 3 df and a δ score of 7.4838, the p-value is 0.057965, which is nearly significant at the $p < 0.05$ level, and implies that the F81 model is on the borderline of being a better model than the JC model for these data.

As a second test, we can evaluate whether introducing rate heterogeneity to the JC model improves the performance of the analysis. In this LRT, the JC model is less parameter-rich, while the JC model plus the γ correction model is more parameter-rich. This makes the simpler JC model the null hypothesis. Again we calculate $-\ln L$ for the two models (JC = 59.7302 and JC plus γ = 57.5932) as $\delta = [2(-57.5932 - (-59.7302))] = 2[2.1370] = 4.2740$. The degree of freedom for this comparison is 1 since the γ correction is a single parameter; if it is removed, the model reduces to the JC model. Examination of a χ^2 table reveals that with 1 df and a δ score of 4.7240, the critical value is 0.019349 (which is significant at the $p < 0.05$ level). Therefore, we can conclude that the simpler JC model and the JC model with γ correction are not significantly different from each other. Comparisons of models that are only one parameter richer than the JC model reveal a lack of statistical significance of improvement over the JC model alone. In this case we use the JC model with the assumption that transitions (TI) and transversions (TV) occur at the same rate for one comparison, and we use the JC model with no invariant sites for the second comparison. When we compare these two models to the JC model alone, the first comparison [JC versus JC + (TI = TV)] has df = 1 and gives a critical value with $p = 0.056771$, and the second comparison (JC versus JC + no invariant sites), also with df = 1, gives a critical value with $p = 0.130970$. Neither of these statistics is significant at the $p < 0.05$ level. In terms of the other 56 models that Modeltest can compare, none of these more complex models improves the statistical significance of the inference at $p < 0.01$; hence, at this significance level, the JC model alone is the most appropriate model for the *CD9* data set we have examined in this example.

Simple Metrics as Measures of Consistency in a Parsimony Analysis

When we generated the trees in Chapter 8, we simply accepted the distance tree, the parsimony result, and the likelihood results as our solution. This acceptance is only one part of systematic analysis. Phylogenomicists also need to assess the strength of the inferences they make with these methods. The simplest measures of inference strength come from measuring the consistency of characters and of the overall inferred tree. These measures count steps on a tree and use various scaling methods to produce a metric that measures either consistency (the consistency index) or retention (the retention index). In addition, there are several measures that use tree length that can be computed to give one an idea of how stable a phylogenetic hypothesis might be. In order to discuss these metrics, we will return to the dog, cat, human, and opossum analysis from Chapter 8. These measures are somewhat qualitative, but they give the researcher some idea of the degree of consistency of the characters being used in an analysis. For instance, suppose two researchers generate two different data sets for the same taxa and obtain phylogenetic trees with a consistency index of 0.999 for the data set of the first researcher and a consistency index of 0.222 for the data set of the second researcher. The first researcher can easily say that his or her data set is more consistent than the second. In fact, the consistency of a data set is inversely proportional to the amount of convergence in a data set, and the first researcher can actually say that their data set has less convergence in it.

Tree length evaluates the degree of parsimony within a given solution

This measure simply counts the number of steps that a given matrix will require to be mapped on a tree. This measure is used to evaluate whether or not a most parsimonious solution has been reached (see Chapter 8).

```
                                1     1
                        5       0     5
cat_CD9         ACTCAACCTTCCAGC
dog_CD9         ACCCAGCCTTCTAGC
human_CD9       AATAATAATTCCAGC
opossum_CD9     AACAACAATTCTAGT
```

The most parsimonious solution to the matrix with opossum as the outgroup was shown in Figure 8.8. Remember there were only three trees to evaluate for this example. The figure shows the number of steps each character lends to the tree. In this case, the (dog (cat, human)) tree is 14 steps long, the (cat (dog, human)) tree is 16 steps long, and the (human (dog, cat)) tree is 12 steps long.

Consistency index estimates convergence in a data set

When maximum parsimony analysis is performed, a simple quick estimation of how much convergence there is in a data set can give the systematist a good idea of how reasonable or appropriate the characters in the analysis are. One index that measures convergence is called the consistency index (abbreviated ci for individual characters and CI for an ensemble of characters). This index can be used to measure the consistency of an individual character (ci) or the overall or ensemble consistency of all characters (CI) over the entire tree. The equation for calculating the character consistency index is:

$$ci = min/step \qquad (9.3)$$

where min = theoretical minimum number of steps the character takes on any tree and step = actual number of steps the character takes. The minimum number of steps is easily calculated by taking the number of character states and subtracting 1 from it. In the above example, we can calculate the 15 character consistency indices as follows:

Characters 1, 5, 9, 10, 11, 13, and 14 all have a single character state, and so min = 1 − 1 = 0.

Characters 2, 3, 4, 7, 8, 12, and 15 all have two character states, so min for each of these characters is 2 − 1 = 1.

Character 6 has four character states and so min = 4 − 1 = 3.

Next we calculate the number of steps the characters have on the tree. The three trees and the steps of the characters on each tree are shown in Figure 8.8. For this example we will use the most parsimonious tree, (human, (dog, cat)), to calculate the character ci values. We could, however, use any of the three possible trees to calculate character ci values (see Discussion Questions at the end of the chapter).

> Characters 1, 5, 9, 10, 11, 13, and 14 each put 0 steps on the most parsimonious tree. Hence the ci for these characters is undefined (0/0 is undefined).

> Characters 2, 4, 7, and 8 take one step on the most parsimonious tree and so their consistency is 1/1 or 1.0.

> Characters 3 and 12 take two steps on the most parsimonious tree and so the ci for these characters is 1/2 or 0.5.

> Character 6 takes 3 steps on the most parsimonious tree and so its ci is 3/3 or 1.0.

The ensemble consistency index is calculated by summing all min values and dividing by steps on the given tree. In this case, characters 2, 3, 4, 7, 8, 12, and 15 each give 1.0 to the min, character 6 gives a 3 and the min summed is 10. The total number of steps is 12 (see Figure 8.8) and so the ensemble CI = 10/12 or 0.83. Either consistency index (character or ensemble) is usually reported to two decimal places and varies between 0.00 and 1.00, with values closer to 1.00 indicating more consistency with the tree for which it was calculated.

Retention index assesses degree of agreement with the maximum parsimony tree

Another measure of the degree of agreement of a single character or an ensemble data set with the maximum parsimony tree is called the retention index. The retention index calculates the relative amount of shared derived characters (synapomorphies) in a tree. Such shared derived characters are defined as being in good agreement with the maximum parsimony topology. The retention index can also be calculated by character (ri) or for an ensemble of characters (RI). The equation for ri is

$$ri = (max - steps)/(max - min) \qquad (9.4)$$

For the dog/cat/human example above, the maximum number of steps is found for mapping each character on a tree in a nonparsimonious way. For characters 1, 5, 9, 10, 11, 13, and 14, $(max - steps)/(max - min) = (0 - 0)/(0 - 0) = 0/0$ or undefined, so we ignore these characters. Character 15 is also interesting because it has a max of 1.0, so $(max - steps)/(max - min) = (1 - 1)/(1 - 1) = 0/0$, which is also undefined, and so it too is ignored. Character 6 has max = 3, min = 3, and steps = 3, so its ri is $(3 - 3)/(3 - 3) = 0/0$, which is also undefined. It turns out, therefore, that ri and RI are calculated by using only phylogenetically informative characters. These characters for our dog/cat/human example are 2, 3, 4, 7, 8, and 12. Each of these characters has a max of 2 and a min of 1. Characters 3 and 12 each cost two steps and characters 2, 4, 7, and 8 each cost one step on the most parsimonious tree. So the ri for characters 3 and 12 is $(2 - 2)/(2 - 1) = 0/1 = 0.00$. Each of these characters contributes 0.00 to the overall synapomorphy on the most parsimonious tree. Characters 2, 4, 7, and 8 each have ri = $(2 - 1)/(2 - 1) = 1.00$, indicating that all four of these characters are fully synapomorphic on the most parsimonious tree. The ensemble RI is simply

$$RI = \Sigma_i (max_i - steps_i)/(max_i - min_i) \qquad (9.5)$$

where i = each character that is phylogenetically informative. In our dog/cat/human example, this sum is $(0 + 0 + 1 + 1 + 1 + 1)/(1 + 1 + 1 + 1 + 1 + 1 + 1) = 4/6 = 0.67$. ri and RI vary from 0.00 to 1.00 and again, the higher the value, the more synapomorphy a character (ri) or ensemble of characters (RI) has with a chosen tree.

It should be pointed out that the CI can be calculated over all characters or using just the informative characters. However, if the uninformative characters are removed from a consistency index analysis, this will inflate the CI value.

Rescaled consistency index addresses some shortcomings of retention and consistency indices

A final measure that can assess the relative quality of characters and ensembles of characters in a maximum parsimony analysis corrects for some of the shortcoming of the retention and consistency indices. Steve Farris, one of the founders of the empirical approach to parsimony analysis, recognized that ci (and CI) is usually deflated because the maximal number of steps of a character can be relatively large, creating a situation where ri and ci are not directly comparable. This is a problem because at most only one of the measures can be accurate, and the other measure (or sometimes both measures) will be flawed. In order to correct for this scaling problem, Farris developed the rescaled consistency index rci for character-by-character measures and RCI for ensemble calculation of the measure. The rci is simply the product of ci × ri for each character, and RCI = CI × RI. So for characters 1, 5, 6, 9–11, and 13–15, rci is undefined because ri is undefined. But for characters 2, 4, 7, and 8, ri = 1.0 and ci = 1.0 on the most parsimonious tree, making rci = 1.0 × 1.0 = 1.0. For characters 3 and 12, rci = 0 and ci = 0, so rci for these two characters is 0. The ensemble RCI is calculated simply by taking the ensemble CI and multiplying by the ensemble RI. For the most parsimonious tree, this is 0.67 × 0.83 = 0.58.

Decay index or Bremer index estimates robustness of a node

A novel method for estimating or visualizing the robustness of a node in a phylogenetic analysis is called the Bremer index (named after Kåre Bremer, a Swedish botanist) and is also known as the decay index. Decay index is a better descriptor of the approach, as the indices generated with this method tell how many steps longer than the most parsimonious solution a tree needs to be for a node to decay or collapse into an unresolved set of relationships. The idea is that if it takes one step longer than the most parsimonious tree for a node to collapse for a particular data set, the Bremer or decay index is equal to 1. Simply put, it only takes looking at a tree that is one step longer than the most parsimonious tree to collapse that node. If it takes a tree 100 steps longer than the most parsimonious tree to collapse a node, the node has a Bremer index of 100. Large Bremer index values suggest that a node is strongly supported by the data: the node with a Bremer index of 100 will require the addition of a huge amount of conflicting data to collapse it. Small Bremer values suggest weak support for a node: for a Bremer index of 1, just adding a small amount of data that conflicts with the node will destroy the node.

Again our dog/cat/human example will be used, this time to calculate a decay index. In the most parsimonious tree (and indeed in all three possible trees for these three ingroups), there is a single node we need to assess for robustness by use of the decay index. In the most parsimonious tree (human (dog, cat)), it is the node that defines dog with cat as a monophyletic group. The number of steps in the most parsimonious tree is 12, and the number of steps in the next shortest tree (cat (dog, human)) is 14 steps. The decay index is therefore 14 − 12 or 2. Decay indices are presented as integers at or near the node they evaluate.

Determining the Robustness of Nodes on a Phylogenetic Tree by Resampling Techniques

Systematists often want to know how well their trees will hold up to further analysis. These approaches take advantage of the resampling approaches we discussed in Chapter 1 to estimate robustness. There are two major kinds of resampling methods that have been used in phylogenetics: bootstrapping and jackknifing.

Bootstrapping analyzes a phylogenetic matrix by resampling with replacement

Bootstrapping was developed by Joe Felsenstein in the 1980s. The initial bootstrapping technique was known as resampling with replacement. In this procedure, the phylogenetic matrix is treated as having removable, duplicatable, and movable columns. In this way, the method can remove columns from an alignment and replace them with duplicated, already existing columns. This approach as it applies to DNA sequences is demonstrated in **Figure 9.7**. Note that some columns (2, 5, 7, 9, and 10) are simply removed, and other columns (1, 3, 4, 6, and 8) are duplicated to create a new matrix with the same number of characters. The next step in a bootstrap analysis is to randomly create a preset number of these resampled matrices. Say for the sake of example here that we will generate 100 randomly resampled matrices where we sample 50% of the columns. The next step is to analyze the 100 resampled matrices, using a phylogenetic optimality criterion (ME, MP, or ML) or a deterministic algorithm (NJ), and keep track of all 100 trees. For each tree, the presence of nodes from the original tree is assessed. This count is tallied for each node, and it is expressed as a percentage indicating the robustness of that node. A node with a 100% bootstrap score would be very highly supported by the data since it is produced in all of the random combinations of the data. In contrast, a node with a 0% bootstrap score would be totally unsupported by the data. In general, though, the thresholds for bootstrap scores are rather arbitrary. Some early examination of bootstrap values and systematic certainty suggested that a bootstrap value greater than 75% was a decent indicator of robustness. However, these early estimates were done on single genes. Phylogenomic-level studies can be expected to have much higher bootstrap values.

For the sequences in Figure 9.7B, there is only a single node that is important for bootstrapping, and this is the most derived node in each of the three trees. The three possible trees that can be generated with species 1, 2, and 3 as ingroups and their Newick formatted trees are shown in Figure 9.7A. A bootstrap analysis simply goes back to all of the trees saved from the resampling procedure in Newick format and counts the number of times a (1,2) tree is found, how many times a (2,3) tree is found, and how many times a (1,3) tree is found. So for instance, if (1,2) format is found in 95 of the trees generated after resampling, and three trees have (2,3) and two trees have (1,3) format, then this means that 95% of the bootstrapped trees contain the node (1,2). It follows that even after resampling, node (1,2) rises to the top as a strongly represented node in the data set. This would be said to have 95% bootstrap "support." When bootstrapping is performed, the number of random matrices generated (usually called replicates) can be altered to any number. It turns out that 100 bootstrap replicates are usually enough for relatively small data sets with respect to number of taxa. Larger bootstrap replicate numbers may be needed for larger data sets. Another aspect of bootstrapping is that the percentage of characters that can be removed and replaced can be set to different values. When small numbers of characters are available in an analysis, perturbing only 30% or 50% of the characters might be desirable. The bootstrap analysis done with 30% character manipulation is less stringent than the one done with say 99% manipulation. Again, phylogenomics-level studies will most likely be held to a higher standard and studies using genome-level information might consider using character perturbation near 100% in bootstrap estimations.

Also, as one might guess, as the number of taxa increases, the number of nodes to consider in a bootstrap analysis gets larger and larger. The general relationship is that there are $n - 2$ nodes to consider for a matrix where n = the number of possible trees. For instance, for four ingroups there are 15 trees but only 13 nodes that are assessed in a bootstrap analysis. For five taxa, there are 105 trees but only 103 nodes that are assessed in a bootstrap analysis. We examine the bootstrap and jackknife approaches in more detail in Web Feature 9, using the phylogenetic analysis programs we have already learned to use.

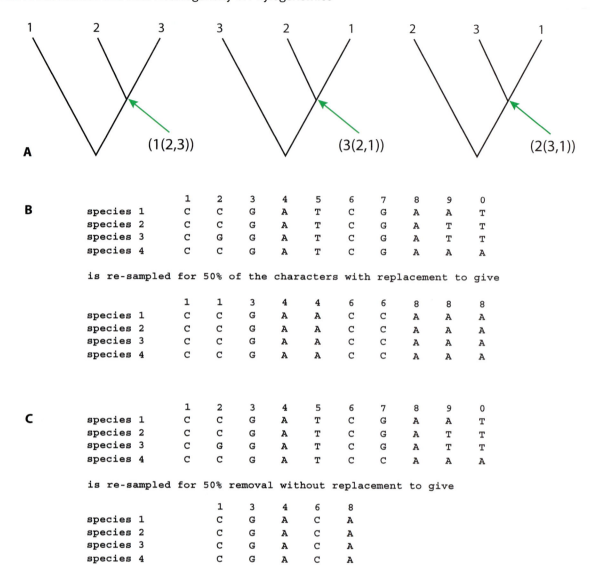

Figure 9.7 Bootstrapping and jackknifing. A: three possible trees for three taxa, showing which monophyletic group is assessed for robustness in the bootstrap and jackknife procedures. B: matrices demonstrating how the bootstrap procedure (or removal with replacement) works. C: demonstration of the jackknife procedure (or removal without replacement). The upper matrix represents the original matrix, and the lower matrix represents one of the matrices produced by the jackknife process.

Jackknifing analyzes a phylogenetic matrix by resampling without replacement

This approach, like bootstrapping, is a resampling technique. The difference is that in jackknifing the columns are removed but not replaced. This procedure is repeated a set number of replicates, say 100, to produce 100 new matrices with randomly removed characters in matrices smaller than the original matrix. As with bootstrapping, one can set the percentage of characters removed without replacing them; for our example, we will use 50%. An example of a jackknifed matrix is shown in Figure 9.7C. As with the bootstrap analysis, three trees are possible. In addition, there is only a single node that is relevant to the jackknife analysis (these nodes are shown in Figure 9.7A). Again, for 100 randomly resampled matrices, a tree is generated and saved in parenthetical notation and the final list of trees is scanned for trees with (1,2), (2,3), or (1,3) formats. Each class of tree is counted and the result is reported as a jackknife percentage. For the matrix in Figure 9.7C, the results would be 95% for (1,2), and 5% for the other two trees combined. Note that neither bootstrap nor jackknife values are treated as statistics. Rather, they are treated as metrics, with values closer to 100% being more reliable and "robust" than values below that percentage.

Parametric bootstrapping applies a distribution model to the data

The methods discussed above for bootstrapping and jackknifing are nonparametric methods. This simply means that the result of the test is not dependent on an underlying distribution of the data. These methods are therefore distribution- and model-free. However, some systematists suggest that models of DNA sequence evolution are important for evaluating trees, so parametric tests, or tests that make assumptions about the distribution of the data, have been developed for bootstrapping in phylogenetics. In parametric bootstrapping, a model is first applied to the data. For phylogenetics, the models can be any of the many likelihood models we have discussed in Chapters 6 and 8. Once the model has been fitted, it can be used to simulate sequence matrices. These matrices will give trees that form a parametric distribution, against which hypotheses can be tested. Soltis and Soltis describe the parametric bootstrap in the following way: "… a single data set can be used to parameterize a model of sequence evolution. This model is then used to simulate new, independent data sets, each of which is analyzed in turn to generate a distribution against which a specific hypothesis can be tested". We present a step-by-step outline of how the procedure works in practice in Sidebar 9.1.

Sidebar 9.1. The parametric bootstrap.

The parametric bootstrap is used to test hypotheses of node existence in phylogenetics in a likelihood context. The following steps are used:

Step 1. Fit a model to the data matrix that is being tested using maximum likelihood.

Step 2. Once the model has been fitted, use it to generate matrices of the same size as the original matrix. This step can be done in sequence simulation programs such as SEQ-GEN. The ML tree is assumed to be the "true" tree and is used as the model tree to simulate the new sequence matrices.

Step 3. Once several hundred new matrices are simulated, separately analyze each matrix. This creates a parametric distribution of trees against which specific hypotheses about the initial tree can be tested. Either likelihood or parsimony methods can be used to accomplish the tree analyses to generate the distribution.

Step 4. The monophyly of specific groups can then be assessed for each tree from each simulated matrix. This results in a distribution of comparisons when carried out over all of the simulated matrices. The significance level of the monophyly of a group is then calculated from this distribution.

Summary

- The trees generated from a particular number of taxa are distributed in three-dimensional tree space, in which trees can be grouped into islands with optimality relationships.

- A computationally rapid search such as neighbor joining can be used to select a starting tree within a fairly optimal location in tree space. Subsequent likelihood or parsimony analysis on this NJ starting tree may produce higher optimality.

- A locally optimal location in tree space can be reached by use of branch swapping methods such as nearest-neighbor interchange, subtree pruning and regrafting, and tree bisection and reconnection.

- The most critical step in exploring tree space is to determine whether there is another peak that reflects a more parsimonious or more likely solution, rather than remaining in a local maximum peak. The ratchet technique is an efficient tool to accelerate this search.

- Rate heterogeneity can be incorporated into likelihood models by use of a γ distribution.

- The likelihood ratio test can be used to determine which models are more likely or probable than others. Programs such as Modeltest can be used to rapidly compare models.

- Simple metrics can be used as measures of consistency in a parsimony analysis. These metrics include tree length, consistency index, retention index, rescaled consistency index, and Bremer or decay index.

- The robustness of nodes on a phylogenetic tree can be analyzed by resampling techniques such as bootstrapping and jackknifing. A distribution model can be added in a technique known as parametric bootstrapping.

Discussion Questions

1. Calculate ci, ri, and rci values for each character in the dog/cat/human example (explained for the most parsimonious tree in this chapter; see Figure 8.8) for the two nonparsimonious trees, (dog (cat, human)) and (cat (dog, human)). Make a table with the values in it. Now calculate the CI, RI, and RCI values for the ensemble of characters for these two trees. Compare these values with those for the most parsimonious tree.

2. In a bootstrap analysis, if one sets the percentage of characters removed and replaced at 15% and then does an analysis of the matrix in Figure 9.7, the node (1,2) is found in 100% of the bootstrap replicates. Next the percentage of characters that are removed and replaced is set at 50% (as in the example in the text), and the node (1,2) is found in 95% of the replicates. Finally the percentage of columns resampled and replaced is set at 95%, and that the node (1,2) is found in only 55% of the replicates. Explain this result.

3. If the same experiment as in problem 2 is performed with jackknifing—that is, removing without replacement 15%, 50%, and 95% of the character columns—would radically different results from the bootstrapping problem be expected? Explain.

Further Reading

Bremer K (1988) The limits of amino acid sequence data in angiosperm phylogenetic reconstruction. *Evolution* 42, 795–803.

Davis JI, Simmons MP, Stevenson DW, & Wendel JF (1998) Data decisiveness, data quality, and incongruence in phylogenetic analysis: an example from the monocotyledons using mitochondrial atpA sequences. *Syst. Biol.* 47, 282–310.

Efron B (1985) Bootstrap confidence intervals for a class of parametric problems. *Biometrika* 72, 45–58.

Farris JS, Albert VA, Källersjö M et al. (1996) Parsimony Jackknifing outperforms neighbor-joining. *Cladistics* 12, 99–124.

Felsenstein J (1985) Confidence limits on phylogenies: an approach using the bootstrap. *Evolution* 39, 783–791.

Holmes S (2003) Bootstrapping phylogenetic trees: theory and methods. *Stat. Sci.* 18, 241–255.

Nixon KC (1999) The parsimony ratchet, a new method for rapid parsimony analyses. *Cladistics* 15, 407–414.

Vos R (2003) Accelerated likelihood surface exploration: the likelihood ratchet. *Syst. Biol.* 52, 368–373.

A Beginner's Guide to Bayesian Approaches in Evolution

One approach to adding a statistical context to phylogenetic analysis is the use of Bayesian methods. This method was applied widely in the late 1990s for systematics and can be used in a wide variety of inferential sciences including phylogenetics, population genetics, and phylogenomics. It is based on a simple theorem developed by the Reverend Thomas Bayes in the 1700s. It was then developed into a more complete statistical approach by Pierre-Simon Laplace in the 1800s. Bayes' rule was first suggested as an approach to phylogenetics in the1960s by Joe Felsenstein at the University of Washington, but he never implemented it. In 1996, several authors resurrected the suggestion and presented it in a more formal manner. It was then actually applied to phylogenetics by John Huelsenbeck in his program called MrBayes (see the Web Feature for this chapter). Huelsenbeck and others developed the technique partly out of the need for a more efficient and rapid search strategy but also because of what he and others felt was a need to impart a statistical aspect to phylogenetic analysis. Bayesian approaches have also been applied to population genetics, and in this chapter we will explain the theorem upon which Bayesian approaches are based. Once we have explained Bayes' theorem, we will show how it can be applied to both population genetics and phylogenomics in this and subsequent chapters.

Bayesian Inference

The Internet abounds with sites explaining how Bayes' theorem works, using examples all the way from hipsters in Williamsburg (New York City) to M&M's to stocks and bonds. We use the following classic example from Chris Westbury of the University of Alberta to get our feet wet with Bayesian inference. We ask the following question about a picnic: "Three tall and two short men went on a picnic with four tall and four short women. What is the probability that a person is tall, given that the person is female?" This is a very silly question and why one might want to know this is not entirely compelling, but here we go. In statistical notation, what we want to know is written as a conditional probability. In such equations a vertical line is used to separate what we want to know from what is given, so that the question we asked above is written as $P(\text{tall}|\text{female})$, where the vertical bar is read as "given that." We can intuitively solve this problem with the following reasoning. There are eight women and four of them are tall, so the probability of being tall, given that the person is female, is $4/8 = 0.5$ or 50%. This intuitive answer can be verified by use of Bayes' rule. Before we get to the theorem, however, we first need to show how Bayesian reasoning is structured.

Westbury points out that a more formal way for solving the question in the situation described above is to restate the conditions in a 2×2 matrix (a bifurcating network that represents the outcomes is also sometimes used to do this):

	female	male
tall	4	3
short	4	2

When we say "given that the person is female" in the statement above, this means we can ignore the "male" column in the 2×2 matrix. The "female" column is all we need to look at, and again in this case we are asking what is the proportion of females on the picnic who are tall. Again our answer is 4/8. There is yet a third way to solve the problem. We can ask what is the ratio of the probability of being both female and tall to the probability of being female. This is a tricky way to reframe the question, but it now places all of our modes of estimation in a probabilistic or statistical framework. So given this third way to ask the question, how do we solve it? The probability of being both female and tall at the picnic is 4/13, and the probability of being female at the picnic is 8/13. So the overall ratio is (4/13)/(8/13) = 4/8 = 0.5. As Westbury states, the reason this is tricky is that "we consider the domain as a whole—all people who went on the picnic—and then take the ratio of two probabilities in that domain." This third way of formulating the question is the way Bayes' rule works. With respect to the 2×2 matrix above, Bayes' rule allows us to pick out the ratio of the probability of being in a cell of interest (tall and female) to the probability of being in a subdomain of interest defined by a conditional probability. The subdomain in Westbury's 2×2 example is being female, a subset of being at the picnic.

This brings us to the formal statement of Bayes' rule. As Westbury points out, Bayes' rule can be formulated in very complex mathematical and statistical notation. Here we will consider its simplest format, because with respect to application of Bayes' rule to systematics and population genetics, the simplest format of Bayes' theorem will suffice. In this simplest formulation of Bayes' rule, we will use two sets of probabilities (call them A and B) and assume that these sum to 1.0 to solidify the math. We start by correctly formulating the question. In a Bayesian analysis, we usually want to know the probability of an outcome given some conditional probability. With the picnic above, we asked what is the probability of being tall for a woman at the picnic. Taking our cue from the above example, we know that this probability [$P(\text{tall}|\text{female})$] is the ratio of two other probabilities. The numerator is the probability of both being a woman and being tall at the picnic considering the entire sample at the picnic, which can be written as $P(\text{female}|\text{tall})$ $P(\text{tall})$. The denominator of the ratio is simply the probability of being a woman, $P(\text{female})$. So with respect to the example above, the probability statement is

$$P(\text{tall}|\text{female}) = [P(\text{female}|\text{tall})P(\text{tall})]/P(\text{female}) \tag{10.1}$$

We can restate this problem as what is the probability of being A given B? We wrote this above as $P(\text{tall}|\text{female})$. To generalize the equation we have developed, we can substitute variable names, A for female and B for tall, to get the generalized form of Bayes' rule derived in Sidebar 10.1.

Generating a distribution of trees is an important application of the Bayesian approach

Bayes' equation is very important for all kinds of statistical reasons, but for now it might not be entirely obvious how it can be used in phylogenetics or in assessing hypotheses. To make the utility of the theorem clearer, we will generalize it further with respect to hypotheses and data, which after all is how we operate in hypothesis testing in science. In the context of scientific hypotheses, what we want to know is the probability of the hypothesis being true given the data. This is written as $P(\text{H}|\text{D})$ and read as "the probability of the hypothesis given the data." By substituting hypothesis (H) into the generalized Bayes' equation for A and data (D) into the equation for B, the result is

$$P(\text{H}|\text{D}) = [P(\text{D}|\text{H})P(\text{H})]/P(\text{D}) \tag{10.2}$$

$P(\text{H}|\text{D})$ is read as the *posterior probability* of the hypothesis given the data because it is computed after, or posterior to, knowing anything about the data. This posterior probability is in essence what we want to know about an experiment or about some scientific data set. $P(\text{H})$ is called the *prior probability* of the possibility

Sidebar 10.1. Deriving Bayes' theorem.

Bayes' rule can be easily derived from the definition of $P(A|B)$ in the following manner:

```
(1)    P(A|B) = P(A&B)/P(B)        [by definition]
(2)    P(B|A) = P(A&B)/P(A)        [by definition]
(3)    P(B|A)P(A) = P(A&B)         [multiply (2) by P(A)]
(4)    P(A|B)P(B) = P(B|A)P(A)     [substitute (1) in (3)]
(5)    P(A|B) = P(B|A)P(A)/P(B)    [Bayes' rule]
```

The derivation is a simple generalization for Bayes' theorem:

$$P(A|B) = [P(B|A)P(A)]/P(B) \tag{10.3}$$

This equation holds a marginal probability, $P(B)$, and a conditional probability, $P(B|A)$, that are needed for Bayes' rule to work. In addition, a prior probability is needed, and this is $P(A)$. So in order to solve for the posterior probability $P(A|B)$, we need to know the likelihood operator, $P(B|A)$, and the marginal probability $P(B)$. We also need to insert our prior probability, but this is a bit trickier than the marginal probability and likelihood operator so we will save describing that for later. The marginal probability $P(B)$ is relatively simple, as it is the probability computed from the law of total probability for B. The conditional probability $P(B|A)$ could pose a problem, but in most cases we actually do know the likelihood operator $P(B|A)$.

For example, in medical diagnostics, we might want to know the probability that someone will have a disease given the symptoms [$P(disease|symptoms)$, the posterior probability]. Bayes' theorem results in

$$P(disease|symptoms) = [P(symptoms|disease)P(symptoms)]/P(disease) \tag{10.4}$$

$P(disease)$ can simply be gleaned from epidemiological data and given as a total probability. The conditional probability that we need to know to compute the posterior probability is $P(symptoms|disease)$, or the probability that someone will have the symptoms given that they have the disease. This latter probability is often knowable in cases of medical diagnostics, and therefore a likelihood operator can be used to compute it.

set H. It is called a prior probability because it is a probability we believe about the hypothesis prior to any knowledge we have about the data (D). The quantity $P(D|H)$ is called the *likelihood operator or function* of D given H, and it can be approximated by using models. $P(D)$ is called the marginal probability of the data. This probability is typically computed from the law of total probability.

Note that once $P(H|D)$ is computed, there is an updated version of knowledge about the hypothesis. What can be done quite easily is to return to Bayes' equation and use this updated value for the probability of the hypothesis as a prior and recompute $P(H|D)$ on the basis of this updated prior. This approach is called recursive Bayesian filtering or iterative updating of the prior. This procedure is very important for employing Bayesian methods in evolutionary biology, and the reader should see a similarity to the Markov chain model (MCM). It turns out that Markov chains have become an important part of Bayesian analysis.

Suppose that now we want to estimate the probability of a node or a clade existing in a given tree. A clade is the group of organisms that emanate from a single common ancestor or, as suggested above, that emanate from a single node in a tree. For instance, if we had a DNA sequence data set with 10 mammals and 10 reptiles as ingroups and fish as an outgroup, the mammals would be one clade and the reptiles would be another clade. The equation that would represent a Bayesian statement about the probability of a clade existing would be written as

$$P(clade|data) = \sum\nolimits_{(any\ tree\ with\ the\ clade)} P(tree|data) \tag{10.5}$$

or, using Bayes' rule to substitute for P(tree|data)

$$P(\text{clade}|\text{data}) = \sum_{\text{(any tree with the clade)}} [P(\text{data}|\text{clade})P(\text{clade})]/P(\text{data}) \qquad (10.6)$$

With this formulation of Bayes' theorem for phylogenetic analysis, there are several components that require further detailed explanation: (1) how models are used and manipulated in Bayesian analysis, (2) how the computation of the posterior is accomplished and controlled to give results that are reliable, and (3) how prior probabilities impact the outcome of a Bayesian analysis.

Equation 10.6 means that the likelihood P(data|clade)P(clade) needs to be estimated. But to do this, we also need to include other parameters like branch lengths, nucleotide change model, site-specific variation, and nucleotide frequency into the P(data|clade) term; hence, a model of evolution needs to be applied. We also need to include the prior probability P(clade). In general, realistic prior probabilities are hard to estimate due to a lack of information, so most analyses will substitute "safe" priors into the equations. These safe priors are usually what are called flat or vague priors. A flat prior for P(clade) is simply where all trees have equal probability, and a vague prior for P(clade) would be where the posterior probabilities being calculated have smaller variance than the prior probability of trees. A flat prior [when P(clade) and P(data) are equal for all estimates in the sum above] means that the posterior probability distribution will then be proportional to the likelihood. So in essence, when you do a Bayesian analysis in phylogenetics, it is approximately like doing a likelihood analysis. In fact, Huelsenbeck has pointed out that "This [Bayesian inference] is roughly equivalent to performing a maximum likelihood analysis with bootstrap resampling, but much faster."

What do we need from a Bayesian phylogenetic analysis? The left side of the Bayes' rule equation above tips us off to this. We are after a distribution of trees in order to calculate the probability distribution of different topologies.

How do we produce this distribution of trees? The real trick to Bayesian phylogenetic analysis and generating the distribution of trees lies in the implementation of a Markov chain Monte Carlo (MCMC) simulation linked to a specific likelihood model to produce a posterior distribution of trees. The MCMC is a simulation trick that is used to explore the distribution of data relevant to a specific question. The MCMC is implemented by using the Markov chain (MC) algorithm, which is a mathematical system that simulates change, as determined by specific rules set out at the beginning of the generation of the chain. Markov chains are known as "memoryless" because they are generated in a linear fashion with the states of next steps in the chain depending only on the current state and not on any states in the past. The chain is generated over several generations of the simulation, and the subsequent large number of steps is then interpreted as a distribution. Since the end product is a distribution, statistical methods can be applied to interpret the distribution. (See Chapter 3 for a full discussion of the MCM.)

In essence, MCMC is used to generate trees to form a probability distribution by use of a Markov chain simulated under specific sets of rules. There are two major problems with this process that can easily be overcome with the MCMC simulation procedure. First, one must make sure that the simulation chains have converged. This problem can be solved by generating a large distribution of trees from the simulations (see description below). The second problem is to take into account the potential for autocorrelation of the simulations. This problem can be overcome by running several simulation chains. All together, this procedure produces a distribution of trees given a model, the priors, and the data (**Figure 10.1**).

Generating a distribution of the trees, therefore, is the essence of the Bayesian approach as it is implemented in phylogenetic analysis programs. The distribution of trees can then be used to determine two properties. First, the single topology with the largest occurrence in the distribution can be found. Second, data can be represented by summarizing the frequency with which particular nodes appear in the posterior distribution of trees. For instance, if A is a sister taxon to

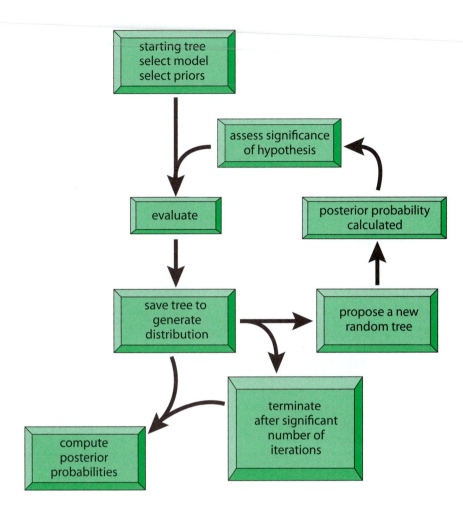

Figure 10.1 Flow diagram of the Bayesian phylogenetic approach.

B in every tree in the posterior distribution, then the posterior probability of that node connecting A with B would be 1.0. If A is a sister taxon to B in 95% of the trees in the posterior distribution of trees, then the posterior probability of that node would be 0.95. Thus a Bayesian phylogenetic analysis can result in a tree, with branch lengths and posterior probabilities assigned to each node in the tree.

MCMC is critical to the success of Bayesian analysis

MCMC is the cornerstone of Bayesian phylogenetic analysis. It is instructive to look at how MCMC works, because its underpinnings will influence the distribution of trees. The MCMC analysis starts by choosing a random tree (Figure 10.1), and by use of a flat prior, a likelihood model, and the marginal probability of the data, the posterior probability of the chosen tree is calculated.

If the product of the likelihood and prior of the newly proposed random tree is greater than that of the current tree, then the new tree is stored along with its parameters. If the posterior probability of the newly proposed random tree is statistically significantly lower than the likelihood of the currently optimal tree, then the tree can be accepted but with a probability that is proportional to the likelihood values of the two trees being compared. MCMC is an efficient way to do these searches, as shown in **Figure 10.2.**

The decisions made by the algorithm to either accept or reject a solution have often been compared to a robot randomly walking around on a topographical

Figure 10.2 The robot metaphor for Bayesian analysis. The curve represents a tree space, and the robots represent generations in the Bayesian analysis. Certain probabilities will always result in the robot taking a step upward; that is, in the context of Bayesian tree searching, accepting the current tree and moving forward up the incline. Another category of probabilities will result in the solution being accepted and stored but in a step downward. A third category of probabilities will result in the robot rejecting the solution; noting that a step forward would be cataclysmic, the robot then steps backward.

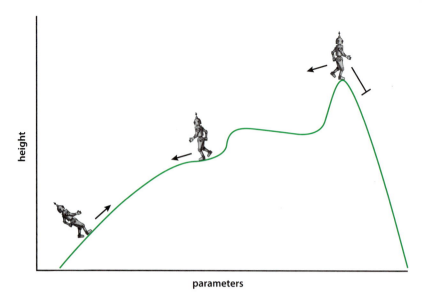

surface. The robot's goal is to generate or collect the full joint posterior probability distribution of tree topologies and branch lengths for the model parameters given the data. This set of probabilities can then be used as a statistical distribution for assessing the credibility of clades in a tree as explained above. It is easy to make the mistake of thinking that the robot is looking for an optimal solution to the tree search, but in fact this is not the case. With an efficient search the optimal tree will probably be in the distribution, but even if not, this does not matter. What matters at the end of the run is the probability distribution of trees and branch lengths of those trees.

MCMC in Bayesian phylogenetics dictates the set of rules for the robot to make decisions that will enhance the speed of exploring the tree space and minimize the amount of energy the robot expends in covering the entire tree space and hence generating the distribution that is so important in Bayesian analysis. Figure 10.2 shows that certain likelihood products will always result in the robot taking a step upward in tree space. Other likelihood products will result in the robot not liking the solution (but storing it) and stepping backward. Still a third category of probabilities will result in the solution being acceptable and stored but in a step downward. This manner of exploring the topology turns out to be incredibly efficient, but it has some quirks to it that can be accommodated by rephrasing the rules and amount of steps the robot is allowed to take.

For instance, given certain rules and a small finite number of steps the robot can take, it will always get stuck on a single topographical peak. It will not, so to speak, get to explore the entire topographical space and escape from local maxima. The simulation avoids getting stuck on peaks by use of coupled MCMC runs. The correct term for this trick is actually Metropolis coupled runs, so we have MCMCMC or MC^3. This approach usually runs four different simulations called chains. One of the chains is called a cold chain (**Figure 10.3**), which essentially behaves and follows the rigorous rules for exploring the tree space described above. The cold chain is the simulation chain from which the sampling distribution is created. The other three chains are called hot or heated chains (Figure 10.3) because the rules are changed for the robot's movement to allow more local exploring of the tree space. The hot chains check out the landscape more efficiently, and the cold chain does a better job of sampling. During the simulation run, the three hot chains are chosen randomly, over and over again, and are compared to the cold chain. Under certain criteria, during a comparison of hot and cold chains, the chains will swap positions in tree space. The comparison and swapping happens throughout

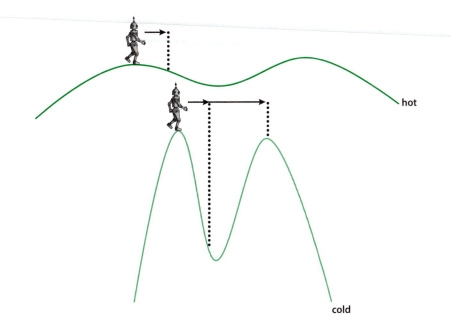

Figure 10.3 Difference between hot and cold chains in Bayesian phylogenetic analysis. The cold chain essentially behaves and follows the rigorous rules for wandering. The hot chains allow visiting local peaks more easily because the rules are changed for the robot's movement. When these approaches are combined, this allows for more efficient discovery of optima. Since the hot chains often get stuck on suboptimal peaks, they are discarded in building the Bayesian tree.

the MCMCMC run, and the trees saved by the cold chain only are used in creating the distribution of trees. This procedure ensures that the cold chain will efficiently jump valleys in the tree space.

One problem with this approach is that it takes a while for the simulation to settle in and to get to higher points in the tree space. As a result the first 10–25% of the trees visited by the cold chain will be very erratic when their likelihoods are examined. Hence, the first 10–25% of the trees simulated are not considered part of the posterior distribution and are discarded. The trees that are discarded are what is called the burn-in. With this kind of simulation, the more generations (ngen) of MCMC that are run, the better the distribution of trees to estimate clade posterior probabilities (also referred to as clade credibilities).

Bayesian Analysis in a Phylogenetic Context

Before we demonstrate how a Bayesian phylogenetic analysis proceeds, first we will focus on the choice of a model, the selection of prior probabilities, and the dynamics of the MCMC simulations.

Model selection can be utilized on any biologically meaningful partition

The matrix for a Bayesian analysis needs to be a set of aligned sequences, either nucleotides or amino acids. The matrix can also be partitioned into genes, gene regions, codons, codon positions, or any other biologically meaningful partition where the researcher can then apply the different models and model parameters to the various partitions. So, for instance, if one desired to apply a different model of sequence evolution to first and second positions versus third positions, the matrix can be partitioned as such. If DNA sequences are used in the analysis, any number of models can be applied as part of the Bayesian approach (see Chapter 8). Likewise, if amino acid sequences are used, several models are available for the analysis of protein-based sequences. The models that can be used in Bayesian analysis can be interchanged easily, and as we pointed out in Chapter 9, a model can be chosen on the basis of likelihood ratio tests of many models tested against one another. Since we are more interested now in demonstrating the mechanics of a Bayesian analysis and we have discussed models at length in Chapter 8, we will not focus further on model selection in Bayesian analysis.

Selection of priors may involve default values, but priors can be adjusted manually

As we described above, the prior distributions of model parameter probabilities ("priors") are critical to a Bayesian analysis. At present in Bayesian phylogenetics, several priors can be set (**Table 10.1**). The settings vary from variable to variable. For instance, the prior settings for transformation probabilities use a Dirichlet distribution. The mathematics of this distribution is beyond the scope of this book, but suffice it to say that Dirichlet distributions have several properties that make them ideal for use in Bayesian analysis. Depending on the number of transformation states (that is, the number of parameters in the 4×4 transformation matrix), the Dirichlet priors will have different characteristics. On the other hand, the branch length priors can be set as either clocklike or unconstrained. What this means is that the change can be modeled as occurring in a regular clocklike fashion or can have large amounts of rate heterogeneity (unconstrained). For the Dirichlet priors, the default and safest setting is a flat distribution. The flat distribution assumes equal probability of variables, and hence the term flat is a good descriptor because no one variable has a higher prior probability than others. For instance, if one is using a generalized time-reversible (GTR) model (with six parameters), the Dirichlet distribution is set at [1.0, 1.0, 1.0, 1.0, 1.0, 1.0]. If one desires to weight the GTR parameters in the analysis, the Dirichlet priors are set higher but kept flat [for example, 100.0, 100.0, 100.0, 100.0, 100.0, 100.0]. The prior settings in Table 10.1 are all set as default values in MrBayes, but they can be changed by using commands in the MrBayes program. As can be seen in the table, they are all set at flat prior distributions. We have discussed the most parameter-rich likelihood model (GTR), but other simpler models can be accommodated with priors as well.

Perhaps the most important prior to be aware of is the branch length prior. This prior seems to have a more extreme impact on the sensitivity of the Bayesian approach. It is an important prior because it is applied to all branches in the tree. With just three taxa, this is not such a daunting set of parameters. However, with larger numbers of taxa, the setting of this prior involves consideration of a large number of independent variables. Not only will the branch length prior impact the inferences made on branch length posteriors, but it will also impact the posterior probabilities on the various nodes in the tree. In general, though, establishing flat priors for most of the other parameters is the safest way to proceed in Bayesian analysis. As we will see when we examine a worked example in MrBayes, while more than 20 priors can be imposed on a Bayesian phylogenetic analysis, it is best to maintain most of them as flat priors.

Setting the MCMC parameters has greater impact for larger data sets

To demonstrate the dynamics of the MCMC simulation in a Bayesian analysis, we will use the simple example in Chapter 8 listed above for the dog/cat/human matrix. In reality, this matrix poses a very simple computational problem, because there are only three topologies to evaluate. Under any model with any combination of priors, the result is always dog + cat with high posterior probability. Changing models also does not drastically change the posteriors, which hover between $p = 0.92$ and 0.97 for the dog + cat topology. The other two topologies fare much more poorly than the dog + cat topology in Bayesian analysis. These results should not be surprising, as we have shown with all other kinds of analyses (NJ, MP, ML, and bootstrap) that the dog + cat topology is by far the preferred topology for this data set.

Table 10.1 Priors for Bayesian analysis.

	Parameter	Options	Default setting
1	Tratiopr	beta/fixed	beta (1.0, 1.0)
2	Revmatpr	Dirichlet/fixed	Dirichlet (1.0, 1.0, 1.0, 1.0, 1.0, 1.0)
3	Aamodelpr	fixed/mixed	fixed (Poisson)
4	Aarevmatpr	Dirichlet/fixed	Dirichlet (1.0, 1.0, ...)
5	Omegapr	Dirichlet/fixed	Dirichlet (1.0, 1.0)
6	Ny98omega1pr	beta/fixed	beta (1.0, 1.0)
7	Ny98omega3pr	uniform/exponential/fixed	exponential (1.0)
8	M3omegapr	exponential/fixed	exponential
9	Codoncatfreqs	Dirichlet/fixed	Dirichlet (1.0,1.0,1.0)
10	Statefreqpr	Dirichlet/fixed	Dirichlet (1.0, 1.0, 1.0, 1.0)
11	Ratepr	fixed/variable = Dirichlet	fixed
12	Shapepr	uniform/exponential/fixed	uniform (0.0, 50.0)
13	Ratecorrpr	uniform/fixed	uniform (−1.0, 1.0)
14	Pinvarpr	uniform/fixed	uniform (0.0, 1.0)
15	Covswitchpr	uniform/exponential/fixed	uniform (0.0, 100.0)
16	Symmetricbetapr	uniform/exponential/fixed	fixed (infinity)
17	Topologypr	uniform/constraints	uniform
18	Brlenspr	unconstrained/clock	unconstrained: exp(10.0)
19	Speciationpr	uniform/exponential/fixed	uniform (0.0, 10.0)
20	Extinctionpr	uniform/exponential/fixed	uniform (0.0, 10.0)
21	Sampleprob	<number>	1.00
22	Thetapr	uniform/exponential/fixed	uniform (0.0, 10.0)
23	Growthpr	uniform/exponential/fixed/normal	fixed (0.0)

However, by examining the results of quick Bayesian analyses of this data set, we can demonstrate some aspects of how the MCMC simulation works. For the K2P run at 100 MCMC generations, we obtain the distribution of trees shown in **Figure 10.4**. The ln L value is given on the x-axis, and the simulation generation number is given on the y-axis. The burn-in used was 25% (that is, the initial 25% of the trees generated were removed). Two cold chains were simulated, and the ln L of each generation was recorded and tabulated. In these plots, there are three major ln L values corresponding to the likelihoods of the three trees that can be generated with three ingroups. Larger numbers of generations for the MCMC simulations do improve the Bayesian posterior for this tiny data set. For the present example, the problem of evaluating the tree space is trivial, but for data sets with larger numbers of taxa and hence immense numbers of trees to examine, the problem is more extreme and the MCMC approach is more useful. Altering priors and

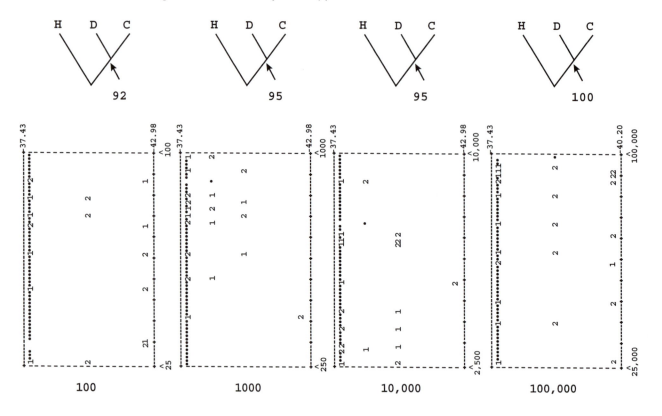

Figure 10.4 Overlay plots for MCMC, ngen = 100, 1000, 10,000, and 100,000, with a simple data set. For this example, three ingroup taxa and one outgroup were used. Trees represent the tree with the highest credibility, and arrows point to the dog–cat node, indicating the posterior probability. Numbers in the plot (1 and 2) refer to the two different runs accomplished by MrBayes.

models in these larger data sets can have a large impact on Bayesian inference, as we demonstrate in the next section.

Increasing ngen in MCMC leads to better distribution of trees but at increased computational cost

In order to examine the impact of altering the number of simulation generations in the MCMC on Bayesian inference, we use a matrix of sequences for the *CD9* gene with 14 taxa in it. This number of taxa (13 ingroup taxa and one outgroup taxon) would generate >10,000,000,000 bifurcating trees. By the same approach as above, an MCMC simulation was generated for this larger data set (**Figure 10.5**). In this example we discarded the burn-in set at the first 25% of the trees. Note that the *x*-axis in these figures varies in range. The ranges become tighter as ngen rises. Note also that for the ngen = 100 simulations (first panel) the ln *L* values range from −5683 to −4971, and that even though we discarded the burn-in, there are still a large number of trees that appear to have low ln *L* values. Even when ngen is increased to 1000 (second panel) and the burn-in is removed, the same pattern emerges, but the range of likelihoods is compressed (−4693 to −4281). Increasing the number of generations of the simulation to 10,000 (third panel) not only increases ln *L* but also compresses further the range of values of likelihoods (−4291 to −4270). Increasing ngen to 100,000 (fourth panel) and to 1,000,000 (not shown) also indicates an even further narrowing of tree distribution based on ln *L* values (−4286 to −4268). The figure demonstrates the importance of setting ngen high enough to avoid the effects of inadequate burn-in. Another important principle demonstrated in the figure is that as ngen rises, the ln *L* values of the two chains converge on each other. This is a desirable property of the distribution of trees for computing posterior probabilities on clades. We will look at convergence in more detail below. One might simply ask, why not just do the MCMC with large ngen to start out with? For some data sets this approach is possible; however, for larger data sets the computation time becomes extremely large. Parallel processing of the problem can help cut computing times, but in general, it is important to determine the limits of an analysis and balance it with time constraints by monitoring how the simulated tree distribution converges.

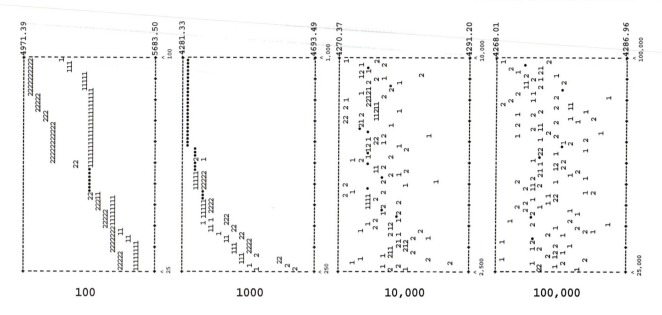

100   1000   10,000   100,000

An Example Using MrBayes

The most broadly available program for Bayesian analysis is a program called MrBayes, developed by John Huelsenbeck and Fred Ronquist. The program can be downloaded for free from the following Website: http://mrbayes.sourceforge.net/download.php. We will use this program as it is freeware and easily accessed for use on Macintosh, Windows, and Linux machines. The program may have been updated (from v3.2.1 at the time of writing); the most recent version should be downloaded so that the examples will work correctly. Download and install the program that runs in a terminal window. This will produce several sample data matrices, a manual, and two programs: one for use on a laptop and the other a parallel version of MrBayes. Once the download is complete, double-click on the MrBayes 3.2.1 (or updated) icon. This should produce a shell that looks like the one in **Figure 10.6**. MrBayes can be exited at any time by typing "q" into the

Figure 10.5 Overlay plots for MCMC, ngen = 100, 1000, 10,000, and 100,000, with a complex data set. For this example, 13 ingroup taxa and one outgroup were used. Numbers in the plot (1 and 2) refer to the two different runs accomplished by MrBayes.

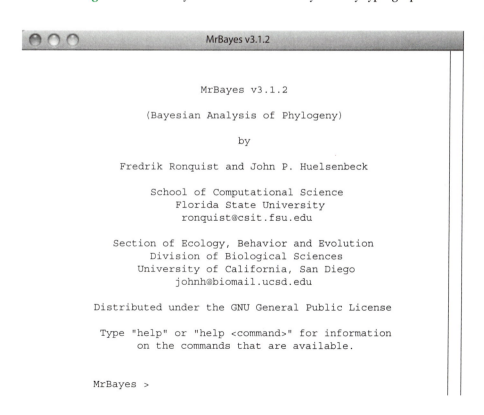

Figure 10.6 Screenshot of the MrBayes console upon opening the program. (Courtesy of F. Ronquist, B. Larget, and J. Huelsenbeck.)

```
MrBayes v3.1.2

(Bayesian Analysis of Phylogeny)

by

Fredrik Ronquist and John P. Huelsenbeck

School of Computational Science
Florida State University
ronquist@csit.fsu.edu

Section of Ecology, Behavior and Evolution
Division of Biological Sciences
University of California, San Diego
johnh@biomail.ucsd.edu

Distributed under the GNU General Public License

Type "help" or "help <command>" for information
on the commands that are available.


MrBayes >
```

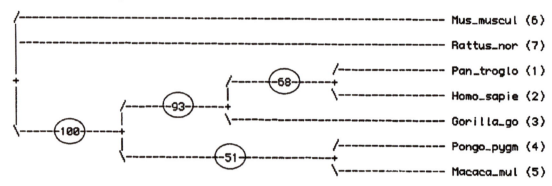

Clade credibility values:

```
/--------------------------------------------------------- Mus_muscul (6)
|
|----------------------------------------------------------- Rattus_nor (7)
|
|                                              /---68---+ /----------- Pan_troglo (1)
|                            /------93------+            \----------- Homo_sapie (2)
+           /----100----+    |              \----------------------- Gorilla_go (3)
|           |                |
\-----------+                |              /----------- Pongo_pygm (4)
            |                \------51------+
            |                               \------------ Macaca_mul (5)
```

Phylogram:

```
/------------------------------------- Mus_muscul (6)
|
|---------------------------------- Rattus_nor (7)
|
|                                                      /-- Pan_troglo (1)
|                                                    /-+
+                                                    | \-- Homo_sapie (2)
|                                                  /--+
|                                                  | \--- Gorilla_go (3)
|                                                /--+
\------------------------------------------------+ /--- Pongo_pygm (4)
                                                 \-+
                                                   \--- Macaca_mul (5)
```

Figure 10.11 Results of MrBayes analysis with generalized time-reversible model settings. Circles indicate posterior probabilities for each node. (Courtesy of F. Ronquist, B. Larget, and J. Huelsenbeck.)

The state frequencies of the Dirichlet priors (entry 10 in Table 10.1) need comment here. Recall from our discussion of how a Bayesian analysis proceeds that we need to include the prior probabilities P(tree) in the estimation procedure. In general, realistic prior probabilities are hard to estimate, so most analyses will substitute safe priors into the equations. These safe priors are usually flat or vague priors. A flat prior for P(tree) is simply one where all trees have equal probability, and a vague prior for P(tree) would be one where the posterior probabilities being calculated have smaller variance than the prior probability of trees. Note that in **Figure 10.12**, point 5, which lists the prior settings for tree topology, says "All topologies equally probable a priori", indicating that the default is set as a flat prior.

Ending a Bayesian analysis involves an assessment of the run's efficiency

Stopping a Bayesian analysis at an appropriate point is important. Huelsenbeck and Ronquist describe the main way in which the efficiency of the run is assessed as follows: "… examine the average standard deviation of split frequencies. As the two runs converge onto the stationary distribution, we expect the average standard deviation of split frequencies to approach zero, reflecting the fact that the two tree samples become increasingly similar. Your values can differ slightly because of stochastic effects." If the average standard deviation is relatively small (very close to zero), then there is no need to do more runs. If it is large, then more runs should be performed.

We will examine this parameter by going back to the original F81 analysis we did with the CD9 dataset. To do this, we need to change the parameters back to the default F81 model by typing "lset nst=1 rates=equal" into the MrBayes prompt. This can be verified by typing "showmodel" into the prompt and inspecting the model parameters. If the MrBayes program is quit and reopened, it will automatically be reset to the F81 model. Once the F81 model is restored, then do the run by

With the basics of E
eral important subj
ulation genomics d

Summary

- Bayes' rule d
  outcome give
  phylogenetic:
  given a mode
  of events imp
- Markov chair
  to a specific l
  a posterior di
  analysis.
- In a phyloger
  on the basis
  Our treatmer
  very little fro
  likelihood fra

Discussion Qu

1. Using a laborat
   a diagnostic tes
   disease, when it
   yield false posit
   disease will be p
   healthy) for 1%
   of the general p
   the people with
   probability the

2. Discuss the role
   How might they
   What might be

Further Readi

Alfaro M & Holder
phylogenetics. *Annu*

Bollback JP (2002) Ba
netics. *Mol. Biol. Evo*

Holder MT & Lewis P
Bayesian approache

Huelsenbeck JP, Ron
inference of phylog
Science 294, 2310–2

Larget B & Simon D
for the Bayesian an
750–759.

```
MrBayes > showmodel

   Model settings:

            Datatype  = DNA
            Nucmodel  = 4by4
            Nst       = 6
                        Substitution rates, expressed as proportions
                        of the rate sum, have a Dirichlet prior
                        (1.00,1.00,1.00,1.00,1.00,1.00)
            Covarion  = No
            # States  = 4
                        State frequencies have a Dirichlet prior
                        (1.00,1.00,1.00,1.00)
            Rates     = Invgamma
                        Gamma shape parameter is uniformly dist-
                        ributed on the interval (0.00,200.00).
                        Proportion of invariable sites is uniformly dist-
                        ributed on the interval (0.00,1.00).
                        Gamma distribution is approximated using 4 categories.

   Active parameters:

        Parameters
        -------------------
        Revmat            1
        Statefreq         2
        Shape             3
        Pinvar            4
        Topology          5
        Brlens            6
        -------------------

   1 --   Parameter  = Revmat
          Prior      = Dirichlet(1.00,1.00,1.00,1.00,1.00,1.00)
   2 --   Parameter  = Statefreq
          Prior      = Dirichlet
   3 --   Parameter  = Shape
          Prior      = Uniform(0.00,200.00)
   4 --   Parameter  = Pinvar
          Prior      = Uniform(0.00,1.00)
   5 --   Parameter  = Topology
          Prior      = All topologies equally probable a priori
   6 --   Parameter  = Brlens
          Prior      = Branch lengths are Unconstrained:Exponential(10.0)

MrBayes > |
```

typing "MCMC ngen=100000" into the prompt. Since MrBayes saves several data files as a result of the run, it will ask if you want to overwrite files. For this exercise, all overwrite questions can be answered yes. If it is desirable to save the output files from the previous run, they should be renamed and saved in the MrBayes folder.

As the analysis proceeds, the average standard deviation of the split frequencies is listed below the likelihoods (see Figure 10.8). When the 100,000 generations have been run, MrBayes will issue the prompt "Continue with analysis? (yes/no)". If the standard deviation has stabilized, the answer is no; if the value is not stable, then the answer is "yes" and more generations should be run. For now, say "no" and stop the run. Next type "sump burnin=250". This command will give a plot like the one in **Figure 10.13**. The graph plots the tree number (that is, the generation number, starting after burn-in trees have been removed) on the *x*-axis and the likelihood of the tree on the *y*-axis. As before, 1s in the figure represent the plots of the first cold chain that MrBayes runs, and 2s represent the second cold chain run. If there is convergence or an overall linear or saturation pattern to the plot, then

Figure 10.12 Model settings for MrBayes with generalized time-reversible + invariant rates with gamma distribution model settings. (Courtesy of F. Ronquist, B. Larget, and J. Huelsenbeck.)

Averag

98100
98200
98300
98400
98500
98600
98700
98800
98900
99000

Averag

99100
99200
99300
99400
99500
99600
99700
99800
99900
10000

Avera

Conti

Figure
betweer
"Continu

Incongruence

Much of modern molecular systematics until the last decade focused on one or a few linked genes. One of the important eye-openers, caused by the generation of information from multiple genes, was that the inferences from these multiple genes individually often gave solutions that were incongruent with each other. In order to deal with this incongruence problem, systematists have developed tools to examine the degree of incongruence between two or more trees in an analysis. There are several reasons the incongruence can exist. The first major reason is that the two data sets simply do not carry enough information to make a strong inference one way or another. In this case, statistical tests have been developed to examine the significance of incongruence among different trees. A second possibility is that the two genes might actually have experienced different evolutionary histories. This has been called the gene tree/species tree problem. In this case of incongruence due to hybridization, lineage sorting, or horizontal transfer, different genes will have experienced different evolutionary histories and hence produced incongruent phylogenetic trees. Coalescence theory was developed to examine these kinds of phenomena during the evolutionary process to help biologists make inferences at the level of populations. Other methods for dealing with the incongruence at higher levels have been developed, such as consensus trees, supermatrices, and supertrees.

Incongruence of Trees

There are two areas of systematics where molecular information promised to be of utility when researchers first started using molecular data. These two areas sit at the opposite extremes of taxonomic divergence. On one end are systematic questions that morphology is unable to address, because of such extreme divergence of forms or because of the lack of morphology, such as with microbes. In these cases, it was difficult to determine which anatomical characteristics were homologous across large evolutionary distances, and hence there was a lack of characters for phylogenetic analysis. On the other end of the spectrum are situations where closely related taxa had changed little anatomically but had indeed diverged, either behaviorally or at the molecular level. Here, homologizing anatomical characters was no problem, but with no variation in the characters, there was no anatomical information that could weigh in on phylogenetic questions.

Certain molecules, on the other hand, were easily homologized, sequenced, and used as characters to weigh in on important and heretofore unexamined phylogenetic questions. Researchers started to use sequences from the small subunit (16S for microbes and 18S for eukaryotes) ribosomal RNA as a tool for understanding the phylogeny of microbes. The reasoning for using this molecule is simple. All organisms have ribosomes, the protein synthesis machinery of the cell. All ribosomes are made up of proteins and an RNA scaffold coded for by genes. The RNA, while highly conserved in some regions of the gene (stems), shows a considerable degree of variation in others (loops). This approach was so successful that it led to the discovery of a brand-new domain of microbial life called Archaea and the discovery that these single-celled nonnucleated organisms were more closely related to eukaryotes than to the other great nonnucleated domain, Bacteria. Subsequent to the microbial studies, the approach was applied to animals and plants with great success. On the other end of the spectrum, researchers started to use sequences of another molecule to look at very closely related taxa. In these studies, the mitochondrial genome was exploited as a phylogenetic tool for examining closely related taxa and for looking at genetic structure within species. Each mitochondrion in the cells of an organism carries a small circular genome that codes for several of the proteins (and structural RNAs like tRNA and rRNA) that are important in the workings of the mitochondrion. Because of the clonal inheritance of the mitochondrion and the biochemistry of DNA replication in the organelle, particular regions of the mitochondrial genome change very rapidly, such that even closely related individuals within species will show variation. This molecule was immediately exploited to examine the relationships of closely related taxa and led to the development of molecular population genetics and the field of phylogeography. This term, which has become embedded in evolutionary biology, was coined by John Avise in the 1990s to encompass the examination of DNA sequence variation and patterns of relatedness of closely related organisms in conjunction with geographic information. Phylogeography usually focuses on the historical patterns of distribution of individuals within a species or between closely related species by use of gene genealogies.

When the sequencing technology advanced enough to result in sequences from more than a single gene for a systematic problem, another form of incongruence in systematics became clear. Application of these genome-level sequencing approaches to a wide array of phylogenetic problems "in between the extremes" led to interesting results and debate in the phylogenetic community because of incongruence. In many cases, the trees constructed from molecular information conflicted with the well-established trees obtained from morphological information. This led to the "molecules versus morphology" controversy that preoccupied many systematists for the last decade of the twentieth century.

Once several genes from the same taxon in phylogenetic analysis were available, incongruence between genes was observed. A classic study that showed this clearly was performed by Rokas et al. This pioneering study used 106 yeast genes for seven ingroup taxa and one outgroup. The study showed that, of the 106 different genes, 24 discrete and well-supported phylogenies could be generated when the genes were analyzed separately. Another, even more extreme, example can be found in diBonaventura et al., where over 2000 genes were used, generating nearly 100 different trees when genes are analyzed separately. More stunning is that only one of the 2000 or so genes actually match the best-supported tree in topology obtained by combining all of the genes (see Figure 1.3).

From a data support perspective, there are two major reasons incongruence can occur. First, incongruent topologies can be obtained simply because the phylogenetic signal from two separate data sets is weak and the tree topologies differ by chance. Second, the differences can be due to the two data sets not sharing the same evolutionary history. We describe the former incongruence as "soft" incongruence and the latter as "hard" incongruence (Maddison and Maddison first used the hard and soft designations to describe polytomies in phylogenetic trees, using the same criteria we use here for congruence). Soft incongruence is a trivial problem that can be assessed with any number of approaches that we describe below, such as the incongruence length difference (ILD) test or the Shimodaira–Hasegawa (SH) likelihood ratio test approach. The problem of hard incongruence is nontrivial and an interesting one; it results when the trees from different genes give significantly incongruent results. We discuss one specific explanation for hard incongruence below in the context of the gene tree/species tree problem. Solving the problem of lack of congruence of trees from two different sources of characters can be looked at in two different ways: taxonomic congruence and character congruence.

Taxonomic congruence achieves consensus by construction of supertrees

Before we discuss the taxonomic congruence approach, we first need to discuss the solution to a similar problem that systematists faced with early phylogenetic analyses. Sometimes with a single data set, multiple equally optimal trees can be obtained. Imagine for the dog–cat–human example that, instead of obtaining the (human (dog, cat)) topology, all three possible topologies had the same number of steps or the same likelihood. In this example, three different trees are equally most parsimonious or have equal maximum likelihoods. These equally optimal trees pose a problem for reporting the results of the parsimony or likelihood analysis. A researcher could simply report all three. A second option is to create a "consensus" of the three equally optimal trees. There are different categories of consensus that can be used here. The first and arguably the most objective is called "strict" consensus. Strict consensus examines the most parsimonious or most likely trees and looks for common nodes in all of the equally optimal trees. In the example there are no nodes in the tree that occur in all three of the equally optimal trees (except for the node that defines the ingroup). Hence a strict consensus of the three equally optimal trees would be a fully unresolved tree, also known as a polychotomy or a star phylogeny.

Another commonly used consensus tree is called a "majority rule" tree. Usually some cutoff for a node to be reported in such majority rule trees is given (most commonly at 50%). What this means is that a node has to exist in at least 50% of the equally optimal trees to be reported. Consider a second example where there are four taxa and an outgroup. Three trees obtained after phylogenetic analysis that are equally parsimonious or have the same maximum likelihood are shown in **Figure 11.1**. In this example, the a-b node exists in two out of three trees, and the c-d node exists in two out of three trees, so both nodes would be included in the majority rule consensus tree. Any node that does not appear in at least 50% of the equally optimal trees is left out of the final majority rule consensus tree. Majority rule trees are usually presented with the frequency among the optimal trees indicated at the node of interest, in this case, two out of three or 0.67 for the c-d and a-b nodes. While there are three additional kinds of consensus trees that researchers use, the strict and majority rule consensus approach are the major ways of reporting multiple optimal trees in phylogenetic analysis.

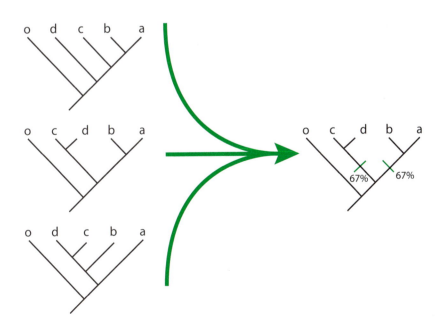

Figure 11.1 Majority consensus rule. Three trees that are equally parsimonious or with equal maximum likelihoods generate a majority rule consensus tree as shown. The percentages on the nodes represent the percentage of trees where that node occurs in the three starting trees. For instance, the node joining a and b is found in two out of three, or 67%, of the starting trees, and the node joining c and d is found in two out of three, or 67%, of the starting trees. All other nodes in the starting trees are found in only one of these trees and these nodes are ignored by the 50% cutoff for a typical majority rule consensus tree.

Consensus approaches make sense as a method for displaying the results of multiple optimal trees generated from a single gene or single kind of data analysis. A consensus approach to reconciling multiple phylogenetic hypotheses generated from different sources has also been suggested by Miyamoto and Fitch. In this approach, also known as "taxonomic congruence," the topologies of the competing hypotheses are treated as multiple optimal hypotheses that can be reconciled through consensus methods. Several algorithms have been developed to implement this approach to solving incongruence. The problem has become to be known as the supertree problem. Suffice it to say here that the supertree approach is at direct odds with the second approach based on total evidence approaches as described below.

Character congruence uses total evidence or concatenation to create supermatrices

The second major approach to dealing with incongruence from different data sets is called the character congruence approach. In a classic paper in systematics, Arnold Kluge called the approach "total evidence." Kluge meant for the name to reflect that all data on a specific systematic or phylogenetic analysis should be brought to bear on the problem at once, and that taxonomic congruence or consensus methods were a poor solution to the incongruence problem. Later, Nixon and Carpenter refined the definition of character congruence and felt the need to rename the approach to "simultaneous analysis". In their terminology they suggest that:

"Simultaneous analysis of combined data better maximizes cladistic parsimony than separate analyses, hence is to be preferred. Simultaneous analysis can allow "secondary signals" to emerge because it measures strength of evidence supporting disparate results. Separate analyses are useful and of interest to understanding the differences among data sets, but simultaneous analysis provides the greatest possible explanatory power, and should always be evaluated when possible."

With the onslaught of molecular data, the terms total evidence and simultaneous analysis seem to have got lost in favor of the term "concatenation," which is, in our opinion, the same as Nixon and Carpenter's "simultaneous analysis." Concatenated analysis, in the context of supertree analyses, is usually referred to as the supermatrix approach. Rokas et al. made a strong argument for the role of concatenation of genome sequences in phylogenomics with their analysis of the yeast data set mentioned above. When a concatenated analysis of the 106 genes for this data set was accomplished, a very reasonable and highly supported hypothesis was generated. The stability and reasonableness of the concatenated hypothesis was used as an argument for concatenation or the supermatrix approach. Barrett et al. offered an interesting argument against consensus based on the fact that concatenation results in the emergence of information that does not emerge when data sets are kept separate (Sidebar 11.1). Other researchers argue strongly against the random concatenation of information in phylogenetics and prefer to focus their attention on sets of genes they feel are "proper" indicators of phylogeny.

Statistical methods assess incongruence to decide whether information should be concatenated

Systematists have developed several methods for determining whether topological incongruence of two trees is soft or hard. The tests that have been developed employ statistical methods, so that a probability statement can be made in the context of the test. Before we go into the tests themselves, we should detail why we would care about testing for incongruence in the first place. Several researchers have contended that if one can recognize significant incongruence between data sets, then some decisions regarding concatenation of the incongruent

Sidebar 11.1 Against consensus?

Barrett et al. made the following argument concerning concatenation by use of the hypothetical data sets shown in **Figure 11.2**. The two data sets support different hypotheses. When a strict consensus tree is drawn (far right), the three taxa b, c, and d are unresolved. However, when the two matrices are concatenated (bottom) and the information in the matrix is analyzed with parsimony, a fully resolved tree is obtained. The gain in resolution is a major argument against consensus.

Figure 11.2 Against consensus. Two genes are examined for phylogenetic information. Matrices on the left represent the primary data. Numbers in parentheses represent the number of columns that have the distribution of characters in the column below. The third matrix is the "concatenated" matrix. The tree at the top right is the consensus tree, and the tree at the bottom right is the supermatrix (combined) tree.

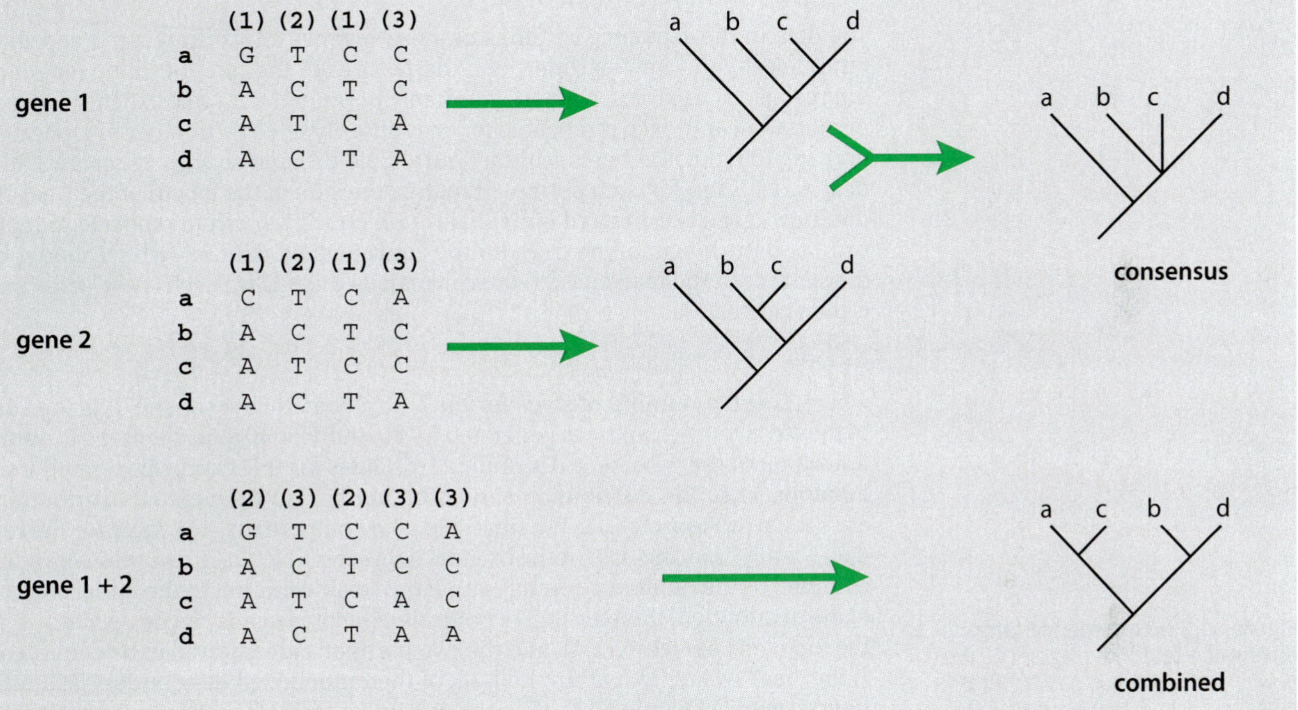

information can be made. Specifically, there are three potential outcomes after a test for incongruence is made. In the best-case scenario, two partitions might be deemed not significantly different in the phylogenetic signal they produce. In this case, most systematists recommend concatenation. The other two outcomes relate to the situation when two data sets are determined to be significantly incongruent. Some researchers have argued that significant incongruence is a good argument for keeping the partitions separate when doing phylogenetic analysis. The final option for incongruent data sets is to concatenate them anyway. Which approach to adopt is a major question in modern systematics.

The incongruence length difference test uses parsimony to determine whether two gene partitions are incongruent

There are two major widely used approaches to assess incongruence of gene partitions. One method is used in conjunction with maximum parsimony (MP) analysis and is called the incongruence length difference (ILD) test. This test relies on the assessment of the length of trees under different constraints and generation of a random distribution of tree lengths under these constraints. If the "real" tree lengths are shorter than 95% of the randomized ones, then the null hypothesis (i.e., that the two gene partitions are incongruent) is rejected. If, on the other hand, the "real" tree lengths are within the 95% distribution, then the null hypothesis cannot be rejected.

Consider the following example: suppose we have two genes that have been sequenced for the same four taxa. The topologies obtained for these two genes turn out to be relatively different from each other in topology. This situation happens frequently when many genes are analyzed. The differences in topology suggest incongruence of the evolutionary histories of the two genes. But are the two trees statistically different from each other? The ILD test is one way to determine whether the two genes are indeed statistically significantly incongruent with each other. This test is calculated by employing a random permutation method, as illustrated in **Figure 11.3**.

The data in the sequence columns of the two genes are concatenated and then randomized, and new partition boundaries are set that are the same length as the real genes to produce a new, randomly permuted data matrix. This random permutation approach is repeated for an arbitrarily selected number of replicates, perhaps 100 (the number is arbitrary but should be large enough to ensure statistical rigor). Then for each permuted matrix, the sum of the length of the trees for the "new" gene is estimated individually (which will vary from replicate to replicate) and the length of the trees for the concatenated data set (which should be the same as in the real data set) are calculated. The ILD for each "new" gene pair is then calculated:

$$\text{ILD} = T_c - T_i \qquad (11.1)$$

where T_c is the number of steps for the concatenated data set and T_i is the sum of the steps for the two trees generated by the individual gene data sets. By randomization of the process, a distribution of values for ILD can be generated from Equation 11.1. The distribution should be normal (two generated distributions are shown in **Figure 11.4**). The final step is to compare the ILD value for the real data (in this case, the ILD for the original data set is 5) to the distribution of values obtained by the random permutations. If the "real" value lies in the upper 95% tail of the distribution, then the null hypothesis of incongruence can be rejected (as in the top panel of Figure 11.4), and the two partitions are inferred to be congruent. If the "real" value lies within the bulk of the randomized distribution (as in the bottom panel of Figure 11.4), then the null hypothesis of incongruence cannot be rejected, and the two partitions are inferred to be incongruent.

Figure 11.3 Incongruence length difference test. The diagram illustrates how a matrix with two partitions (dark green and light green) is randomized in this test.

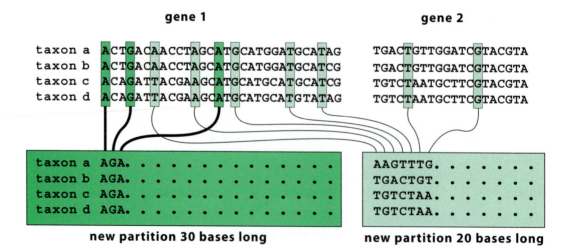

new partition 30 bases long   **new partition 20 bases long**

Likelihood ratio tests compare likelihoods to determine whether two gene partitions are incongruent

The second method is used in conjunction with maximum likelihood (ML) analyses and is called the Shimodaira–Hasegawa (SH) test. The first step in this test is to generate tree topologies for the two gene partitions by likelihood methods. The topology of the second gene tree is then enforced on the data for the first gene and vice versa, and likelihoods are calculated for the trees. Then a likelihood ratio test of tree likelihoods is used to determine if the two gene partitions are congruent (Sidebar 11.2). A significant likelihood ratio will result in rejection of the null hypothesis that the two gene partitions are congruent. The KH test is only valid for pre-specified trees, so if one wants to compare trees generated by ML, it is an invalid approach. The SH test was developed to allow for direct comparisons of two ML-generated trees.

Fork indices provide measures of tree similarity

There are several measures available in the literature that can measure the similarity of the branching order and topology of phylogenetic trees. These measures are rough indicators of incongruence of trees. Several of these measures are automatically generated by the phylogenetic programs that exist, and some are stand-alone programs. See http://bioinfo.unice.fr/biodiv/Tree_editors.html for a complete listing of tree manipulation and comparison programs.

The simplest tree comparison measure is to count the number of nodes that two trees have in common. This can be done by constructing a strict consensus tree of the two trees being compared and counting the nodes in the strict consensus tree. Another measure that can be used in a broader comparative way is the consensus fork index (cfi). This measure is calculated by constructing a consensus tree of the two trees being compared. The cfi is then found by dividing the number of bifurcating nodes on the consensus tree by the maximum number of possible nodes. For instance, if we have two trees with 50 nodes that we want to compare, we first construct a consensus tree of the two trees. Three nodes are lost in this consensus tree as a result of difference in topology; this means that 47 of the possible nodes are still resolved and bifurcating. The cfi is then estimated as 47/50 or 0.94, which means that 94% of the nodes are common to the two trees. These two measures, node counting and cfi, do not take into consideration bias caused by tree symmetry, nor do they consider polytomies in the calculation of the metric. Polytomies are simply nodes that have more than two branches coming from them. A node with three branches emanating from it is called a trichotomy; one of the more famous trichotomies before molecular data were applied to the problem was the human/chimp/gorilla relationship. With respect to the metrics discussed above, this means that sometimes a polytomy near the base of the tree will severely affect the tree comparison. A related measure called the Rohlf CI1 was developed to overcome these problems and basically addresses the problem of polytomies in trees. This metric gives greater weight when polytomies that are more basal (closer to the root of the tree) occur in the tree comparison. The Rohlf CI1 therefore corrects for biases introduced by tree topology.

The Gene Tree/Species Tree Problem

The "hard" incongruence problem has also been referred to as the gene tree/species tree problem. In other words, not all gene trees reflect the species tree, due to the genes having different evolutionary histories. There are three basic ways that a gene tree can be incongruent with a species trees, and all three relate to the idea that genes can have different evolutionary histories. First, specific genes might be under different kinds of natural selection, and this might result in incongruent

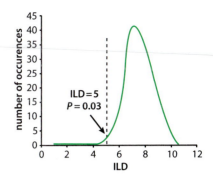

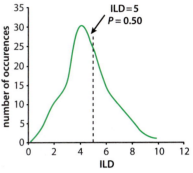

Figure 11.4 Two hypothetical distributions of incongruence length difference (ILD) tests. Top: the distribution implies that the null hypothesis can be rejected; thus the inference is that the partitions are congruent. Bottom: this distribution implies that the null hypothesis (that is, that the two matrices are incongruent) cannot be rejected (p = 0.50); thus the inference is that the partitions are incongruent.

histories of genes. Second, some genes might be horizontally transferred from one group of organisms to another. Such horizontal history disrupts the vertical history of species. This problem is thought to be so rampant in the microbial world that some researchers have decided that a bifurcating tree of life in Bacteria and Archaea is not a possibility. Third, by pure chance and a process called lineage sorting, some genes can diverge with different patterns than others. We will discuss

Sidebar 11.2. Kishino–Hasegawa and Shimodaira–Hasegawa tests in action.

Consider the foxp2 example used in Chapter 10. The tree topology that we obtained from nearly all analyses of the data set gave a reasonable topology, except for the inference of orangutan with macaque as a monophyletic group. The Shimodaira–Hasegawa (SH) test, adapted from the Kishino–Hasegawa (KH) test, can assess the statistical robustness of the ML tree versus large numbers of alternative hypotheses such as the one obtained from the foxP2 data. The KH test is very specific in that it allows one to examine the relationship of two prescribed trees. The KH and SH tests are implemented in most available phylogenetic packages.

The items needed for a KH test are the data, a likelihood model, and alternative trees to test. Since we are interested in the orangutan + macaque grouping, we will compare the parsimony and likelihood topology with the accepted topology for the taxa in our foxP2 matrix (**Figure 11.5**). The test is very simple. Likelihoods for both hypotheses are estimated, and the difference of log likelihoods is evaluated for significance by use of a normal distribution. For our example, the results are

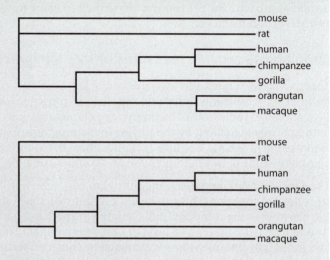

Figure 11.5 Two alternative trees are compared with the Kishino–Hasegawa test for the foxP2 data set. The tree on the bottom is the tree that is accepted by most primatologists.

```
                         diff      KH-test
tree     -ln L          -ln L         P
------------------------------------------
  1    4301.13956      0.14024      0.00*
  2    4300.99932      (best)
```

*P < 0.05.

The KH test indicates that we can reject the null hypothesis that the two trees differ in their likelihood support from the data; thus the orangutan + macaque grouping is no more likely than the accepted topology.

The major steps of the SH test are listed here:

Step 1. Generate a sample of N bootstrap samples of the sites in the sites in the original data set. For each bootstrap replicate, compute $\ln L$.

Step 2. For each replicate, evaluate each candidate tree (T_{ML}, T_2, T_3, T_4, ..., T_j, where T_{ML} is the maximum likelihood tree) and center likelihood scores from those trees.

Step 3. For each bootstrap replicate, compute for the tree how much the centered value is below the maximum across all trees for that replicate. In this way a distribution can be generated.

Step 4. For each tree, the likelihood values are compared to the distribution.

The results of the SH test are

```
                     SH-test
tree                    P
--------------------------------
TML                   0.49
```

The results of the SH test indicate that we can reject the null hypothesis that our likelihood tree generated from the data is significantly better than the specified accepted tree. Again, the orangutan + macaque grouping is not significantly more likely than the accepted topology.

the ramifications of natural selection and horizontal transfer in subsequent chapters, but we focus here on the examination of genealogy of closely related individuals within species and in closely related species and the phenomenon of lineage sorting. Lineage sorting occurs in ancestral populations that have yet to diverge. Consider an ancestral population that will eventually diverge into three new species called a, b, and c. In the ancestral population of these three species, some genes will be transmitted such that the gene in a and b is more closely related than either one is to c. For other genes, a different pattern of divergence will result in the gene in b and c being more closely related to each other than to a. And finally, in some genes, the divergence process will allow for the gene in a and c to be more closely related to each other than either is to b. Hence, depending on the overall pattern of divergence, different genes will have different histories (**Figure 11.6**).

Phylogeography adds a new dimension to population-level analysis

A major advance in modern evolutionary biology in the latter part of the twentieth century was the development of a research paradigm called "phylogeography." In the late 1970s and the 1980s, researchers interested in the genetic relationships of closely related organisms and the genetic dynamics of populations began to use mitochondrial DNA (mtDNA) as an indicator of variation and of relatedness of individuals (see below). John Avise, then at the University of Georgia, pioneered this work, using a wide array of organismal systems. It is interesting to note that, up to that time, genetic information obtained from proteins or chromosomes was

Figure 11.6 Diagrammatic representation of the process of lineage sorting. A: the light green tree shows the species tree; in this case, A and B are more closely related to each other than each is to C. The oval marks the part of the interacting system of these species where they share a common ancestor. This region is where gene systems "sort" through recombination and segregation of chromosomes. B (expansion of the oval in panel A): the three different genes (dotted line, dark green line, and light green line) will sort in three different ways. C: the gene trees for each of the three possible sorting events are shown. Note that each one has a different topology even though the genes eventually sort into their respective species.

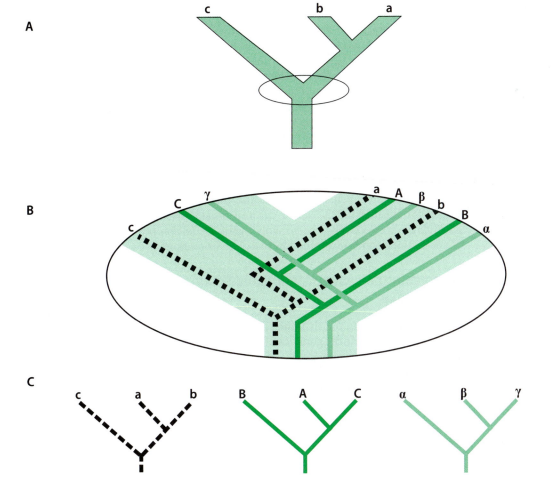

converted to genetic similarities, and demographic hypotheses about the organisms were tested using a statistic called the F statistic (F_{ST}), which we will discuss in detail later. F_{ST} was used to determine the degree of gene flow between populations. Avise realized that the genealogies obtained from mtDNA analysis could be related to the geographic distributions of the organisms.

In fact, the lack of correlation of geography with genetic patterns was used as a null hypothesis in this new area of population genetics and evolutionary biology. Essentially, the mtDNA data, when converted to genealogies, could add a new dimension to population-level analysis. Avise called this new paradigm "phylogeography" to indicate the importance of the geographical component of organismal history with their genetic patterns. Avise intuited that this powerful null hypothesis could lead to inferences about gene flow across geographic areas, about hybridization between species, and other demographic aspects of the evolution of organisms. Other authors subsequently developed algorithms to take advantage of the extra added genealogical aspect of mtDNA data. In this context they developed the nested clade analysis (NCA) approach, which could examine demographic aspects of the evolution of populations such as restricted gene flow, range expansion, and past fragmentation of populations. NCA has recently been criticized for its potential to generate an alarmingly large false-negative inference rate.

Lineage sorting was first observed in mtDNA

Many of the first examples of lineage sorting involved the different phylogenetic patterns researchers obtained when comparing mtDNA versus anatomical or protein electrophoretic results. The very earliest studies of organisms at the population level and the species boundary utilized mtDNA because of its rapid rate of change and its ease of manipulation. Early on, researchers realized that the patterns of relationships determined from mtDNA were often at odds with the patterns obtained from other sources. Most researchers attributed these differences to the unique biology of mitochondrial DNA. Remember that mtDNA is inherited as a clonal maternal marker in most organisms. Its clonality ensures that a negligible amount of recombination occurs in the divergence of the molecule, and its maternal inheritance can lead to interesting and counterintuitive patterns of divergence. When sequencing of nuclear genes became commonplace in the late 1990s, the incongruence of gene trees constructed from nuclear genes were at odds with the trees constructed from mitochondrial DNA. More recently, patterns obtained from mtDNA and Y chromosomal DNA are commonly compared with the patterns obtained from microsatellite studies to examine the phylogenetic patterns that differ between mtDNA and nuclear DNA. The idea behind these studies is that since mtDNA and Y chromosomal DNA follow maternal and paternal lineages, respectively, issues involving the sexual components of the evolution of populations can be examined by use of these markers.

Three examples illustrate lineage sorting in studies of closely related taxa

Researchers studying the dynamics of species formation frequently utilize DNA sequences to examine the genetics of closely related populations or closely related species. There are many examples of this problem ranging from plants to animals. Here we present two empirical examples and one simulation example that are highly cited in the literature as emblematic of lineage sorting.

The first example comes from Craig Moritz's lab at the University of California, Berkeley. Moritz and colleagues examined the genealogical patterns of seven nuclear genes for eight closely related species in two subgenera of *Thomomys* (geomyid rodents). The interesting aspect of the species in this group is that the population sizes of these species are presumably larger than the divergence times of the species in the group would suggest. This interesting aspect of *Thomomys* species in the light of coalescent theory suggests that there should be a high degree of incongruence among gene trees constructed from the seven nuclear loci used in the study. In this study, none of the seven gene trees gave the same topology. Moritz and colleagues then use two approaches to establish the phylogenetic relationships of the individuals in the data set they collected. The first approach is the standard concatenation approach, and the second is an approach called BEST (Bayesian estimation of species trees) that simultaneously analyzes species and gene trees. We examine the BEST program in Web Feature 8. Concatenation resulted in strong support for six of the eight recognized species in the study, while BEST resulted in strong support only for the split between the two subgenera in the study. The upshot of the study, according to Moritz et al., is that concatenation ignores the conflicting signal from the different gene genealogies and overestimates the speciation events in the group. They caution against overreliance on the concatenation method.

The second empirical example comes from Lacey Knowles' lab at the University of Michigan. This study focused on several individuals each from four species of *Melanopus* grasshoppers and utlilized five nuclear-encoded gene regions and one mitochondrial gene. With four species, the number of strictly bifurcating trees possible is 15, so the researchers could infer fairly precisely the probabilities for the potential trees. The upshot of this study is that each gene region delivered a different topology that was best interpreted in a coalescent context. The authors coined the term ESP to refer to the estimation of species phylogeny through coalescent methods. They shy away from the concatenation approach, criticizing it as forcing congruence when none exists.

Our third example is a simulation study from James Degnan's lab at Harvard. Using simulations, these researchers "examined the performance of the concatenation approach under conditions in which the coalescent produces a high level of discord among individual gene trees." They simulated data to produce phylogenetic matrices for four ingroup taxa for varying numbers of genes from 10 to 6000. For four taxa, there are 15 possible trees, and so they could simply count how many times these 15 trees are obtained that are incongruent with the species tree. Trees generated from matrices that matched the species tree were called matching trees (MT). Trees that did not match were called swapped (ST) or symmetric (S1, S2, and S3) trees. Gene trees that are more probable than the MT topology are referred to as anomalous. They were able to show that, for simulations with short branches at the base of the tree, there was a high probability of obtaining an anomalous tree. They conclude from this simulation study that concatenation can lead to inferences that depart from the species tree.

Genome-level examples of lineage sorting have also been documented

Some recent examples of lineage sorting in the literature concern closely related species and genome-level analysis (**Figure 11.7**). One of the first major reports from the *Drosophila* 12 Genomes consortium concerned a claim of the overwhelming existence of lineage sorting among three species of flies in the data set (*Drosophila melanogaster*, *Drosophila yakuba*, and *Drosophila erecta*). The distributions of genes and proteins that support the three possible arrangements of the three taxa are shown in Figure 11.7B. It is clear from this figure that the frequency of conflicting topologies for these three taxa is quite high. Whether or not this pattern is produced entirely by lineage sorting is an important issue. Another case of lineage sorting in closely related lineages concerns the three closely related subspecies of mouse: *Mus musculus musculus, Mus musculus domesticus*, and *Mus musculus castaneus* (Figure 11.7C). White showed that the three possible topologies occur with high frequency for the various genomic regions examined by those authors. A third genome-level example of the lineage sorting problem can be

Figure 11.7 Four recent genome-level studies where massive amounts of lineage sorting have been implied. A: yeast data set of Rokas et al., showing the three topologies considered in that paper. B: *Drosophila* study of Pollard et al. C: mouse study of White et al. D: primate study of Hobolth et al. The percentage of genes in the genome of the organism giving that topology is shown. The asterisk indicates that the percentages for two of the topologies in the primate study were combined rather than computed individually.

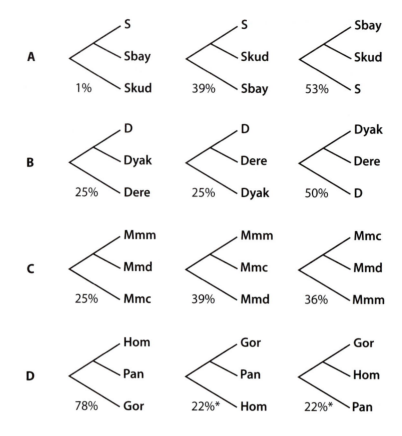

found in the yeast data set of Rokas et al. (Figure 11.7A). Finally, Hobolth has used human, chimpanzee, and orangutan genome sequences to suggest that incomplete lineage sorting has occurred in the divergence of these three species (Figure 11.7D). The question with these studies is not if lineage sorting occurs. Lineage sorting is a logical extension of coalescent theory (see Chapter 3). The problem really is how frequently lineage sorting impacts vertical evolution and hence how strongly it might influence phylogenomic inference. If the levels of lineage sorting seen in flies, mice, and yeast are any indication of the frequency, then there are indeed a large number of genes affected by this coalescent phenomenon.

Coalescence offers a partial solution to the gene tree/species tree problem

As more and more genes have been sequenced for phylogenetic problems, it quickly became obvious that the existence of incongruence in the phylogenetic signal is more the norm than the exception. What this means is that not all genes are telling the same evolutionary story. As we discussed above, the reasons for the incongruence of genes with each other are complicated, and the problem has been called the gene tree/species tree problem. Some phylogeneticists have suggested that consideration of the coalescent process can partially solve the problem of gene tree incongruence. Scott Edwards of Harvard University points out that, prior to the availability of large numbers of gene sequences, systematists and phylogeneticists basically equated gene phylogenies with species phylogenies. Edwards points out the fallacy in doing this and suggests that the coalescent approach (see Chapter 15) is a valid one for solving the gene tree/species tree problem. Edwards' group has developed a coalescent approach for estimating species trees. The approach utilizes a likelihood function that involves the probability distribution of gene trees given a species tree. Students interested in this approach should consult the BEST program developed by Liu et al., which is discussed below.

Horizontal Transfer

Microbiologists have long recognized the fact that microorganisms can swap DNA quite easily and have called this horizontal gene transfer (HGT). This process is relatively simple for bacteria and has been known in *Escherichia coli* since the origins of phage research several decades ago. Microbiologists have also recognized that HGT would have an impact on phylogenetic signals. If an HGT event occurs in a particular gene, then the relationships of organisms determined by use of that gene will show incongruence because of the interruption of vertical history caused by horizontal transfer. The molecular mechanisms by which HGT can occur are transformation, conjugation, and transduction. Some microbiologists have suggested that HGT is so prevalent that reconstructing phylogenies of domains such as Bacteria and Archaea is not possible. Microbiologists like Ford Doolittle and Eric Bapteste have suggested that HGT forces phylogeneticists to accept a comblike phylogeny for microbes. They argue for pattern pluralism (a pluralistic or multifaceted understanding of the patterns involved) because in some cases nonvertical evolution (HGT) "may be the relevant historical process, and nets or webs are the appropriate way to represent what is a real but more complex fact of nature." Basically, they ask that researchers be open-minded about the possible patterns that can be hypothesized to explain evolution at this level.

As Doolittle and Bapteste suggest, one way to accommodate the presence of large amounts of HGT in reconstructing phylogeny is to use nets to represent relationships. Victor Kunin, of the Joint Genome Institute, and colleagues have utilized a three-dimensional way of viewing HGT in phylogenies that they call the "net of life." They use classical approaches to phylogeny reconstruction to generate a three-dimensional tree. The tree is flat (in the *x*- and *y*-directions) until horizontal transfer occurs, and then branches extend into the *z*-direction. Any horizontal transfer is denoted by the connection of branches in the *z* direction with lines (see **Figure 11.8**). The width of the lines represents the number of gene families involved in the HGT events. In this way, both HGT and the vertical history of the microbes can be displayed.

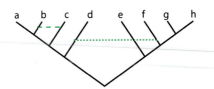

Figure 11.8 Tree showing how horizontal transfer is depicted. In this example, horizontal transfer events are detected between two species pairs on the tree: species b and c (dashed line) and species d and f (dotted line).

Programs That Consider Nonvertical Evolution and Lineage Sorting to Infer Phylogeny

Because of the large number of genes in the genomes of organisms and because coalescence is analyzed via simulations, this approach has become very computer-intensive. In addition, the detection of horizontal transfer and its characterization is also computationally intensive. Therefore, to conclude this chapter, we briefly discuss the various programs that exist to infer phylogeny when problems like lineage sorting and horizontal transfer are prevalent. Some programs use input from both gene trees and species trees to resolve the gene tree/species tree problem. Other programs accommodate horizontal gene transfer by producing nets or webs as output.

Coalescence programs use both gene trees and species trees as input

As more and more genomic information is collected, the number of gene partitions that can be included to address a phylogenetic problem grows larger and larger. As we mentioned above, the probability that topologies of the trees generated by single genes is not the species tree is quite high. To approach this phenomenon, COAL (http://www.coaltree.net) was developed by Degnan and Salter. COAL computes the probabilities of tree topologies for genes when a species topology is known. Lengths of the branches in the tree are computed with coalescence units (number of generations divided by effective population size). The program uses three input files: a species tree file, a gene tree file, and a command file. Important output options include generation of probabilities of gene trees given a specific species tree and generation of the number of coalescent histories for a data set. One limitation of the program is that, in its current form, only one gene tree can be processed at a time, but this can be overcome by performing multiple separate runs for gene trees. Another program that accomplishes similar operations is BEST. The assumptions that the program uses are that "discrepancies between the gene and the species tree are due exclusively to lineage sorting with free recombination between genes." BEST will perform a Bayesian analysis that results in the topology of the species tree and also gives divergence times and population sizes. BEST is integrated into the MrBayes program.

Programs that consider horizontal gene transfer generate nets and webs

Several programs have been developed to construct nets or webs from data where researchers suspect that HGT has been prevalent. All are based on the premise that there is a need for a richer visualization of the results of a phylogenetic analysis than a strictly bifurcating tree. Huson and Bryant have developed a program called SplitsTree that constructs phylogenetic diagrams using split networks. SplitsTree can construct networks using distances, characters, or even a collection of trees to obtain the network. Lumbermill is a program developed to keep track of HGT. The program uses an arbitrarily chosen tree against which vertical and horizontal events are measured. The program uses a tree and a set of partitioned gene sequences as input. The information in the gene sequences is then displayed, showing support on the tree for vertical events and horizontal events. Conflict at nodes can be identified and hence possible points of horizontal transfer can be inferred using the software. T-Rex (Tree and Reticulogram Reconstruction; http://www.trex.uqam.ca/index.php?action=trex&menuD=2&project=trex) is yet another program that detects conflicting signals between phylogenetic trees. It produces a reticulated phylogeny and identifies all possible positions in the reticulated tree where horizontal events may have occurred.

Summary

- Relationships among microbes are difficult to address by morphology. Instead, sequences from ribosomal RNA can be used as a tool for understanding the phylogeny of microbes because the "stem" portions are highly conserved, even among divergent organisms.

- To look at very closely related taxa, the mitochondrial genome was exploited as a phylogenetic tool. Particular regions of the mitochondrial genome change very rapidly, such that even closely related individuals within species will show variation.

- Out of necessity, the tools for dealing with incongruence of trees obtained from different sources have grown since the explosion of genome-level data. This is because, as more and more genes were sequenced, the incongruence of inferences made from these genes became evident.

- There are two major strategies when faced with multiple gene partitions: to combine (concatenate) or not. The subject of whether to concatenate or not is complicated.

- Statistical methods assess incongruence to decide whether information should be concatenated. The incongruence length difference test uses parsimony and SH and KH likelihood ratio tests assess likelihoods to determine whether two gene partitions are incongruent.

- Not all gene trees reflect the species tree, owing to the genes having different evolutionary histories. This is known as the gene tree/species tree problem, and it can arise from natural selection, horizontal gene transfer, and/or lineage sorting.

- Coalescence offers a partial solution to the gene tree/species tree problem. Coalescence programs use both gene trees and species trees as input.

Discussion Questions

1. In a system of four closely related species (a, b, c, and d) with a species tree as in **Figure 11.9**, how many gene trees are possible that can be placed over the species tree? Draw two of the possible trees and show how they would fit onto the species tree.

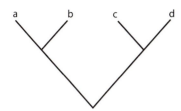

Figure 11.9 Species tree for four taxa.

2. Contrast how phylogenetic hypotheses would be impacted by lineage sorting and horizontal transfer.

3. The following rooted trees with 10 taxa (a–i) are in Newick format (see Chapter 8). What are the cfi values for these two trees in each part?

(A)
```
(((((((((a, b), c), d), e), f), g), h), i)
(((((((((i, h), g), f), e), d), c), b), a)
```

(B)
```
(((((((((a, b), c), d), e), f), g), h), i)
(((((((((a, b), c), d), e), f), g), i), h)
```

(C)
```
(((((((((a, b), c), d), e), f), g), h), i)
(((((((((a, b), e), c), d), f), g), h), i)
```

Further Reading

Avise JC, Arnold J, Ball RM Jr. et al. (1987) Intraspecific phylogeography: the mitochondrial DNA bridge between population genetics and systematics. *Annu. Rev. Ecol. Syst.* 18, 489–522.

Bonaventura MD, Lee E, DeSalle R, Planet P (2010) A whole-genome phylogeny of the family Pasteurellaceae. *Mol. Phylogenet. Evol.* 54, 950-956

Doolittle WF (2009) The practice of classification and the theory of evolution, and what the demise of Charles Darwin's tree of life hypothesis means for both of them. *Philos. Trans. R. Soc. B* 364, 2221–2228.

Doolittle WF & Bapteste E (2007) Pattern pluralism and the Tree of Life hypothesis. *Proc. Natl. Acad. Sci. U.S.A.* 104, 2043–2049.

Farris JS, Källersjö M, Kluge AG, & Bult C (1995) Constructing a significance test for incongruence. *Syst. Biol.* 44, 570–572.

Kishino H & Hasegawa M (1989) Evaluation of the maximum likelihood estimate of the evolutionary tree topologies from DNA sequence data, and the branching order in hominoidea. *J. Mol. Evol.* 29, 170–179.

Kluge A (1989) A concern for evidence and a phylogenetic hypothesis of relationships among Epicrates (Boidae, Serpentes). *Syst. Zool.* 38, 7–25.

Knowles L (2009) Statistical phylogeography. *Annu. Rev. Ecol. Evol. Syst.* 40, 593–612.

Miyamoto MM & Fitch WM (1995) Testing species phylogenies and phylogenetic methods with congruence. *Syst. Biol.* 44, 64–76.

Rokas A, Williams BL, King N, Carroll SB (2003) Genome-scale approaches to resolving incongruence in molecular phylogenies. *Nature* 425, 798–804.

Adapting Population Genetics to Genomics

Population genetics has a long history, as it was invented concomitant with the rediscovery of Mendel's laws in the early 1900s. The theoretical underpinning of population genetics was discussed briefly in Chapter 3, and this chapter extends that discussion to include the use of phylogenomic high-throughput data such as microsatellites and single-nucleotide polymorphisms (SNPs). The nature of DNA variation at the population level has become extremely important in human genomics, and recent work has expanded our knowledge of how variable we are as a species. More importantly, this variation can be used in studies of human disease and to scan whole genomes for disease-related genes by the genome-wide association study (GWAS) approach. Classic approaches to population genetics, such as F statistics, and more recent novel approaches, such as STRUCTURE, have broadened the view that population geneticists have of the natural world. These approaches are at the foundation of how we understand natural populations in an evolutionary context.

Modernizing Population Genetics by Use of High-Throughput Methods

As we saw in Chapter 3, the new synthesis of evolution in the 1940s provided cohesion and a strong research focus for evolutionary biology for several decades, well into the 1970s. What happened to change things? One word might explain it: molecules. Blood groups and early protein sequencing were used prior to exact DNA sequence characterization. In the late 1960s scientists discovered that they could visualize proteins by a process called protein electrophoresis. This advance led to a huge leap in population studies in evolutionary biology. In this technique, researchers were able to isolate proteins from blood, tissue, or internal organs. Proteins that migrated different distances in gels were assumed to have different amino acid sequences. Since the central dogma of molecular biology (DNA → RNA → Protein) was known at that time, the link to genetics was clear. Proteins are the products of genes, and if differences in proteins can be traced via protein electrophoresis, the genes of organisms can be followed by proxy. Protein electrophoresis is fast and efficient in comparison to the techniques used prior to the advent of molecular biology. For instance, an evolutionary geneticist studying individuals within a species before the discovery of protein electrophoresis would look for behavioral, chromosomal, or anatomical differences that might have a genetic basis. They would then characterize the organisms on the basis of this genetically based attribute. This work was painstaking and required massive amounts of time and resources because the individuals needed to be subjected to many anatomical and physical tests.

When protein electrophoresis was invented and used widely, a group of individuals could be examined very rapidly, not just for a single attribute but for as many proteins as there were assay systems available. In essence, this was the beginning of the genetics data onslaught, otherwise known as high-throughput evolutionary studies, which has culminated in the genomics-level analyses that are now routinely seen in evolutionary biology.

Kimura and Lewontin contributed important new ways to think about genes in nature

Two important ways of thinking about genes in nature arose from the protein electrophoresis deluge. The first approach was proposed by geneticist Motoo Kimura in 1968 and was articulated in his book *The Neutral Theory of Evolution*. Kimura pointed out that the majority of evolutionary change in proteins was neutral or, in other words, did not occur by natural selection. He introduced the notion that genetic drift and neutral evolution were the major movers of evolution in nature. These ideas have become very important in many fields of biology but particularly in genomics, where the search for functional portions of the genome is the equivalent of trying to find locations that are not evolving neutrally.

The second approach came from Richard Lewontin and was articulated in his book entitled *The Genetic Basis of Evolutionary Change*. Lewontin described the paradigm changes that were necessary to cope with the onslaught of protein data. In many ways, he foresaw the data deluge of genomics and argued forcefully that evolutionary biologists should prepare themselves for it by expanding their thinking about how to analyze organisms in nature. His book opened the way for a whole new approach to population genetics and understanding of the evolutionary process in nature. These two new ways of thinking foreshadowed the bioinformatics age as they both offered new methodological approaches and theoretical vantages for viewing populations in nature.

The Hardy–Weinberg theorem has been extended in modern population genetics

As we saw in Chapter 3, most of the workings of modern population genetics rely on a simple mathematical formulation called the Hardy–Weinberg theorem. Because modern population genetics has utilized genes and alleles as the currency of analysis, a rich theoretical framework for applying population genetics to genomics has developed. One of the more important advances that aided computational approaches to population genetics is the development of model-based analysis, upon which much of the current analytic literature on population-level processes is based. For a particular population, a distribution of probabilities yielded by the genetic process can be estimated without incorporating predetermined model parameters. Instead, by use of a data set and a model of evolution, probabilities can be calculated for the parameters of the model via maximum likelihood methods, and statements using posterior probabilities can be made by approaches with Bayesian methods. The keys to this way of thinking about populations are a data set, a model, and some idea of the distribution of the parameters without model restrictions. In essence, any genetic parameter within a population, such as mutation rate, recombination rate, migration, selection, and drift, can be manipulated in this way.

There are many assumptions and models that are used in population genetics. We have already discussed the assumptions of the Hardy–Weinberg principle. The models themselves are not so important for our purposes. However, these models incorporate parameters that are relevant to our discussion of population genetics. Current models of population genetics and the programs that are used to manipulate population-level information reveal the parameters listed in Sidebar 12.1 that can be estimated by model-based approaches. We will explore some of these parameters and models in more detail in Chapter 13.

DNA Variation among Individuals

Phylogenomics can be applied both to species and to individuals within a species. For the two different levels of analysis, the characters used to denote variation between the groups must be appropriate for the data set being studied.

Sidebar 12.1. Parameters incorporated into modern population genetics models: after marjoram and Tavare.

Mutation and recombination rates: among the first parameters to be approached by these new techniques.

Demographic parameters: population size, population substructure and mating patterns, and migration.

Selection: regions of the genome under selective pressure. In particular, the HapMap project has benefited from this approach in identifying episodic selection at the genome level. In addition, a phenomenon called selective sweep can be examined via this process.

Ancestral inference: time to the most recent common ancestor, also known as TMRCA. TMRCA identifies the divergence time of a population and is also used to infer the age of certain mutations.

Genome-level phenomena: genomic changes within a population, such as identification of recombination hotspots, reconstruction of haplotypes, and linkage disequilibrium.

Human disease association studies: identification and mapping of disease genes mostly used in human population genetics. Genome-wide association studies (GWAS) are used extensively in disease association studies and are reliant on population genetic modeling.

General population description: estimates of genetic diversity, genetic distances between populations, estimates of population subdivision, tests of selective neutrality within populations, estimates of gametic phase, allelic richness, kinship and relatedness at individual level. Many of these parameters and estimates can be obtained from statistical packages and programs that we will discuss in the next few chapters. These kinds of analyses have become an important part of conservation genetic research.

Individual-centered analysis: estimates of recent migration rates, assignment of individuals to populations, detection of genetic structure among individuals in populations, and hybridization in populations, as well as detection of hybrids in populations.

When relationships between species are investigated (interspecies variation), the presence or absence of genes or gross physiological characteristics (number of fingers, presence of an opposable thumb, etc.) can be used as diagnostic characters. However, when all the individuals being investigated are of the same species (intraspecies variation), more specific markers of variation need to be used. This is because all humans (except for those with severe diseases) generally have the same number of genes, 10 fingers, and an opposable thumb. Because they are uniform throughout the species, these attributes are not useful as diagnostic characters.

In order to investigate intraspecies variation, physiological markers such as height, weight, and hair color could be used. In order to obtain more fine-grained information, variation at a molecular level needs to be obtained. While there are a great variety of variations that can occur between the genomes of individuals, the two most commonly used markers of variation are SNPs and microsatellites, which will be discussed in the following sections. After we describe these markers, we will explain how they are used in population genetics.

Single-nucleotide polymorphisms can be used to differentiate members of the same species

Single-nucleotide polymorphisms, or SNPs, occur when the nucleotide at a particular position in the genome of one individual differs from the nucleotide at that position in another individual, while the surrounding sequence is identical (see Chapter 2). As an example, consider these two sequences that are from the same genomic location in two individuals:

ATTGTGTA**T**ATTGTATGTA

ATTGTGTA**G**ATTGTATGTA

A SNP is found in these two sequences at the nucleotides marked in boldface type. The individual with the top sequence has a T, while the individual with the bottom sequence has a G at the same locus.

For human studies, SNPs and other variations are discussed in relation to the human genome reference sequence that was produced by the Human Genome Project. In relation to this reference sequence, on average, a human genome will have several million SNPs. Since the human genome contains 3 billion base pairs, the several million SNPs will amount to ~0.1% of the genome varying on average between individuals. To state things another way, there is on average one SNP per 1000 base pairs (some estimates indicate even one every 300 base pairs) between any two randomly chosen humans. While this is the average rate, there are great variations between regions of the genome in their level of SNPs. For example, genes and their promoters have a low level of SNPs because their sequence is under negative selection. In contrast, heterochromatic and repetitive regions of the genome have very high levels of SNPs. All SNPs that have been identified in humans are contained in the dbSNP database (http://www.ncbi.nlm.nih.gov/SNP). As of the summer of 2012, there were over 30 million SNPs in the database. This number is expected to greatly increase as the number of genomes that have been sequenced increases.

The main reason that SNPs have become a major force in biological research is due to the fact that it is easy to profile them in an individual. Companies, mainly Affymetrix and Illumina, have produced microarrays (see Chapter 18) that can profile millions of SNPs in the human genome. These SNPs were selected by studying many individuals from different populations and are thought to be a good representation of variation among humans. Each microarray contains probes matching the two known alleles of each SNP, and it can therefore be used to determine which allele a person's genome contains. Since the human genome is diploid (that is, we have two copies of each chromosome), the array can also tell whether a person is homozygous for an allele or heterozygous, with a different allele on each copy of the chromosome. Running one of these SNP arrays on a person is a relatively inexpensive procedure (by scientific standards), costing much less than a thousand dollars.

Microsatellites provide another analytical tool for species where SNPs are less abundant

Used correctly, SNPs can be extremely important in understanding the population genetics of organisms. Unfortunately for the population geneticist interested in plants or animals, the number of SNPs identified in other species is much lower than in humans. Consequently, plant and animal population geneticists have turned to another tool that can rapidly and concisely identify polymorphisms that can be used in the population genetics of other organisms. This tool is the analysis of microsatellites, and it has many uses (**Table 12.1**). Microsatellites are highly polymorphic regions of genomes that consist of short repeats. These repeats most often consist of two to four nucleotides, but they can be of any length.

A variation of the approach actually was pioneered in the 1980s by Alec Jeffreys as a forensic tool to identify humans involved in crimes. Jeffreys' approach was a bit different from the approaches that are now used, because polymerase chain reaction (PCR) did not exist when he first applied the use of repeated regions. His approach focused on larger repeat regions in the genome that were variable and hence was called the variable number of tandem repeat region or VNTR approach. Detection of these VNTRs required that they occur over quite long stretches of the genome, and this approach was greatly enhanced by the introduction of PCR. The approach employed by most population geneticists today uses smaller regions where more precise repeats occur; these are called microsatellite approaches, and

Table 12.1 Uses of microsatellites in evolutionary biology

Population genetics
Markers for gene flow and migration studies
Markers for determining stock structure in agriculture and conservation genetics
Genetic probes for difficult tissues
Markers to determine the species origin of unidentifiable larval forms
Markers to determine genetic variability and identification of unidentifiable biological samples (gut contents, commercial samples, etc.)
Markers for determining genetic identity of scat samples
Pedigree maps
Markers for pedigree analysis in disease studies
Paternity tests
Understanding diseases, forensics, or quantitative trait loci (QTLs)

they have both advantages and disadvantages (**Table 12.2**). Both VNTR and microsatellite approaches take advantage of the co-dominant inheritance (co-dominance is a situation where both alleles of a locus are expressed in equal amounts) of the repeated regions. Three major kinds of repeat variants are used in microsatellite studies: di-, tri-, and tetranucleotide repeats (**Figure 12.1**). The utility of microsatellites lies in their high degree of variability produced by length variation. The length variation is produced by slippage in DNA during replication, called

Table 12.2 Advantages and disadvantages of microsatellites

Advantages
Polymorphism levels are high for the most part
Most are co-dominant genetic markers (unlike RFLP or AFLP, which are mostly dominant)
Present in every organism examined to date
Very abundant; about one every 6 kb of sequence in human genome
Show random spacing in the genome
Can use identical primers to find same loci in closely related species
Easier to interpret from primary gel information, in contrast to RAPD, AFLP, or RFLP
Neutral markers; di- and tetranucleotide repeats are neutral because they are usually found in noncoding regions; however, trinucleotide repeats can change number of amino acid repeats in a protein (Huntington's chorea)
Disadvantages
Development of primers is both expensive and time-consuming
To attain sufficient statistical power, several markers are needed
Different rates of evolution at different loci can skew results

dinucleotide	5′-GATCGACG**ATATATATATATATATATATATATATAT**AGCATCGAGCAATGAC-3′	15 repeats
trinucleotide	5′-GATCGACG**CAGCAGCAGCAGCAGCAGCAGCAGCAG**AGCATCGAGCAATGAC-3′	9 repeats
tetranucleotide	5′-GATCGACG**GAATGAATGAATGAATGAATGAATGAAT**AGCATCGAGCAATGAC-3′	7 repeats

Figure 12.1 Three most commonly used microsatellite formats: di-, tri-, and tetranucleotide.

slip-strand mispairing (SSM). Such slippage usually either increases or decreases the repeat length by one unit. A second cause of variation in microsatellites is recombination, either unequal crossing over or gene conversion.

Since the whole genome sequences of many study organisms in population genetics are not known, a microsatellite study of the population of organisms usually starts with the discovery of microsatellites in the genome by the following approach. First, genomic DNA is obtained from an individual of the study organism. The DNA is fragmented, either by use of restriction enzymes or through some mechanical process like sonication. Next small fragments of the genomic DNA are shotgun-cloned into a cloning vector to create a random genomic library of the study organism. Shotgun cloning is explained in Chapter 2 in the context of whole genome sequencing. The approach used to generate a library for microsatellite screening is similar. The library is then screened using a di-, tri-, or tetranucleotide like 5′-ATATATATATATATATATA-3′. The screening process with the aforementioned probe will detect clones of the genomic DNA with TA repeats in it. Several di-, tri-, and tetranucleotides are used to screen the library, and any clone that appears to contain a repeat is then sequenced to verify the existence of a repeat in the clone. Any genomic sequence that shows the existence of a repeat is then scanned to design oligonucleotide primers that can be used to amplify the region of the genome of the study organisms where the repeat exists. Usually, a researcher hopes to discover 20 or 30 of these regions. These primers are then used to screen a population of individuals of the study organisms. Other approaches to primer design have been developed that use programs that scan whole genomes for di-, tri-, and tetranucleotide repeats. Since the whole genome is often times known for several species, the regions 3′ and 5′ to these repeats in their respective genome data bases can be used to design primers.

If there is variation in the number of repeats in the amplified regions, this will manifest itself by differences in the length of the amplified region. An example of how a dinucleotide microsatellite might vary among three individuals is shown in **Figure 12.2**. Any region amplified by the designed primers that shows variation in the initial population is then used to detect variation of these new co-dominant markers in larger population genetics studies. Most microsatellite studies that are focused on classical population genetics problems start off with at least eight of these microsatellite markers, but larger numbers of markers are needed for studies that hope to identify individuals using microsatellites. Note also that the markers for genetic variability that are analyzed as a result of a microsatellite study are not simple nucleotide changes but rather length differences. This latter fact means that the data input into microsatellite studies are not simple strings of Gs, As, Ts, and Cs but rather numbers that represent the different lengths of the microsatellites. These initial fragment length data are then converted into frequencies of the various variants for the microsatellites examined in the target populations.

Extending Fundamental Population Genetics

Single-nucleotide polymorphisms and microsatellites are utilized in population genetics calculations to characterize populations and to compare different populations. Before explaining the application of population genetics techniques to these types of data, we will explain some basics of mathematical population genetics as an addendum to what was presented in Chapter 3.

1-atcgagcgacatgctacgcaATATATATATATATATATATATATATATATATATcgacgatcgatcgatcgac(...)

2-atcgagcgacatgctacgcaATATATATATATATATATATATATATATcgacgatcgatcgatcgac(...)

3-atcgagcgacatgctacgcaATATATATATATATATATATATATATATATATATATATATcgacgatcgatcgatcgac(...)

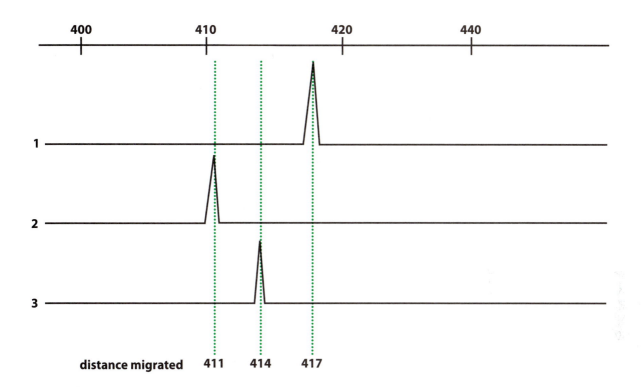

Tajima's *D* distinguishes between sequences evolving neutrally and those evolving non-neutrally using allele frequencies

Often one of the first things a population geneticist wants to do is determine whether or not a gene or locus under study is evolving neutrally or is under any one of or combination of several factors such as *directional* or balancing selection, range expansion or contraction, genetic hitchhiking, or introgression. Fumio Tajima, a population geneticist, developed a test statistic called *D* in 1989 to analyze this phenomenon. Tajima used a very intuitive view of genetic drift to formulate *D*. Genetic drift is the most common result of neutral evolution and it depends strongly on population size in which the genomes of organisms are evolving (see Chapter 3). Populations of constant size with constant mutation rate will reach equilibrium with respect to gene allele frequencies. At equilibrium, the characterization of allele frequencies has important properties. One is the number of segregating sites in the genes under study. Another is the quantity of nucleotide differences between pairs of sequences. The latter are called pairwise differences and are usually one of the first parameters calculated when comparing DNA sequences at the population level. Tajima noted that these pairwise differences needed to be standardized and developed a metric called π (pi) which is the sum of pairwise differences measured for a gene or allele in a population divided by the number of pairs measured. The metric π and the number of segregating sites are then used to estimate *D*, which can identify sequences that are not behaving neutrally at equilibrium.

To compute *D*, one needs at least three sequences of the same gene from individuals in a population. The test becomes more powerful with greater numbers

Figure 12.2 Detection of microsatellites. The sequences are derived from three individuals, the first having 11 AT repeats, the second having 14 AT repeats, and the third having 17 AT repeats (top). The grid represents a typical result of amplifying the AT repeat region and running on a gel (bottom). The gel is scanned for DNA and the amplification patterns are depicted as peaks. The distance fragments migrate is shown on the scale, which begins at an arbitrary length of 400 base pairs.

of individuals in the sample. *D* compares the total number of segregating sites (that is, sites that are polymorphic) and the average number of mutations between pairs of genes in the sample. Under neutrality, these values should be the same. If the two values differ by more than expected by chance, then the null hypothesis of neutrality is rejected. Since Tajima's *D* is not computationally intense, it can be used to scan through whole genome sequences for genes and regions of the genome that are evolving in non-neutral fashion.

F statistics measure the degree of isolation of entities

The most common measure used in population biology is called the *F* statistic. It has been used to characterize populations based on the degree of isolation certain study populations have from each other. Because populations can be defined hierarchically, the measure was eveloped so that it works at different hierarchical levels: individuals (subscript I), subpopulations (subscript S), and total populations (subscript T). Sewall Wright was among the first to develop statistical methods in population genetics to characterize this hierarchical variation. He and Gustave Malécot independently developed three statistics that are at the heart of most population genetic comparative studies. These measures have been described concisely by Holsinger and Weir:

"F_{IT} which is the correlation between gametes within an individual relative to the entire population; the F_{IS}, which is the correlation between gametes within an individual relative to the subpopulation to which that individual belongs; and the F_{ST}, which is the correlation between gametes chosen randomly from within the same subpopulation relative to the entire population."

Later, Clark Cockerham developed a mathematical and statistical framework for these *F* statistics; he pointed out that a statistical approach and a consideration of sampling was needed to enhance their utility. These statistics were developed to characterize the genetic diversity of populations at different hierarchical levels. In essence, these statistics measure the impact genetic cross talk (migration, mutation, and drift) on heterozygosity at the different hierarchical levels of subdivision as explained above.

As an example, consider F_{ST}. When two large populations are compared and extensive gene flow between them is observed, we can state that these two populations will have little if any differentiation between them. This results in a very small F_{ST} that can be tested for its statistical difference from 0.0. On the other hand, small populations with little or no migration between them will have much lower heterozygosity as a result of their isolation, and hence higher degrees of differentiation. In this latter case, F_{ST} will be closer to 1.0. Wright had a very subjective way of interpreting these *F* statistics: 0.0–0.25 represented little or no differentiation at the hierarchical level, 0.25–0.50 represented a significant small degree of differentiation, 0.50–0.75 represented a moderate degree of differentiation, and 0.75–1.0 represented a high degree of differentiation. Modern approaches have refined the methods used to generate the statistics and also the interpretation of these statistics.

As stated above, Cockerham realized that the sampling properties used to compute *F* statistics were an important consideration. Masotashi Nei, an influential evolutionary geneticist at Pennsylvania State University, recognized that there are really two kinds of distributions or samples that can be considered when computing statistics about populations. The first is a distribution of allele frequencies in populations (**Figure 12.3A**). However, each population is also a subsample of the overall sample (Figure 12.3B), and each of these subpopulations has an allele distribution (Figure 12.3C). To explain this further, consider that scientists can decrease the amount of variance in their samples by simply increasing the

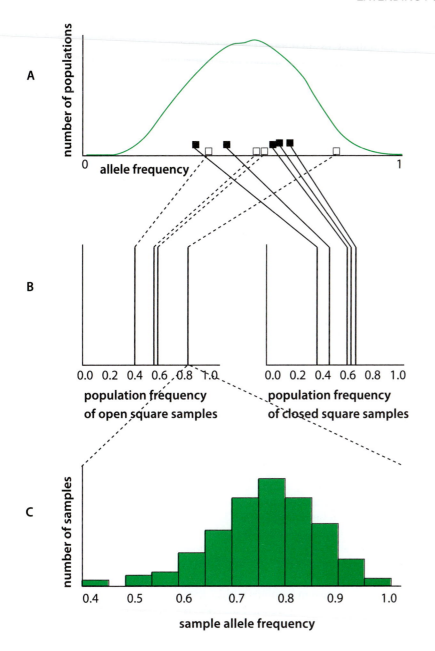

Figure 12.3 Distributions that are used in population genetics. A: allele frequency distribution. B: evolutionary sampling. Each population is a subsample of the overall sample. C: statistical sampling. Each of these subpopulations has an allele distribution or distributed sample. (Adapted from K.E Holsinger & B.S. Weir. *Nature Reviews Genetics* 10:639–650, 2009. Courtesy of Nature Publishing Group.)

number of individual data points collected. Variance decreases as sample size increases. Population geneticists, on the other hand, cannot control the genetic sampling variance. This sampling variance is a property of the evolutionary process and is what contributes to the differentiation of populations.

Nei developed a different kind of statistic that measures differentiation of populations, only it takes advantage of the kind of distribution seen in Figure 12.3B. He called the statistic G_{ST}, which is used to compute the genetic differentiation among populations. It is defined by use of population frequencies and not the allele frequency as typical F statistics are computed. G_{ST} allows for some interesting things to be computed. Because typical population genetics studies are based on the types of distributions in Figure 12.3A, they are based on real measures of heterozygosity at nuclear loci. The type of distribution seen in Figure 12.3B is the focus of G_{ST} measures. Because the distributions in Figure 12.3B can be derived from both haploid and diploid data, these kinds of comparisons can be useful with clonal markers such as mtDNA or Y chromosomes or even bacterial genome-level polymorphisms. It should be emphasized that F_{ST} and G_{ST} measure different things.

Because haplotype (mitochondrial DNA) data and microsatellite data contain information on the evolutionary distance between populations, various kinds of information can come from them. G_{ST} is one statistical method that can be used to estimate differentiation among populations. The Φ_{ST} statistic was developed as a means to obtain better insight into the patterns of differentiation among populations too, and it complements G_{ST}. For microsatellite data, because they constitute a different kind of data, a third statistic has been developed, called R_{ST}. We demonstrate how these statistics are calculated in the Web Feature for this chapter.

There are two approaches to estimating population-level statistics

Holsinger and Weir describe two major approaches to estimating F_{ST}: method of moments analysis (MOM) and merged likelihood Bayesian methods. MOM analysis is relatively simple since it is based on algebraic concepts. The parameters important in MOM analysis are

$$F = \frac{\sigma^2_P + \sigma^2_I}{\sigma^2_P + \sigma^2_I + \sigma^2_G}$$

$$\theta = \frac{\sigma^2_P}{\sigma^2_P + \sigma^2_I + \sigma^2_G}$$

$$f = \frac{\sigma^2_P}{\sigma^2_I + \sigma^2_G}$$

where the σ values are variances that can easily be calculated via the statistical principles outlined in Sidebar 1.1 in Chapter 1. The subscripts refer to the following: P is population, I is individual, and G is genotype. θ is interpreted as the effect that allele frequency among populations has on genetic diversity or F_{ST}. f is the correlation between gametes within an individual relative to the subpopulation to which that individual belongs (equivalent to F_{IS} as defined above), and F is simply the correlation between gametes within an individual relative to the entire population (equivalent to F_{IT} as defined above). The variances (σ) in the above equations can be estimated by a statistical approach called ANOVA (analysis of variance).

Since we have discussed Bayesian methods for phylogenetic inference previously (Chapter 10), the Bayesian approach should be familiar. The likelihood approach uses a probability distribution that utilizes a model characterizing the variation in allele frequencies among populations, by use of a preset distribution (a multinomial distribution) that is a descriptor for genotype samples within populations. θ and f are estimated in this approach by maximizing the likelihood value as a function of θ, f, and the allele frequencies found in the populations. Several programs are available that perform these estimations. We will introduce two programs—GENDIVE, which is a stand-alone program, and GenePop, which is a Web-based approach—in the Web Feature for this chapter. As mentioned, Bayesian analysis of population-level genomic data uses the very same likelihood function and preset prior distributions of f and θ and allele frequencies. As with the phylogenetic approach, Markov chain Monte Carlo (MCMC) methods are used to sample from the posterior distributions of θ and f. As with likelihood approaches, there are several programs that implement Bayes' rule in population genetic estimates.

FST and related measures have four major uses in evolutionary biology

Holsinger and Weir list four major uses of F_{ST} measures in evolutionary biology:

- **Estimating migration rates.** Attempts to use F_{ST} to determine migration rate center around an equation that Wright derived:

$$F_{ST} \approx \frac{1}{4N_e m + 1}$$

which says that F_{ST} is inversely proportional to effective population (N_e) size and migration rate (m). However, any estimate of migration via this approach has to be scrutinized heavily, because this equation assumes symmetrical exchange of genes between populations and this is rarely the case in nature. One way to overcome the difficulties inherent in the F_{ST} approach to estimating migration rates is to use the coalescent approaches we described in Chapter 3.

- **Inferring demographic history.** Evolutionary phenomena such as sex-biased dispersal or selection on loci can be detected by the F_{ST} approach.

- **Identifying genomic regions under selection.** In Chapter 13, we will discuss sequence-based methods for detecting natural selection in protein coding genes. The approach mentioned here is to use F_{ST} to detect selection and is different from the methods discussed in Chapter 13. In the present case, F_{ST} values can be determined locus by locus (here SNP by SNP) for clusters of genes, and the F_{ST} values can be correlated to degree of selection at a locus. A distribution of F_{ST} values over a genomic region with more than 10 gene regions of over 20,000 SNPs is shown in **Figure 12.4.** In this example, the majority of SNPs were inferred to be not under selection, while about 50 showed signs of selection as inferred from the F_{ST} estimates made with Bayesian approaches. A few of these showed extremely large F_{ST} values and hence strong degrees of statistically significant selection (these are the extreme "outliers" in Figure 12.4).

Figure 12.4 Hypothetical distribution of FST values over a genomic region with more than six gene regions and over a large number of single nucleotide polymorphisms. In this hypothetical example, the majority of single nucleotide polymorphisms (SNPs) were inferred to be not under selection (closed circles), while approximately 50 SNPs (open circles) showed signs of selection as inferred from the F_{ST} estimates calculated with Bayesian approaches. Seven of the SNP regions showed extremely large F_{ST} and hence strong degrees of statistically significant selection. The solid rectangles at the bottom of the figure refer to gene regions on a chromosome. (Adapted from K.E Holsinger & B.S. Weir. *Nature Reviews Genetics* 10:639–650, 2009. Courtesy of Nature Publishing Group.)

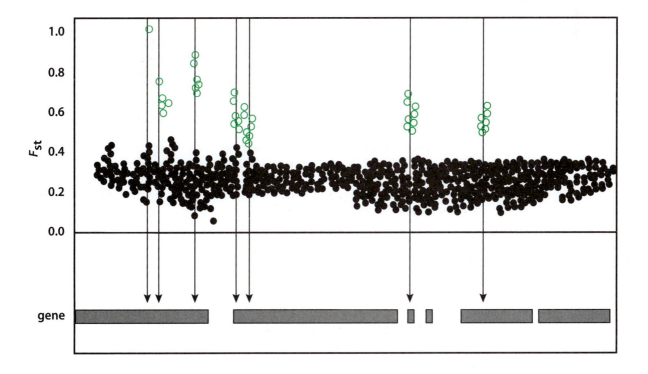

- **Forensic science and association mapping.** This method uses a Bayesian approach to determine the probability that a collected matched genotype of a potential perpetrator of a crime is unique. In association studies, the approach is similar.

Imputation

Knowledge about the genetics of populations has been used in many studies to allow for the imputation of untyped variants of individuals. In a GWAS study, hundreds of thousands of genotypes for SNPs are usually obtained, but there is often a desire to know the genotype at additional markers. This desire is generally motivated by the requirement of comparing data between different arrays (Affymetrix and Illumina) that interrogate different SNPs. The determination of genotypes for untyped markers is known as imputation.

Basically, imputation is based upon the fact that, within a population, nearby loci on a chromosome are linked, and thus the genotype at one locus should be highly predictive of a genotype at a nearby locus. The HapMap project and the more recent 1000 Genomes project have determined the frequencies of SNPs across the genome in many different populations, and these data are used as the resources for imputation. As an example, we will illustrate the imputation of data into a Caucasian American sample by using the HapMap data and Illumina genotype data. First, all of the study samples are run on the Illumina chip and their SNP calls at the designated loci are determined. Then the HapMap data for Caucasians (CEU), which includes both Affymetrix and Illumina data, is downloaded. Next, for each location on the Affymetrix array that is not typed on the Illumina array, nearby SNPs are selected that appear to be linked to each other. This linkage will be evident by a clear correlation of genotypes at one locus to those at another. For instance, all individuals with a G at locus 1 have a T at locus 2, and all individuals with an A at locus 1 have a C at locus 2. By use of this linkage information, the genotypes can be imputed from the HapMap reference onto the Illumina data to give a bigger picture of the genome of study individuals.

The reason population information is so important is that the linkage of SNPs decreases greatly as populations become more heterogeneous. While among Caucasians there is a strong link between locus 1 and locus 2, this may not be the case among other populations where the loci are not linked due to erosion of linkage as a result of lineages being older.

Population-Level Techniques: Mismatch Distribution Analysis and STRUCTURE Analysis

There are many novel approaches that have been developed recently that use genomic information to make inferences about the evolution of populations. We will examine some of these in the Web Feature to this chapter. Here we mention two approaches that are particularly useful in examining the differentiation of populations.

Mismatch distribution analysis compares haplotype data of populations

Mismatch distribution analysis (MDA) allows for the comparison of populations by use of haplotype data. It is a relatively simple approach in that the technique simply quantifies the pairwise differences between haplotypes and uses these data to characterize demographic events such as range expansion and spatial expansion of populations in a species. MDA uses a parametric bootstrap to test hypotheses of range expansion and spatial expansion. Results of simulations

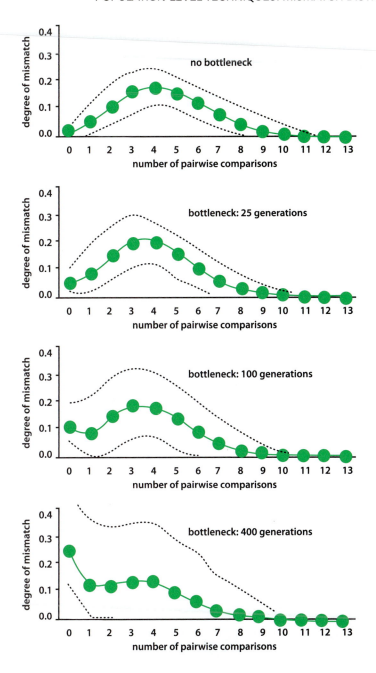

Figure 12.5 Mismatch distribution analyses. The four distributions are produced from a model with a bottleneck except for the first one. The second distribution models a bottleneck for 25 generations, the third for 100 generations, and the fourth for 400 generations. Dotted lines are 95% confidence intervals.

showing the shapes of curves under specific conditions are shown in **Figure 12.5**. The calculated numbers of pairwise differences are on the x-axis, and the frequencies among all the comparisons that can be made are on the y-axis. The dotted lines refer to confidence intervals (95%). The four different cases in this figure demonstrate that the shape of the curve changes under different demographic models. In addition, the simulations show the sensitivity of the approach. Hence the approach with MDA is to obtain plots from actual data and compare them to simulated plots of the same population size for various models of evolution of the populations. If the plots can be shown to be significantly different, then certain demographic models can be rejected. The figure also shows that, under the assumptions of the analysis, a recent bottleneck (decrease in population size) can be detected by the shape of the plot only after at least 100 generations post-bottleneck. One shortcoming of this approach is that it cannot be used on microsatellite data but only on haplotype data.

STRUCTURE analysis reveals substructure and genetic cross talk

This approach allows one to examine population-level data to see if there is significant substructuring in the data set. Substructuring refers to the footprint of genetic isolation or alternatively to interconnectedness of populations as discovered by use of genetics. The discovery of substructuring and its impact on various populations is important because it allows the population geneticist to better characterize the genetic cross talk of subdivided populations. The approach can be used with a wide variety of techniques that result in allele frequency data, such as restriction fragment length polymorphism (RFLP) studies, AFLP (amplified fragment length polymorphism), microsatellites, and SNPs. A model is assumed where one can set the number of populations (K), or on the other hand, one can set K as unknown. Individuals in the sample are assigned to a population. The model also allows for the assignment of an individual to two or more populations jointly. The way the approach works is to assume that within population variation is in Hardy–Weinberg equilibrium (HWE; see Chapter 3). Individuals are assigned to populations by use of the model to maintain HWE and LE. The approach uses Bayesian methods to do the assignments. It has been used in many high-profile publications and has been established as a major way to assign individuals to populations and to generate hypotheses concerning the number and composition of populations. Two hypothetical STRUCTURE analyses are shown in **Figure 12.6**. In these diagrams, individuals are plotted on the *x*-axes and each individual is represented by a column. The *y*-axes represent the probability of inclusion in a population. The individual identified with the black arrow in Figure 12.6A is assigned with a probability of 1.0 to population 2, which is represented by a dark gray column. The individual identified with the green arrowhead is assigned to population 3 (dark green) at a probability of 0.4 and to population 4 (light green) at a probability of 0.6. Figure 12.6B shows an idealized situation in which the assignments of all individuals to their respective populations have probabilities of 1.0.

One critical aspect of the analysis is the number of MCMC generations used to make the assignments. The impact of number of generations on the precision of the assignments is demonstrated in **Figure 12.7**. In this example, there are three populations represented by the three vertices of the triangle. The individuals are

Figure 12.6 STRUCTURE diagrams. A: STRUCTURE analysis where the assignments to populations are "imperfect." The different shades (black, gray, dark green, and light green) represent assignment to a particular "population." Dotted lines indicate boundaries of the populations. Each column represents an individual in the study. The green arrow points to an individual that was assigned to at least two populations, in this case to populations 3 and 4. The black arrow points to an individual that was assigned to a single population (gray), in this case to population 2. B: STRUCTURE analysis where the assignments to the four populations are "one to one." Dotted lines indicate boundaries of the populations.

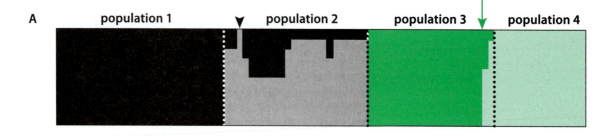

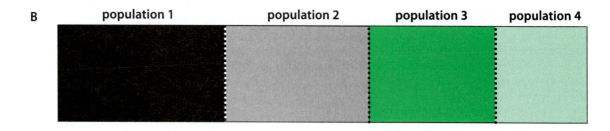

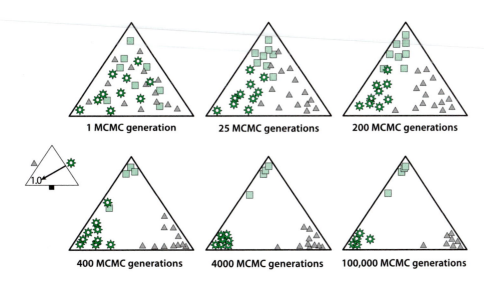

Figure 12.7 Triangle plots from STRUCTURE showing how the number of generations in the Bayesian simulation step can impact the clustering of individuals. Three populations are depicted by the vertices of the triangle; individuals, who are differentiated by their localities, are represented by sunbursts, green squares, and gray triangles. The inset shows how the triangle plots are read. Each side of the triangle represents the 0.0 probability assignment for the subset of individuals depicted on that side. As the points on the 0.0 axis move further away from the side of the triangle, the probability of assignment rises, as shown by the open sunburst. If any open sunburst resides at the vertex opposite that side of the triangle, then the probability of assignment of that individual is 1.0. MCMC, Markov chain Monte Carlo. Image generated with STRUCTURE. (Derived from J. K. Pritchard, M. Stephens, & P. Donnelly. *Genetics* 155:945–959, 2000.)

assigned by the MCMC approach at 1, 25, 200, 400, 4000, and 100,000 generations. The individuals are differentiated by the locality where they were found. The figure demonstrates that assignment becomes more precise with increased number of generations in the MCMC simulations.

Summary

- Population genetics has been modernized by use of high-throughput methods. Kimura and Lewontin contributed important new ways to think about genes in nature, and the Hardy–Weinberg theorem has been extended.

- DNA variation among individuals can be quantified by examining single-nucleotide polymorphisms or microsatellite DNA.

- *F* statistics measure the degree of isolation of a population. These statistics can be estimated by method of moments analysis and merged likelihood Bayesian methods.

- Statistical measures have four major uses in evolutionary biology: estimating migration rates, inferring demographic history, identifying genomic regions under selection, and forensic science and association mapping.

- Imputation, or the determination of genotypes for untyped markers, can be accomplished by analyzing SNP linkage in microarrays.

- Mismatch distribution analysis uses a statistical approach to compare populations by use of haplotype data.

- STRUCTURE analysis examines population-level data to see if there is significant substructuring in the data set. This allows the population geneticist to better characterize the genetic cross talk of subdivided populations.

Discussion Questions

1. The following microsatellite is discovered in a screen of the genome of polar bears:

5′-agctacgatcgacacgtactagcagcatgctacatcgatcgaATATATAT
ATATAATATATATATATAATATATATATATAagctacgatcgatcga
tcgtacgatcgatcgaga-3′

Design a primer pair with both primers 18 base pairs in length for screening other polar bear genomes for the microsatellite.

2. The sequences below for two loci are obtained in a microsatellite study of a population for two loci:

Make a table showing the microsatellite profiles for the seven individuals (I1–I7) in this population. You can simply use the number of dinucleotide repeats as an indicator of how far they will migrate in a sequencing gel.

Locus 1

I1   ...atcgagcgacatgctacgcaATATATATATcgacgatcgatcgatcgac...

I2   ...atcgagcgacatgctacgcaATATATATATATATATATATATATATATcgacgatcgatcgatcgac...

I3   ...atcgagcgacatgctacgcaATATATATATATATATATATATATATATATATcgacgatcgatcgatcgac...

I4   ...atcgagcgacatgctacgcaATATATATATATATATATATATATATcgacgatcgatcgatcgac...

I5   ...atcgagcgacatgctacgcaATATATATATATATATATATATATATATATATATATcgacgatcgatcgatcgac...

I6   ...atcgagcgacatgctacgcaATATATATATATATATATATATATATATATATATcgacgatcgatcgatcgac...

I7   ...atcgagcgacatgctacgcaATATATATATATATATATATATATATATATcgacgatcgatcgatcgac...

Locus 2

I1   ...atcgatcgatgcatcatcGTGTGTGTGTGTGTGTGTGTGTGTGTGTagctagctactacatcga...

I2   ...atcgatcgatgcatcatcGTGTGTGTGTGTGTGTGTGTGTGTGTGTGTGTGTGTGTGTagctagctactacatcga...

I3   ...atcgatcgatgcatcatcGTGTGTGTGTGTGTGTGTGTGTGTGTGTGTGTagctagctactacatcga...

I4   ...atcgatcgatgcatcatcGTGTGTGTGTGTGTGTGTGTGTGTGTGTGTGTGTGTGTGTGTGTagctagctactacatcga...

I5   ...atcgatcgatgcatcatcGTGTGTGTGTGTGTGTagctagctactacatcga...

I6   ...atcgatcgatgcatcatcGTGTGTGTGTGTGTGTGTGTGTGTGTGTGTGTGTGTGTGTGTagctagctactacatcga...

I7   ...atcgatcgatgcatcatcGTGTGTGTGTGTGTGTGTGTGTGTGTGTGTGTGTGTGTGTGTagctagctactacatcga...

3. The following table was obtained for a mismatch distribution analysis. It was generated by computing the number of differences between each individual in a population of 500 individuals for a small stretch of mtDNA. Once each of the pairwise distances (for 500 individuals, how many comparisons were made?) was measured, the frequency of the distances was recorded over the whole population.

calculated number of differences	frequency of occurrence
1	2%
3	8%
7	20%
9	50%
13	10%
15	5%
17	3%
19	2%

Graph the mismatch distribution of this population.

4. The mismatch distributions obtained for three different populations are shown in **Figure 12.8.**

The frequency of occurrence of a particular genetic distance is graphed on the *y*-axis and the genetic difference is plotted on the *x*-axis. The solid line represents the theoretical curve for a population expansion model with conservative assumptions. Characterize each of the three populations shown in the figure with respect to population expansion.

5. Discuss the merits and drawbacks of being able to set the population number (K) as a variable in doing STRUCTURE analysis.

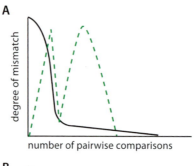

A

degree of mismatch

number of pairwise comparisons

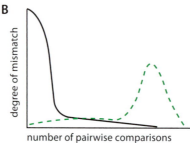

B

degree of mismatch

number of pairwise comparisons

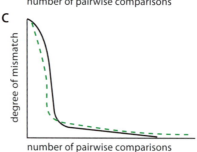

C

degree of mismatch

number of pairwise comparisons

Figure 12.8 Three hypothetical mismatch distribution analysis curves. The solid line represents the theoretical curve for a conservative model of population expansion and the dashed line represents the hypothetical population.

Further Reading

Holsinger KE & Weir BS (2009) Genetics in geographically structured populations: defining, estimating and interpreting FST. *Nat. Rev. Genet.* 10, 639–650.

Hudson RR (1990) Gene genealogies and the coalescent process. *Oxford Surv. Evol. Biol. 7,* 1–44.

Kuhner MK (2009) Coalescent genealogy samplers: windows into population history. *Tree* 24, 86–93.

Marjoram P & Tavaré S (2006) Modern computational approaches for analysing molecular genetic variation data. *Nat. Rev. Genet.* 7, 759-770.

Pollard KS, Salama SR, King B et al. (2006) Forces shaping the fastest evolving regions in the human genome. *PLoS Genet.* 2: e168 (DOI 10.1371/journal.pgen.0020168).

Pritchard JK, Stephens M & Donnelly PJ (2000) Inference of population structure using multilocus genotype data. *Genetics* 155, 945–959.

Rogers AR & Harpending H (1992) Population growth makes waves in the distribution of pairwise genetic differences. *Mol. Biol. Evol.* 9, 552–569.

Detecting Natural Selection: The Basics

As we saw in Chapter 3, the products of natural selection manifest themselves at many levels: the level of amino acids and the level of phenotype of the proteins are just two. Ultimately though, DNA sequences are where the action is when thinking about natural selection. Fortunately, because much is known about the genetic code (see Chapter 2), scientists realized that the positions in amino acid codons behaved differently in the face of natural selection, and quantifying these differences could then allow researchers to estimate levels of natural selection. To do this, researchers needed to be able to establish a null hypothesis for the lack of selection. In this chapter, we will see that codon positions where a change does not alter the identity of the amino acid provide the basis for this null hypothesis. Researchers have been able to take advantage of this peculiarity of the genetic code to survey whole genomes for the occurrence of natural selection at the molecular level.

Analyzing DNA Sequences for Natural Selection

As discussed in Chapter 12, population genetic methods such as Tajima's D can be used to detect whether or not a gene or allele is evolving neutrally. However, while evolution is generally discussed with regard to traits and alleles, the real elemental changes occur at the level of individual DNA molecules, where mutations appear. These mutations occur either through random errors in the copying of DNA or through the action of an external mutagen that causes DNA changes. In most cases, when a mutation occurs, it will occur only at a single nucleotide position, and the surrounding sequence will be unaffected. The result of a mutation is highly influenced by its location within the genome. Because genes are the most important and most well-characterized portions of the genome, the discussion here will focus on mutations within genes.

DNA sequences can be examined for silent and replacement changes

As discussed in Chapter 2, there is redundancy in the genetic code. For many amino acids, there are multiple three-letter codons that code for the same amino acid. For instance, there are four codons for proline (green rectangle in **Figure 13.1**) that all begin with the sequence CC. The third position in the codon can be G, A, U, or C, and the amino acid produced will still be proline. This position is thus considered "silent," because changing it will not change the identity of the amino acid at that position. However, if the second position of the proline codon is changed to any nucleotide other than C, it will produce a different amino acid in the resulting protein. These sites are known as "replacement" sites.

Silent substitutions do not affect the function of an enzyme or a protein, because they do not change amino acids in the protein. However, as we shall see in Chapter 14, these silent sites may alter the amount of an enzyme or protein that is produced as a result of codon bias. Replacement changes, by definition, result in the

Figure 13.1 The genetic code. The green rectangle indicates the codons that encode proline. (Courtesy of the National Institutes of Health.)

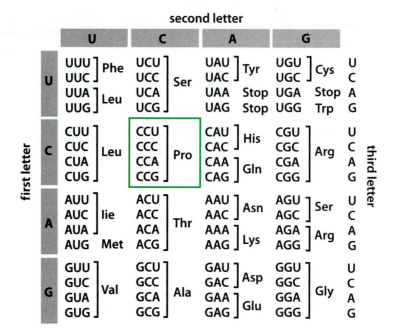

replacement in a protein of one amino acid for another. There are three outcomes of replacement substitutions as they affect the function of a protein: neutral, purifying, and positive selection. A neutral replacement has no impact on the function of the protein. Such changes are usually changes from one amino acid to another that is similar in size and biochemical properties, such as a leucine to isoleucine change. This will not affect the fitness or the function of the protein that is produced. A purifying change is where the substitution produces a protein that lacks the function or is very poor at accomplishing the function of the original protein. Hence such a change will have very low fitness and will eventually be eliminated from the population. A classic example of this kind of change is the sickle cell anemia allele. This allele simply is a change in the nucleotide in the second position of the sixth amino acid in the β-globin protein. The nucleotide change causes a change in this position from a glutamic acid in the usual β-globin protein to a valine in the sickle cell protein. This change from a hydrophilic to a hydrophobic amino acid alters the three-dimensional structure of the β-globin proteins of individuals with two copies of the mutation. This altered three-dimensional structure change causes a cascade of effects in the red blood cells, the most obvious of which is the sickling or bending of the red blood cells.

Positive selection is where the change actually increases the fitness or functionality of the protein as a result of changing the structure of the protein. We will illustrate several of these kinds of replacement changes in Chapter 14. Using sequence information, we can actually quantify whether the evolution of a protein is neutral, purifying, or positive.

In order to begin this discussion and simplify the calculations for detecting natural selection at the molecular level, we need to make some assumptions. First, we assume there is a diploid population of N individuals in which a mutation occurs in the gene of one individual. If events are random, a new mutation has a $1/2N$ chance of making it to the next generation, due to the fact that each individual in the population has two copies of an allele at each locus. This means that in large populations, $2N$ will be large and the probability of the mutant allele surviving will be small. But in a small population, $2N$ will be small, and the probability of the mutant allele surviving will be relatively large. This equation also tells us that if a protein is evolving neutrally, a silent mutation would have as good a chance as a replacement change to survive or "go to fixation" in a population. Hence, if evolution is neutral for this new mutation, the probability of fixation of a silent mutation should be equal to the probability of fixation of a replacement mutation. To quantify the role of selection in a protein, then, we need to be able to count the number of silent changes and compare it to the number of replacement changes.

Simply getting raw counts of silent changes and replacement changes is not enough, though, because there are many more possible replacement changes than silent ones. A correction is needed for the skew of silent versus replacement sites caused by the degeneracy of the genetic code. After this correction, the ratio of silent to replacement sites can be used to test whether populations are behaving in a neutral fashion or whether natural selection is at work. The rate of silent mutations provides a background measure or null hypothesis of neutrality.

The ratio of the rate of appearance of silent sites versus replacement sites can be calculated with a codon substitution model, which clearly shows that the ratio under neutrality is 1.0. In this model, the silent substitution rate is labeled dS (synonymous) and the replacement rate is labeled dN (nonsynonymous). If no selection is occurring, then there is no preference for either type of substitution, and they will both occur at the same rate, giving a ratio of 1. The ratio of dN to dS is known as ω:

$$\omega = dN/dS \qquad\qquad (13.1)$$

If $\omega = 1.0$ is the null hypothesis when testing for natural selection, what is the significance when $\omega \ll 1.0$? This scenario occurs when the replacement rate is extremely low relative to the silent rate; for example, in a situation where natural selection eliminates all variants that change the amino acid sequence of a protein. In other words, natural selection purifies the pool of variants to eliminate any changes that might alter the function of the protein. This is likely to occur in a critical location of a protein, such as the active site of an enzyme, where any amino acid change would drastically alter the protein function.

If $\omega \gg 1.0$, the replacement rate is much greater than the silent rate, meaning that natural selection prefers the new variants over the existing ones. In this case, the background silent rate is less than the replacement rate. These new changes presumably make the protein functionally different in ways that are beneficial for the organism. This kind of selection is known as positive Darwinian selection. The range of values for dN/dS depends on the method used, but all methods allow for comparison to the null hypothesis. We will show several examples of positive Darwinian selection in the reproductive proteins of *Drosophila* in Chapter 14.

Several variables affect the detection of natural selection at the genomic level

DNA sequence analysis can effectively describe natural selection at the genomic level. However, in order for this approach to attain greater validity, three phenomena need to be considered. The first factor involves accurately calculating the dN and dS rates. Consider the following example. If we have a codon that codes for histidine (H), it can be either CAU or CAC. The other codons that start with CA, CAA and CAG code for glutamine (Q). Now recall the transition (Ti) to transversion (Tv) phenomenon that was described in Chapters 2 and 8: transitions (changes between pyrimidines, C and U, or between purines, A and G) occur more frequently than transversions (changes from pyrimidine, C or U, to purine, A or G, or vice versa). The higher rate of transitions results in more silent changes (see Chapter 8). This issue needs to be considered in calculating ω in order to get an accurate estimate. If the rate difference between transitions and transversions is tenfold, then for these codons mentioned above we would see the third position mutating from C to U back to C and then to U again and so on. We would not see many transversions, and an accumulation of silent changes will be observed. This pattern is because transitions occur more frequently than transversions in comparisons among closely related species. In this case the role of neutral processes

marine lifestyle. In total, myoglobin consists of 154 amino acids in mammals, but for this example, we will use only the first 20 amino acids as shown below:

```
          M   V   L   S   D   A   E   W   Q   L   V   L   N   I   W   A   K   V   E   A
                          *                               *                       *
whale   atg gtg ctc agc gac gca gaa tgg cag ttg gtg ctg aac atc tgg gcg aag gtg gaa gct
sheep   atg ggg ctc agc gac ggg gaa tgg cag ttg gtg ctg aat gcc tgg ggg aag gtg gag gct
             #           #                           ##      #
          M   G   L   S   D   G   E   W   Q   L   V   L   N   A   W   G   K   V   E   A
```

The amino acid and DNA sequences from whale are shown on the top, while the equivalent sheep sequences are on the bottom. These sequences were aligned with each other by use of an alignment program such as CLUSTALW (see Chapter 6).

Basic dN and dS calculations begin with counting the observed number of changes

Begin by counting the actual number of silent and replacement changes between the two sequences. In the sequences above, the number of silent or synonymous changes (marked by *) between the two sequences is three. Note that all these changes occur in the third codon position. The number of amino acid replacements or nonsynonymous changes (marked by #) between the two sequences is five. On the basis of these results, would it be possible to calculate dN/dS as 5/3 = 1.67? If we could simply count these differences as being an adequate indicator of selection, then the inference would be that these two sequences have shown a large amount of positive natural selection. It then could be stated that this segment of myoglobin has experienced positive Darwinian selection possibly because whales need altered myoglobin molecules to transport oxygen more efficiently. However, we cannot use this calculation because not all sites in the above sequences are equal with respect to silent and replacement events.

Scaling for redundancy and getting the number of potential substitutions is necessary for determining dN/dS

Each site in the sequence needs to be scaled by its redundancy with respect to the genetic code, which involves calculating the number of potential changes that can occur for each codon. But remember that there are two kinds of changes that can occur: synonymous and nonsynonymous. For a nondegenerate amino acid (methionine and tryptophan), this task is simple: a change in any position of the codon will code for a different amino acid. The first amino acid in both sequences is a methionine (M), coded for by ATG as shown below:

```
111
ATG
```

where the numbers above the base pairs indicate the degeneracy of the position. Each position in the codon is onefold degenerate, and a change in any position will be a nonsynonymous change. Since there are three sites, the number of potential nonsynonymous sites in this codon is three, and there are zero sites in this codon that are synonymous. The silent and replacement counts for each codon should sum to three. We have summarized the counts that need to be made for all 61 codons that code for amino acids (the three stop codons are not counted) in Sidebar 13.2.

Returning to the whale and sheep sequences, we can now count the number of potential substitutions for each codon. For synonymous sites, the numbers of potential changes for the two sequences in the comparison are

	M	V	L	S	D	A	E	W	Q	L	V	L	N	I	W	A	K	V	E	A
whale	atg	gtg	ctc	agc	gac	gca	gaa	tgg	cag	ttg	gtg	ctg	aac	atc	tgg	gcg	aag	gtg	gaa	gct
	0	1	1	2/3	1/3	1	1/3	0	1/3	2/3	1	4/3	1/3	2/3	0	1	1/3	1	1/3	1

$$S_{whale} = 3(0) + 7(1) + 1(^4/_3) + 3(^2/_3) + 6(^1/_3) = 12.33 \qquad (13.2)$$

Sidebar 13.2. Potential number of silent and replacement events for the 20 amino acids.

For codons that are one-, two-, three-, and fourfold degenerate, we have listed the potential events by category. For the three amino acids that are sixfold degenerate (that is, L, R, and S have six codons each), the silent and replacement events vary for each codon, so we have listed these counts per individual codon.

degeneracy	amino acids	silent	replacement
onefold	M, W	0	3
twofold	F, Y, H, Q, N, K, D, E, C	1/3	8/3
threefold	I	2/3	7/3
fourfold	V, P, T, A, G	1	2
sixfold	L, R, S	variable	variable

codons	silent	replacement
L		
TTA	2/3	7/3
TTG	2/3	7/3
CTG	4/3	5/3
CTA	4/3	5/3
CTT	1	2
CTC	1	2
R		
AGA	2/3	7/3
AGG	2/3	7/3
CGA	4/3	5/3
CGG	4/3	5/3
CGT	1	2
CGC	1	2
S		
AGT	2/3	7/3
AGC	2/3	7/3
TCT	4/3	5/3
TCC	4/3	5/3
TCG	1	2
TCA	1	2

```
sheep  atg  ggg  ctc  agc  gac  ggg  gaa  tgg  cag  ttg  gtg  ctg  aat  gcc  tgg  ggg  aag  gtg  gag  gct
        M    G    L    S    D    G    E    W    Q    L    V    L    N    A    W    G    K    V    E    A
        0    1    1   2/3  1/3   1   1/3   0   1/3  2/3   1   4/3  1/3   1    0    1   1/3   1   1/3   1
```

$$S_{sheep} = 3(0) + 8(1) + 1(^4/_3) + 2(^2/_3) + 6(^1/_3) = 12.67 \tag{13.3}$$

To obtain the number of potential synonymous sites (S), the average for the two sequences is calculated as follows:

$$S = (S_{whale} + S_{sheep})/2 = (12.33 + 12.67)/2 = 12.5 \tag{13.4}$$

For potential nonsynonymous sites, the numbers of potential changes are counted for the two sequences:

```
        M    V    L    S    D    A    E    W    Q    L    V    L    N    I    W    A    K    V    E    A
whale  atg  gtg  ctc  agc  gac  gca  gaa  tgg  cag  ttg  gtg  ctg  aac  atc  tgg  gcg  aag  gtg  gaa  gct
        3    2    2   7/3  8/3   2   8/3   3   8/3  7/3   2   5/3  8/3  7/3   3    2   8/3   2   8/3   2
```

$$N_{whale} = 3(3) + 7(2) + 1(^5/_3) + 3(^7/_3) + 6(^8/_3) = 47.67 \tag{13.5}$$

```
sheep  atg  ggg  ctc  agc  gac  ggg  gaa  tgg  cag  ttg  gtg  ctg  aat  gcc  tgg  ggg  aag  gtg  gag  gct
        M    G    L    S    D    G    E    W    Q    L    V    L    N    A    W    G    K    V    E    A
        3    2    2   7/3  8/3   2   8/3   3   8/3  7/3   2   5/3  8/3   2    3    2   8/3   2   8/3   2
```

$$N_{sheep} = 3(3) + 8(2) + 1(^5/_3) + 2(^7/_3) + 6(^8/_3) = 47.33 \tag{13.6}$$

To obtain the number of potential nonsynonymous sites (N), the average for the two sequences is calculated as follows:

$$N = (N_{whale} + N_{sheep})/2 = (47.67 + 47.33)/2 = 47.5 \tag{13.7}$$

Note that $S_{whale} + N_{whale}$, $S_{sheep} + N_{sheep}$, and S + N all equal 60. Since the number of potential changes for synonymous and nonsynonymous sites totals 60 for any 20-amino-acid sequence (20 amino acids × 3 positions), we can compute S by the counting approach and simply subtract S from 60 to get N.

We can now calculate the proportion of observed synonymous (pS) and nonsynonymous (pN) substitutions (where proportion = actual number observed/ potential number calculated):

$$pS = 3/12.5 = 0.24 \tag{13.8}$$

$$pN = 5/47.5 = 0.105 \tag{13.9}$$

The ratio pN/pS, which we will take as a proxy for dN/dS, would therefore be 0.105/0.24 = 0.438. Since this ratio is less than 1.0, it would imply that these sequences are evolving as a result of purifying selection, which is a different inference than if we had simply calculated the raw synonymous and nonsynonymous changes that have occurred since the divergence of these two species.

Pathways of codon change are an important element in calculating dN/dS

Although the above corrections for redundancy yield a partial solution for a proxy for dN/dS, we need to take into account the different pathways that can occur in analyzing each codon before we can settle on the final dN/dS. For each codon in an alignment, there are three positions that need to be considered. For a given pair of codons, there can either be zero changes (the two codons are identical at the DNA sequence level), one change (only one position in the codon is different between the two sequences), two changes (two positions are different), or three changes (all three nucleotide states are different). In terms of the possible pathways that could be followed to arrive at these differences, the following observations apply.

- If there is one difference between the sequences, there is one pathway.

- If there are two differences between sequences, there are two pathways.

- If there are three differences between the sequences, there are six pathways.

The best way to account for these pathways is to apply a maximum likelihood (ML) approach to correct for the potential occurrence of multiple pathways. In the ML approach, certain pathways are weighted differently. To demonstrate this approach, we examine the problem using a simpler counting method developed by Nei and Gojobori that does not weight the kinds of changes; this is called an unweighted counting method.

Look at the sixth position in the whale and sheep sequences:

```
            A
            *
whale       gca
sheep       ggg
              #
            G
```

In order for the two species to possess their codon states, one needs to consider the identity of their common ancestor. The nucleotide state of the common ancestor of whale and sheep is inferred through character reconstruction as explained below. In the codon shown above, there are differences in two positions (marked by * and #). Therefore, two pathways need to be examined.

```
A       G       G
gca  →  gga  →  ggg
```

```
A       A       G
gca  →  gcg  →  ggg
```

Each pathway requires one nonsynonymous (N) change and one synonymous (S) change:

```
                        S    N
A       G       G
gca  →  gga  →  ggg      1    1

A       A       G
gca  →  gcg  →  ggg      1    1
```

In the comparison of sheep and whale myoglobin, there are two potential synonymous replacements and two potential nonsynonymous replacements for the two pathways. The adjustment for this codon is simple. The proportion of potential changes that are synonymous is $2/4$ or $1/2$, and likewise $2/4$ or $1/2$ of the potential changes are nonsynonymous. If these proportions are applied to the two actual changes, $1/2(2) = 1$ should be synonymous and $1/2(2) = 1$ should be nonsynonymous. Because this matches the proportion of observed changes, no adjustment is needed by the multiple pathways correction.

The other codon with two or more changes is the 14th amino acid in our sequence:

```
            I
whale       atc
sheep       gcc
            ##
            A
```

Again, there are two differences and therefore two pathways of change between these two codon states:

```
I     T     A
atc → acc → gcc

I     V     A
atc → gtc → gcc
```

In this case, each pathway leads to two nonsynonymous (N) changes and zero synonymous changes (S):

```
                    S   N
I     T     A
atc → acc → gcc     0   2

I     V     A
atc → gtc → gcc     0   2
```

Thus, among the two pathways, there are four potential nonsynonymous changes and zero potential synonymous changes. The correction for this codon would be $^0/_4$ synonymous changes and $^4/_4$ nonsynonymous changes. If these proportions are applied to the two actual changes, $^0/_4(2) = 0$ should be synonymous and $^4/_4(2) = 2$ should be nonsynonymous. Because this matches the proportion of the observed changes, no correction is needed over the raw count approach.

This example demonstrates that, for sequences that evolve slowly and incur delayed nucleotide modification, the raw count method is an effective approach for calculating sequence changes. Other sequences, in which more significant change has occurred, require consideration of the pathways and implementation of nucleotide substitution model corrections.

Codon change pathways can be used to account for redundancy

To demonstrate how the approximate approaches to estimating natural selection are accomplished, we will use the carboxy terminus of the myoglobin gene of sheep and whale as an example. This sequence is shown below.

```
      R   H   P   G   D   F   G   A   D   A   Q   A   A   M   N   K   A
                  *       *       *   *   *           *
whale agg cat cct ggg gac ttt ggt gcc gac gcc cag gca gcc atg aac aag gcc
sheep aag cat cct tca gac ttc ggt gct gat gca cag ggc gcc atg agc aag gcc
      #           ##                                  #           #
      K   H   P   S   D   F   G   A   D   A   Q   G   A   M   S   K   A
```

Scoring individual positions in each codon for their degeneracy in the whale sequence yields

```
      R    H    P    G    D    F    G    A    D    A    Q    A    A    M    N    K    A
whale agg  cat  cct  ggg  gac  ttt  ggt  gcc  gac  gcc  cag  gca  gcc  atg  aac  aag  gcc
      214  112  114  114  112  112  114  114  112  114  112  114  114  111  112  112  114
```

For the sheep sequence, scoring individual positions for their degeneracy results in the following:

```
sheep aag  cat  cct  tca  gac  ttc  ggt  gct  gat  gca  cag  ggc  gcc  atg  agc  aag  gcc
      K    H    P    S    D    F    G    A    D    A    Q    G    A    M    S    K    A
      112  112  114  214  112  112  114  114  112  114  112  114  114  111  214  112  114
```

The raw counts of synonymous and nonsynonymous potential sites for the two sequences are calculated as follows:

The whale sequence has 34 positions of onefold degeneracy, eight positions of twofold degeneracy, and nine positions of fourfold degeneracy.

$$S_{whale} = 34(0) + 8(^1/_3) + 9(1) = 11.67 \qquad\qquad (13.10)$$

$$N_{whale} = 34(1) + 8(^2/_3) + 9(0) = 39.33 \qquad\qquad (13.11)$$

The sheep sequence has 33 positions of onefold degeneracy, nine positions of twofold degeneracy, and nine positions of fourfold degeneracy.

$$S_{sheep} = 33(0) + 9(^1/_3) + 9(1) = 12 \qquad\qquad (13.12)$$

$$N_{sheep} = 33(1) + 9(^2/_3) + 9(0) = 39 \qquad\qquad (13.13)$$

The number of potential synonymous sites is therefore $[(11.67 + 12)/2] = 11.835$, and the number of potential nonsynonymous sites is $[(39.33 + 39)/2] = 39.165$. Since there are six synonymous differences (marked by *) between the two sequences, dS is $6/11.835 = 0.507$; dN is $5/39.165 = 0.128$ (there are five nonsynonymous differences, marked by #); and dN/dS is 0.128/0.507, or 0.252. These results are clearly in the purifying selection zone. However, there is one codon in the comparison that requires further examination because of the pathway problem:

```
G
  *
ggg
tca
##
S
```

There are three changes in this codon: two are nonsynonymous (marked by #) and one is synonymous (marked by *). These three changes could have arisen by six possible pathways, which requires that we recalculate the potential synonymous (S) and nonsynonymous (N) changes that can occur in this codon.

The six pathways are as follows:

```
                               S   N
(1)   G     G     A     S       1   2
      ggg → gga → gca → tca

(2)   G     A     A     S       1   2
      ggg → gcg → gca → tca

(3)   G     W     stp   S       -   -
      ggg → tgg → tga → tca

(4)   G     W     stp   S       -   -
      ggg → gga → tga → tca

(5)   G     A     S     S       1   2
      ggg → gcg → tcg → tca

(6)   G     W     S     S       1   2
      ggg → tgg → tcg → tca
```

Pathways 3 and 4 are ignored because they require stop codons as intermediates in the reconstructions, and such stop codons would disrupt the protein and be highly disadvantageous. Since this evolutionary scenario would be highly unlikely compared to other pathways, these pathways can be ignored. Therefore, this analysis results in four pathways of three changes each, for a total of 12 potential changes, with 4 of the 12 being synonymous ($^4/_{12} = ^1/_3$) and 8 of the 12 being nonsynonymous ($^8/_{12} = ^2/_3$). If these proportions are applied to the three actual changes, $^1/_3(3) = 1$ should be synonymous and $^2/_3(3) = 2$ should be nonsynonymous. This matches the results found, so no corrections are needed in this case. However, in some cases not all codon pathway changes can be accommodated by simple counting as we will see in the next chapter.

Summary

- Natural selection can be detected by comparing sequences at the nucleotide and amino acid level.
- Some changes in DNA codons will change the structure of proteins (nonsynonymous or replacement changes), while other changes in DNA codons will have no effect on protein structure (synonymous or silent changes).
- The ratio of nonsynonymous to synonymous changes, dN/dS or ω, can be used to quantitate whether a protein experiences purifying natural selection, is under neutral evolution, or is experiencing positive Darwinian selection.
- The dN/dS ratio may be determined by approximate or model-based methods. Approximate approaches determine the actual and potential numbers of synonymous and nonsynonymous changes between two sequences and their proportionality.
- Redundancy in the genetic code, and the different pathways by which codon change can occur, also must be taken into account in calculating dN/dS.

Discussion Questions

1. Calculate the exact number of DNA sequence changes that are silent relative to all replacement changes. Leave the three stop codons out of the calculation.

2. What is the molecular reason that there are more potential replacement sites than silent sites?

3. Show that the following values are true for the potential number of silent and replacement events for the codons for the amino acid R (arginine).

```
R
          silent       replacement
AGA       2/3          7/3
AGG       2/3          7/3
CGA       4/3          5/3
CGG       4/3          5/3
CGT       1            2
CGC       1            2
```

4. Given the discussion of ω in this chapter, what would be a good null hypothesis to test when ω is used? How would one establish a statistical test for this null hypothesis?

5. Using the approximate methods described in this chapter, compute S, N, pS, pN, dN, and dS for the following sequences:

| L | S | H | V | W | P | R | S | S | L | P | F | P | H | R | H | L | W | H | V | P |

ttg agc cac gtg tgg ccg aga agc tcc ttg cct ttc cca cac aga cat ctt tgg cat gtg cca

tgg agc cac gcg ttg cag agc tgc tgg atg agc atc cca ctc aga cga caa ctc aat gtt ccc

| W | S | H | A | Q | S | C | W | M | S | I | P | L | R | R | Q | L | N | V | P |

Further Reading

Comeron JM (1995) A method for estimating the numbers of synonymous and nonsynonymous substitutions per site. *J. Mol. Evol.* 41, 1152–1159.

Hurst LD (2002) The *Ka/Ks* ratio: diagnosing the form of sequence evolution. *Trends Genet.* 18, 486.

Li W-H (1993) Unbiased estimation of the rates of synonymous and nonsynonymous substitution. *J. Mol. Evol.* 36, 96–99.

Li W-H, Wu CI & Luo CC (1985) A new method for estimating synonymous and nonsynonymous rates of nucleotide substitution considering the relative likelihood of nucleotide and codon changes. *Mol. Biol. Evol.* 2, 150–174.

Nei M & Gojobori T (1986) Simple methods for estimating the numbers of synonymous and nonsynonymous nucleotide substitutions. *Mol. Biol. Evol.* 3, 418–426.

Refining the Approach to Natural Selection at the Molecular Level

In Chapter 13, we examined the nuances of the genetic code that have allowed researchers to examine molecular sequences for the footprint of natural selection. By establishing the null hypothesis that lack of selection means there should be equal amounts of silent and replacement substitutions in molecular sequences, we established an important working model for the detection of selection. However, simply counting the number of silent and replacement substitutions in a molecule will not give an accurate estimate of dN/dS. There are a number of aspects of DNA sequence evolution that need to be taken into account to arrive at an accurate estimate for dN/dS (ω). First, we need to account for some of the factors that produce a disproportionate number of silent and replacement changes in DNA sequences (see Chapter 13). The next step is to take into account the evolutionary processes that govern sequence change, such as multiple hits to sequences. In addition, the models of sequence evolution that we studied in previous chapters can be applied to the process of detecting selection. In addition to these problems, we also need to consider that specific regions of proteins might be under different selective constraints than others and that, in different parts of phylogenies, episodes of selection might be different. One of the best examples of natural selection at the molecular level also needs to be addressed, and this is the phenomenon of codon usage bias.

Accounting for Multiple Hits in DNA Sequences for dN/dS Measures

An important assumption underlying the calculations discussed in Chapter 13 is that we see all of the sequence changes that have occurred throughout evolutionary history. In other words, when we see an A in a position in the whale sequence and a T in the same position in the sheep sequence, we assume that the evolutionary history of that change is a simple A to T change between the two species. But the evolutionary changes at the site can also be A → T → C → T, or A → C → T, or any number of other scenarios. This assumption is known as the multiple hits problem and needs to be accommodated in the calculation of dN/dS ratios.

The Jukes–Cantor conversion corrects for multiple hits

The major problem in calculating dN and dS is the multiple hits phenomenon, which can be addressed with the Jukes–Cantor conversion. The problems that are resolved by this method are similar to those encountered in reconstructing phylogeny with respect to estimating nucleotide substitutions. The Jukes–Cantor conversion for the number of synonymous and nonsynonymous changes in two sequences is a natural logarithmic conversion using the following equations (see Chapter 8):

$$dS = -(^3/_4) \ln [1 - (^4/_3)pS] \tag{14.1}$$

$$dN = -(^3/_4) \ln [1 - (^4/_3)pN] \tag{14.2}$$

For the example from the amino terminus of the myoglobin gene, ds = $-(^3/_4)$ ln $[1 - (^4/_3)(0.2446)] = 0.2990$ and dn = $-(^3/_4)$ ln $[1 - (^4/_3)(0.1045)] = 0.1126$. Thus the dN/dS ratio = 0.3765. This value is smaller than 1; therefore, we cannot come to the conclusion that positive Darwinian selection is at work. In fact, this value for dN/dS suggests that there is strong purifying selection to maintain the sequences in the myoglobin protein.

Estimating natural selection requires adjusting the calculation of sequence changes

In the examples provided above and in the previous chapter, we have moved from simple counting of nucleotide changes to adjusting the method for calculating synonymous and nonsynonymous changes by the following protocol:

- Align sequences.

- Count the number of "actual" synonymous and nonsynonymous sites in each sequence.

- Obtain the total of synonymous and nonsynonymous changes by averaging for sequences in the comparison.

- Count the number of "potential" changes by summing all possible pathways between each pair of codons. For the majority of changes in closely related species and for slowly evolving sequences, the correction for pathways will be trivial.

- Correct for the substitution model.

Expanding the Search for Natural Selection at the Molecular Level

Model-based approaches to estimating natural selection accommodate four problems that occur when scientists examine natural selection at the sequence level and need to be assessed: (1) statistical significance of the calculations; (2) nonuniform selection over different parts of the protein; (3) nonuniform selection over temporal evolution of the protein; and (4) comparisons of sequences for natural selection in a phylogenetic context. We describe the nuances of each of these issues below.

Statistical tests of significance are required at various levels

Over the past two decades, evolutionary biologists have realized that there are two taxonomic levels where natural selection can be examined and that the statistical problems encountered at each level are different. The first level concerns comparisons of the sequences of different species. The significance of these kinds of calculations is examined with a variety of approaches, as discussed below. If sequences within a species from different populations are evaluated, Fisher's exact test is used to compare dN/dS within species versus across species. The classical example of this approach is the MacDonald–Kreitman test for natural selection.

The MacDonald–Kreitman test (MK91) is an improvement over the Hudson–Kreitman–Aguade (HKA) test that was developed in 1987. For the sequences of two alleles in a population, the HKA test used the level of polymorphism of these alleles to test for neutral versus adaptive sequence evolution. MacDonald and Kreitman refined this approach to include polymorphic *and* fixed sites to estimate the role of neutrality versus selection in sequence evolution. Their approach utilizes DNA sequence information from large numbers of individuals from at least two species. This large number of individuals from each species allows one to classify each site in the sequence as "fixed" or "polymorphic." Next, from knowledge of the genetic code in codons, the researcher can classify each position in the sequence as "synonymous" (silent) or "nonsynonymous" (replacement). Hence there are four possible ways a variable site can be classified: fixed and synonymous; polymorphic and synonymous; fixed and nonsynonymous; or polymorphic and nonsynonymous. If the evolution of the sequences is neutral, one would expect equal frequencies of all four categories. If not, then there should be a statistically significant departure from this expectation.

The following example shows how the sites are scored. This example is a short stretch of 12 nucleotides of DNA coding for four amino acids (as shown by the spaces dividing the sequences into codons) for two species and four individuals from each species.

```
Species 1
            --$ *-- --- #-%
ind 1    GGG TGA CTT GAT
ind 2    GGG TGA CTT AAT
ind 3    GGG TGA CTT AAC
ind 4    GGG TGA CTT AAT


Species 2
            --$ *-- --- #-%
ind 1    GGA CGA CTT GAT
ind 2    GGA CGA CTT AAC
ind 3    GGA CGA CTT GAT
ind 4    GGA CGA CTT GAC
```

In the above example, the dashes (-) indicate invariant sites and are ignored in the rest of the example. (*) represents a site that is fixed and nonsynonymous. Note that it is fixed and different in the two species and that it occurs in the first position of the second codon of the sequence. The change from T to C is a replacement change, as it causes a shift in amino acid identity. The dollar symbol ($) indicates a site that is fixed and synonymous, as the change from G to A in the third position of this codon does not change the amino acid in this position. The hash symbol (#) indicates a site that is polymorphic and nonsynonymous, as it occupies the first position of a codon and a change results in a different amino acid, and the percent symbol (%) indicates a site that is polymorphic and synonymous. In this short sequence, there would be one of each category of change that is important for estimation of the role of natural selection in the evolution of this sequence.

MacDonald and Kreitman examined the alcohol dehydrogenase (*ADH*) gene, a favorite gene of population geneticists for natural selection. The product of this gene helps the fly break down alcohol it ingests. They used 12 *Drosophila melanogaster* individuals, 6 *Drosophila simulans* individuals, and 24 *Drosophila yakuba* individuals to do the test. After sequencing all of the *ADH* genes from the 42 individuals, they found 68 sites that were variable among the sequences. They then classified these 68 sites in the way shown above and obtained the results in **Table 14.1**. The table holds all the information we need to do a statistical test to see if there is a departure from neutral expectations. The statistical test employed is a common one when there is a 2 × 2 table of data, called Fisher's exact test. The specific expectation of the test is that the ratio of fixed nonsynonymous (N_f) to fixed synonymous (S_f) sites should be equal to the ratio of polymorphic nonsynonymous (N_p) to polymorphic synonymous (S_p) sites:

$$N_f/S_f = N_p/S_p \tag{14.3}$$

```
UUU 22.2 (35846)    UCU  8.7 (14013)    UAU 16.5 (26648)    UGU  5.2 ( 8458)
UUC 14.9 (25565)    UCC  8.9 (14420)    UAC 12.3 (19766)    UGC  6.4 (10285)
UUA 13.8 (22316)    UCA  8.1 (13117)    UAA  2.0 ( 3163)    UGA  1.1 ( 1751)
UUG 13.0 (20904)    UCG  8.8 (14220)    UAG  0.3 (  435)    UGG 14.3 (24656)


CUU 11.4 (18366)    CCU  7.2 (11657)    CAU 12.8 (20631)    CGU 20.2 (32590)
CUC 10.5 (16869)    CCC  5.6 ( 8961)    CAC  9.4 (15116)    CGC 20.8 (33547)
CUA  3.9 ( 6257)    CCA  8.4 (13507)    CAA 13.7 (23703)    CGA  3.8 ( 6166)
CUG 51.1 (82300)    CCG 22.4 (36178)    CAG 29.4 (47324)    CGG  6.2 ( 9955)


AUU 29.7 (47838)    ACU  9.1 (14639)    AAU 19.2 (30864)    AGU  9.4 (15123)
AUC 23.9 (38504)    ACC 22.8 (36724)    AAC 21.7 (34907)    AGC 16.0 (25800)
AUA  5.5 ( 8835)    ACA  8.1 (13030)    AAA 34.0 (54723)    AGA  2.9 ( 4656)
AUG 27.2 (43846)    ACG 14.0 (24122)    AAG 11.0 (17729)    AGG  1.8( 2915)


GUU 18.1 (29200)    GCU 14.4 (24855)    GAU 32.8 (52914)    GGU 24.2 (38983)
GUC 13.8 (23870)    GCC 25.2 (40571)    GAC 19.2 (30953)    GGC 28.1 (45226)
GUA 10.9 (17561)    GCA 20.7 (33343)    GAA 39.3 (63339)    GGA  8.9 (14286)
GUG 26.2 (42261)    GCG 32.3 (52091)    GAG 18.7 (30158)    GGG 11.8 (18947)
```

Further Reading

Almeida FC & DeSalle R (2009) Orthology, function, and evolution of accessory gland proteins in the *Drosophila repleta* group. *Genetics* 181, 235–245.

Bennetzen JL & Hall BD (1982) Codon selection in yeast. *J. Biol. Chem.* 257, 3026–3031.

Dean AM & Thornton JW (2007) Mechanistic approaches to the study of evolution: the functional synthesis. *Nat. Rev. Genet.* 8, 675–688.

Gouy M & Gautier C (1982) Codon usage in bacteria: correlation with gene expressivity. *Nucleic Acids Res.* 10, 7055–7074.

Hughes AL (2007) Looking for Darwin in all the wrong places: the misguided quest for positive selection at the nucleotide sequence level. *Heredity* 99, 364–373.

Ikemura T (1981) Correlation between the abundance of *Escherichia coli* transfer RNAs and the occurrence of the respective codons in its protein genes. *J. Mol. Biol.* 146, 1–21.

Kosakovsky Pond SL & Frost SDW (2005) Not so different after all: a comparison of methods for detecting amino acid sites under selection. *Mol. Biol. Evol.* 22, 1208–1222.

McDonald JH & Kreitman M (1991) Adaptive protein evolution at the Adh locus in *Drosophila*. *Nature* 351, 652–654.

Nielsen R & Yang Z (1998) Likelihood models for detecting positively selected amino acid sites and applications to the HIV-1 envelope gene. *Genetics* 148, 929–936.

Plotkin JB & Kudla G (2011) Synonymous but not the same: the causes and consequences of codon bias. *Nat. Rev. Genet.* 12, 32–42.

Powell JR & Moriyama EN (1997) Evolution of codon usage bias in *Drosophila*. *Proc. Natl Acad. Sci. U.S.A.* 94, 7784–7790.

Sharp PM & Li W-H (1987) The codon adaptation index: a measure of directional synonymous codon usage bias, and its potential applications. *Nucleic Acids Res.* 15, 1281–1295.

Wright F (1990) The effective number of codons used in a gene. *Gene* 87, 23–29.

Yang Z (2007) PAML 4: phylogenetic analysis by maximum likelihood. *Mol. Biol. Evol.* 24, 1586–1591.

Genome-Level Approaches in Population Genetics

The nature of genome-level information has changed the face of population genetics in several areas. This chapter will delve into some of these special subjects. Because high-throughput approaches can generate whole genome-level profiles for individuals, this information can be used to scan genomes for associations with genetic disorders. There are several approaches that have been used to address genome-level questions about populations. The first we address here are called genome-wide association studies (GWAS). These studies take advantage of the ease with which DNA sequences for large numbers of individuals from different populations can be obtained. The second concerns the coalescent. So far we have only touched upon coalescent theory in the context of gene tree/species tree problems. Coalescent theory is also a valid approach to evaluating the impact of drift on populations, as well as to help detail migration and recombination. Analysis of an important population genetic phenomenon that has benefit from genomics is the study of natural selection. As we saw in Chapters 13 and 14, analysis of sequences by use of the genetic code to set up a null hypothesis has been a fruitful approach to understanding natural selection. Phylogenomicists have also developed other techniques using genome-level data to detect selection. These areas include examination of patterns of variation to detect selective sweeps and phylogenetic shadowing methods that pinpoint genes and noncoding regions that have experienced strong conservation or acceleration.

Genome-Wide Association Studies

The most valuable use of single-nucleotide polymorphism (SNP) arrays has been for disease studies that are known as genome-wide association studies (GWAS). In a GWAS study, a researcher will collect hundreds or thousands of DNA samples from individuals having a particular disease (cases) and from nondiseased control individuals (controls). The DNA of each person will then be profiled by use of the SNP array or next generation DNA sequencing in order to determine their allelic state at each of the SNPs. Then a statistical test will be performed for each SNP to determine if it is represented to a greater level in either the case or the control individuals. After a multiple testing correction is applied for the millions of markers tested, any significant SNPs are reported to be associated with the particular disease.

A simple example illustrates the association technique

Here is a simple example of the association technique. There are eight samples, four from case individuals and four from control individuals. For each individual at each locus, we have their SNP status. The SNPs for each position are T/G, A/T, C/T, A/T, G/T, and A/T. The cases have the SNP statuses of TACTGT, TATAGA, GATAGT, and TACATA, while the control individuals are TTCTTA, TTTAGT, GTCTGT, and GTTAGA. If we line these all up and look at the pattern of SNPs for diseased and control individuals, we have

```
cases
T-A-C-T-G-T
T-A-T-A-G-A
G-A-T-A-G-T
T-A-C-A-T-A
controls
T-T-C-T-T-A
T-T-T-A-G-T
G-T-C-T-G-T
G-T-T-A-G-A
```

Note that the only SNP where there is a distinct difference between the two groups is in the position (underlined and bold) where all the cases have an A and all the controls have a T. All other positions in the compared sequences have both nucleotide possibilities in both the cases and the controls. Because there is a straightforward correlation between disease and SNP status at the second position, this position would have a very strong *p*-value and would be correlated with the disease. Of course this is a simplified example, and in a real GWAS study there are millions of SNP loci and thousands of individuals. It is also very unlikely that in a real experiment there will be such a "smoking gun" where there is a complete distinction between the cases and the controls at a particular SNP. In most cases, there will be a mixture of alleles in both cases and controls, and statistical tests will be required to determine which SNPs are significantly differentiated between the two groups.

Such GWAS studies have been performed for hundreds of diseases, and the degree of success has varied. The overall approach of a GWAS study is shown in **Figure 15.1**. For some diseases, such as age-related macular degeneration, a GWAS study has identified a very penetrant causative variation. For many other diseases, though, the results of GWAS studies have been less robust and they are able to explain only a small number of disease cases. Even so, disease-related GWAS studies continue to be performed because they are a rapid and relatively cheap method to identify causative alleles for diseases.

On the basis of the results of these GWAS studies, companies such as 23andMe, Navigenics, and deCODE genetics market testing kits to determine an individual's propensity for a wide variety of diseases. As discussed in Chapter 2, microarrays can be used to "resequence" human genomes. These companies use the microarray resequencing approach to obtain SNP information. Each of the companies takes an individual's DNA, and hybridizes it to an Affymetrix array (1.3 million SNP markers on the chip), an Illumina array (currently 2.5 million SNP markers on the chip), or an array that is proprietary to the company. The results of the microarray analysis are then compared to the database of SNPs compiled from GWAS studies to determine which diseases a person may be prone to. One major criticism of these companies is that the GWAS studies underlying the disease predictions for many of the SNPs are not robust.

More in line with the phylogenomic focus of this book, SNPs have also been used to investigate human genealogy. Since the genome of an individual is a product of the genomes of their ancestors, along with spontaneous mutations, the genomes of individuals with a common ancestor should be similar. Using this idea, SNPs have been found that are suggested to be associated with individuals whose ancestry can be traced to different geographical locations. Ancestry is a complicated issue, though, and the limitations of using this approach to determine ancestry of humans and human populations is controversial. As with microsatellites, SNPs can also be used for forensic purposes to determine the degree of genealogical separation between two individuals. This approach would be used in a paternity test or other situations where the biological link between two individuals is being questioned.

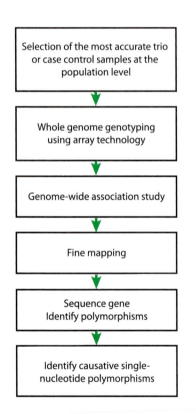

Figure 15.1 Flow chart for the application of human population genetic approaches to genome-wide association studies. In a genome-wide association study, samples are collected from individuals having a particular disease (cases) and from nondiseased control individuals (controls). The DNA of each person is profiled to determine their allelic state at each of the single-nucleotide polymorphisms (SNPs). A statistical test is performed for each SNP to determine if it is represented to a greater level in either the case or the control individuals. Any significant SNPs are reported to be associated with the particular disease.

The National Human Genomics Research Institute maintains a database of genome-wide association studies

One tangential result of the GWAS studies is the large amount of SNP data that have been made available for diverse populations. A researcher interested in the population genetics of a certain ethnic group can utilize the data that were obtained for GWAS in that group. This reuse of data greatly decreases the costs of the research. In addition, SNP data have been used to investigate the admixture of different populations and the historical background of ethnic groups.

The National Human Genomics Research Institute (NHGRI) has kept tabs on the hundreds of papers published using GWAS. The increase in the number of papers using the genome-wide association approach is shown in **Figure 15.2**. As of August 2011, there were 1520 publications and greater than 10 million disease associated SNPs catalogued by the NHGRI.

Role of the Coalescent in Population Genetics

We briefly discussed the coalescent in Chapter 11 in the context of the gene tree/species tree problem. However, the broadest use of the coalescent has been in population genetics, so in this chapter we will explore the coalescent in more detail. Most of the results obtained from coalescence theory pertain directly to the process of genetic drift and sampling error. In essence, what population geneticists would like is a simulation of gene genealogy, making some basic assumptions about the kind of population they are dealing with. These basic assumptions are that the population is diploid and is modeled on the Fisher–Wright model. The Fisher–Wright model makes some of the very same assumptions as the Hardy–Weinberg principle, such as no mutation, no selection, and random mating. For ease of computation, the Fisher–Wright model also assumes nonoverlapping generations. Note that infinite population size is *not* an assumption of the Fisher–Wright model, and so this parameter can be focused on in the coalescent approach.

As pointed out in Chapter 3, the linchpin of coalescence analysis is the backward-looking approach taken to examine the present-day distribution of alleles for a gene. To develop the theory further, we will assume that the population size is

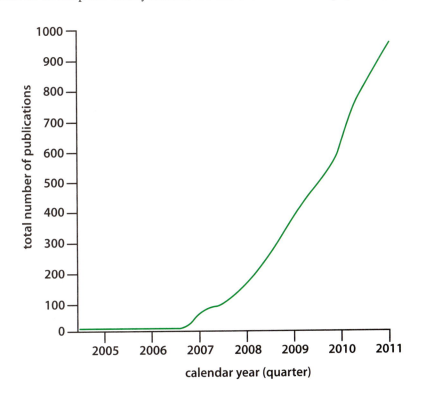

Figure 15.2 Published genome-wide association reports, 2005–2011. The graph shows the gradual but significant increase in genome-wide association studies since 2006.

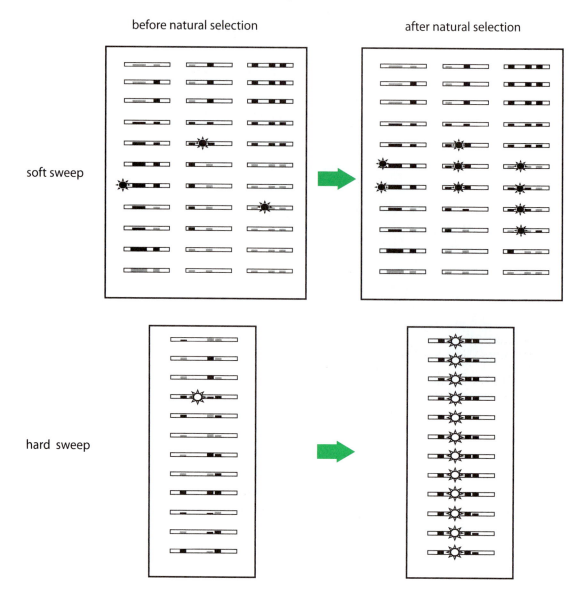

before natural selection   after natural selection

soft sweep

hard sweep

Figure 15.7 Difference between a hard sweep and a soft sweep. Top: polygenic process (soft sweep). Three chromosomes are shown for 11 individuals. Three mutations or variants are involved in a polygenic trait that is beneficial to the population. The impact of natural selection on the population for this trait is shown in the right panel. Note that the strong signal of homozygosity around the selected loci is missing. Bottom: hard sweep process as explained in Figure 15.3.

genome (but not necessarily in regions coding for proteins). Allan Wilson and his colleagues, at the University of California at Berkeley, pioneered the examination of the role of regulatory changes in phenotypic evolution in primates. Mary Claire King and Allan Wilson, using protein allozyme information, estimated that the divergence between humans and common chimpanzees (*Pan troglodytes*) was about 1%. This result means that one out of every 100 amino acids differs between humans and chimpanzees. However, there is extreme anatomical and behavioral divergence of humans and chimps. Wilson and his colleagues then asked how the small amount of structural protein differences could account for the extreme anatomical differences among organisms. King and Wilson hypothesized that gene regulation was the key to understanding this problem. Not until gene sequences from the genomes of humans and chimpanzees were compared in detail could the actual DNA sequence divergence of coding and noncoding regions between chimps and humans be estimated. King and Wilson were not far off with their allozyme estimate, as the human to chimpanzee difference turned out to be 1.23% at the DNA sequence level. More recent work has focused on pinpointing the regions of the human genome that might be involved in these regulatory regions. The methods that we will discuss here for detecting conservation and accelerated evolution are all based on a principle initially developed for comparing sequences across larger phylogenetic distances, known as phylogenetic shadowing, so we will discuss this approach first.

Phylogenetic shadowing identifies regulatory elements in DNA sequences

Phylogenetic shadowing matches sequences with their phylogenetic placement, with the objective of identifying regulatory elements in DNA sequences. The process begins by aligning the DNA sequences with any of the available alignment tools (see Chapters 5 and 6). Distance-based methods are most often used to detect regions of the alignment with significant similarity and these regions are mapped on the consensus sequence of the alignments or on the longest sequence in the set of sequences being examined. The logic is that if a region of regulatory DNA shows significant conservation, it should be involved in conserved regulation of the gene with which it is associated. Phylogenetic shadow mapping for the APOA gene in humans and other vertebrates is depicted in **Figure 15.8**. The likelihoods for sequence conservation are shown in the graph over the length of the promoter region of the Apo(a) protein. The peaks represent regions that are not conserved, and the valleys represent regions that are highly conserved. The regions of high conservation would be candidates for involvement in conserved regulatory processes.

Evolutionary phylogenetic shadowing of closely related species is a Web-based tool that performs the calculations for phylogenetic shadowing. eShadow (discussed in the Web Feature for this chapter) uses two distinct methods for finding functional elements: the divergence threshold (DT) scan and the hidden Markov model islands (HMMI) approach. In DT, multiple (or pairwise) sequence alignments are scanned. The computational design and specific algorithms that are incorporated in eShadow can detect coding exons, noncoding elements, and protein domains.

Figure 15.8 Phylogenetic shadowing comparison of the Apo(a) protein over several mammalian taxa. The graph shows where in the length of the gene strong conservation occurs across the large number of mammal species (represented by bars under the x-axis). The differences in length of the gray and green bars refer to the differences in length of the signal detected. The position in the gene is given in base pairs along the x-axis, and the degree of conservation is given as log likelihood on the y-axis. bp, base pair position on the chromosome. (Adapted from I. Ovcharenko, D. Boffelli, and G. Loots. *Genome Research* 14:1191–1198, 2004.)

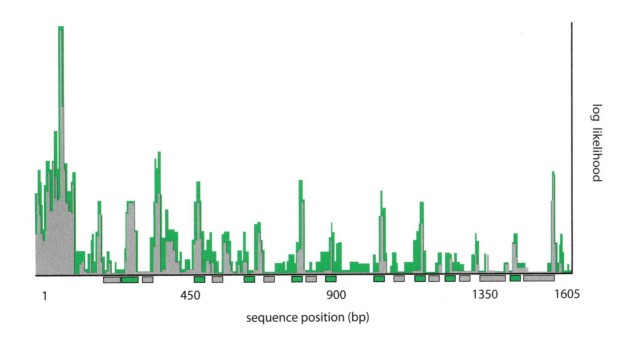

Regions of the human genome experience accelerated evolution

Pollard et al. first described regions of the human genome that have undergone extensive accelerated change relative to other organisms. They devised a ranking system based on a likelihood ratio test, which allowed them to pinpoint certain genes that showed extreme acceleration in human populations. They named these regions HARs (human accelerated regions). In their initial study, they pinpointed one gene that had previously been uncharacterized that they called *HAR1F*. Pollard et al. were then able to demonstrate that this gene was expressed in neural tissues and more specifically co-expressed with an important neural gene called *reelin*. Using similar approaches, Prabhakar et al. examined noncoding regions of the human genome to determine if some of these regions that do not code for proteins showed accelerated evolution. They identified nearly 1000 regions of the human genome that showed evolutionary acceleration. They concluded that "the strongest signal of human-specific noncoding sequence evolution that we detected was near genes specifically involved in neuronal cell adhesion." Bush and Lahn scanned the genomes of primates and humans to determine similar patterns. They also obtained a list of genes and gene regions that experienced acceleration in the human lineage. They also addressed the problem of how well the three studies (Pollard, Probhakar, and Bush) agree in the genes they identify as HARs. A Venn diagram demonstrating the overlap of the three studies is shown in **Figure 15.9**. While there are some scaling problems with the comparison (the Probhakar study identifies 10 times more regions than the other two studies), the result is that there are very few regions of the human genome that consistently show acceleration. Perhaps the best comparison to discuss is the Pollard/Bush comparison, where six regions of the human genome show acceleration. The upshot of acceleration in this context is that it is rare and most likely correlated to adaptive evolution.

Regions that are both strongly conserved and rapidly deleted are of interest

McLean et al. have used an interesting approach that combines the examination of sequences for conservation and for their rapid deletion in the human lineage. The approach was to compare the genomes of chimpanzees with other organisms other than humans and to identify regions of strong conservation. Next these highly conserved regions were aligned to the human genome to detect where they resided *and* more importantly to determine if they were deleted; hence the name condel (for "conserved" and "deletion") was coined for these regions.

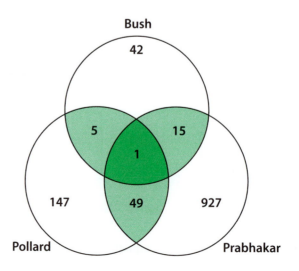

Figure 15.9 Overlap of identified human accelerated region genes as depicted in Venn diagrams. Three groups of researchers (Bush and Lahn, Pollard et al., and Prabhakar, Noonan, Paabo, and Rubin) used different methods to identify regions of the human genome, known as human accelerated regions, that have undergone accelerated change relative to other organisms. The diagram illustrates how many regions were identified by one, two, or all three of the studies. (Adapted from E. Bush & B. Lahn, *BMC Evolutionary Biology* 8:17, 2008. Courtesy of BioMed Central.)

Surprisingly, nearly 600 regions of the human genome, of average size about 100 base pairs, could be identified as condels. Even more surprising is that only one of these condels actually occurs in a coding region. When the functions of the genes near the deletions were examined, many of the deletions occur near neural development genes, indicating some functional significance to the discovery and location of "condels" in brain development and function in our lineage.

Summary

- Genome-wide association studies take advantage of the ease with which DNA sequences for large numbers of individuals from different populations can be obtained. They use single-nucleotide polymorphism (SNP) arrays to detect SNPs associated with disease.

- The coalescent addresses the time for an allele to coalesce and the variation in populations under drift. It explores a large number and broadly representative sample of plausible genealogical scenarios.

- High-quality DNA sequence data from a random sample constitute the best input for a coalescence analysis. Importance sampling or correlated sampling can be used to generate a collection of simulated genealogies.

- If a locus is under intense positive selection, loci closely linked to it will also be affected; these loci are called hitchhiking loci. The hitchhiking loci can be neutral or even detrimental, and yet they will behave as if positive Darwinian selection is acting upon them.

- Selective hard sweeps can be detected by the extended haplotype approach, the site frequency spectrum (SFS) approach, by subdivisioning, and by reduced variability in regions of the genome. Soft sweeps are more difficult to detect.

- Phylogenetic shadowing identifies regulatory elements in DNA sequences. Regions of the human genome that have experienced accelerated evolution can be detected and compared. Regions that are both strongly conserved and rapidly deleted are also of interest.

Discussion Questions

1. The following series of SNPs is found for a group of diseased and control individuals:

```
1 2 3 4 5 6      SNP number
disease
T-A-T-A-T-T
T-T-T-A-T-A
A-A-T-A-T-T
T-A-T-A-A-A
control
T-T-T-T-T-A
T-A-T-T-T-T
A-A-A-T-A-T
A-T-A-T-A-A
```

Which SNP shows an association with the disease? Look more closely at SNP 3. Write a description of what is happening with this SNP.

2. Describe what might be expected if a GWAS approach is applied to two groups of people with a full complement of SNPs for both groups. Group A has a long evolutionary history, while group B is an offshoot of group A and has been isolated from group A for some time. Which group would be a better one to use GWAS to detect an association of disease with SNPs? Which group would it be difficult to determine an association for, and why?

3. Discuss the major similarities and differences between a hard sweep and the polygenic model of selective sweeps.

4. What factors might influence a positive result for the HAR approach?

5. Discuss the impact of the overlap of positive results from the three HAR studies mentioned in the text (Pollard, Probhakar, and Bush) and illustrated in Figure 15.9.

Further Reading

Bush EC & Lahn BT (2008) A genome-wide screen for noncoding elements important in primate evolution. *BMC Evol. Biol.* 8, 17.

Carr SM, Marshall HD, Duggan AT et al. (2008) Phylogeographic genomics of mitochondrial DNA: highly-resolved patterns of intraspecific evolution and a multi-species, microarray-based DNA sequencing strategy for biodiversity studies. *Comp. Biochem. Physiol. D: Genomics Proteomics* 3, 1–11.

Hernandez RD, Kelley JL, Elyashiv E et al. (2011) Classic seductive sweeps were rare in recent human evolution. *Science* 331, 920-924.

Hindorff LA, Junkins HA, Hall PN et al. (2004) A Catalog of Published Genome-Wide Association Studies. Available at www.genome.gov/gwastudies.

King M & Wilson A (1975) Evolution at two levels in humans and chimpanzees. *Science* 188, 107-116.

Kuhner MK (2009) Coalescent genealogy samplers: windows into population history. *Trends Ecol. Evol.* 24, 86–93.

Maynard Smith J & Haigh J (1974) The hitch-hiking effect of a favourable gene. *Genet. Res.* 23, 23–35.

Pollard KS, Salama SR, King B et al. (2006) Forces shaping the fastest evolving regions in the human genome. *PLoS Genet.* 2, e168.

Prabhakar S, Noonan JP, Paabo S, & Rubin EM (2006) Accelerated evolution of conserved noncoding sequences in humans. *Science* 314, 786–790.

Presgraves DC, Gerard PR, Cherukuri A & Lyttle TW (2009) Large-scale selective sweep among segregation distorter chromosomes in African populations of *Drosophila melanogaster. PLoS Genet.* 5, e1000463.

Pritchard JK, Pickrell JK & Coop G (2010) The genetics of human adaptation: hard sweeps, soft sweeps, and polygenic adaptation. *Curr. Biol.* 20, R208–R215.

Quilez J, Short AD, Martínez V et al. (2011) A selective sweep of >8 Mb on chromosome 26 in the Boxer genome. *BMC Genomics* 12, 339–345.

Slatkin M (2001) Simulating genealogies of selected alleles in populations of variable size. *Genet. Res.* 145, 519–534.

Yi X, Liang Y, Huerta-Sanchez E et al. (2010) Sequencing of 50 human exomes reveals adaptation to high altitude. *Science* 329, 75–78.

Genome Content Analysis

Over 2000 whole genomes of living organisms have been sequenced by various laboratories around the world, and this number is expected to expand exponentially over the next several years. In order to make sense of the onslaught of data, a new field of biology called comparative genomics has developed, where the complete genomic information of different species is compared and contrasted. This endeavor requires the storage of large amounts of information, as the typical bacterial genome is on the order of 2 million base pairs and the typical eukaryotic genome is two–four orders of magnitude larger. In previous chapters, we explored the use of GenBank and other databases to observe how single genes can be compared by use of informatics. This chapter discusses the analysis of whole genomes and the extraction of information via evolutionary approaches.

Genome Biology Databases

In Chapters 2 and 7, we discussed the technologies involved in DNA sequencing and data storage. It is becoming clearer that once the cost of genome sequencing becomes reasonable, we will be able to sequence the entire genome of any organism or specimen inexpensively and rapidly.

The various kinds of evolutionary projects that use genome-level information are shown in **Figure 16.1**, which outlines where the studies are in the DNA sequencing space and how new technology will allow them to expand. On the basis of this scenario, it can be theorized that if technology allows for the production of whole genomes at a reasonable cost, all categories of studies will converge on comparative genomics, including analyses such as DNA barcoding (see Chapter 17) and population-level studies. DNA barcoding requires only a sequence from a single gene, the cytochrome oxidase I gene (*coxI* or *COXI*), and microsatellite assays, which are used in population-level studies, need only 10–20 microsatellite markers to be examined (see Chapter 12). If the cost of a whole genome sequence can be reduced to the same order of magnitude as these two approaches, then it stands to reason that these other techniques might be abandoned and whole genome sequences used in their stead. Hence it is important to explore some of the techniques and approaches that have been developed to look at fully sequenced genomes.

Viral genomes are frequently the target of whole genome sequencing. Given the impact of the illnesses caused by influenza, human immunodeficiency virus (HIV), human papillomavirus (HPV), and other viruses, the sequencing and storage of viral genomes is important. In addition, since prokaryotic and eukaryotic cells possess either plastid or nuclear genomes, different databases are required to catalog the information in each type of genome. The Genomic Biology Website, which can be accessed via either the National Center for Biotechnical Information (NCBI) home page (see Chapter 4) or the "Genome" site on the NCBI home page (http://www.ncbi.nlm.nih.gov/sites/genome), offers several resources by which these genomes can be analyzed (**Figure 16.2**). On the left side of the Genome home page, links to several important databases are provided, which

Figure 16.1 Progression of different kinds of "high-throughput" sequencing projects. Ovals represent the position of the approach in the sequencing space. The number of taxa in the study is shown on the *y*-axis and the number of base pairs utilized is shown on the *x*-axis (the numbers on both axes represent powers of 10). The Barcode of Life (BoL) project is forecasted to obtain a 600 base pair fragment of sequence for the 1.7 million named species on the planet. The Tree of Life (ToL) project analyzes at least 10,000 base pairs to determine the phylogenetic relationships of the living species on this planet. Both BoL and ToL will be discussed in detail in Chapter 17. Mitochondrial genomics (mt) and chloroplast genomics (cp) approaches focus on the mitochondrial and chloroplast genomes as sources of information for phylogenetics. The animal mtDNA genome typically consists of approximately 16,000 base pairs and the cpDNA genome is believed to consist of approximately 100,000 base pairs. Over 1000 animal mtDNA and over 250 cpDNA genomes have been sequenced to date. Bacterial and archaeal genomics (pg) approaches yield about 400,000 base pairs in intracellular symbiotic bacteria such as *Wollbachia* and *Buchneria* and approximately 10,000,000 base pairs in some of the larger bacterial genomes. To date, nearly 2000 bacterial and archaeal full genomes have been sequenced, and many more are expected within the next five years. The eukaryotic genomics (eg) approach is intended to study genomes that vary widely in size from tens of millions to several billions of base pairs. Because of their large size, fewer genomes of eukaryotes have been sequenced, but next-generation sequencing techniques promise to increase this number greatly. The solid arrows represent the expected increases in sequencing power for each of the approaches. The dashed arrow represents the desired trajectory of ToL studies as a result of next-generation sequencing.

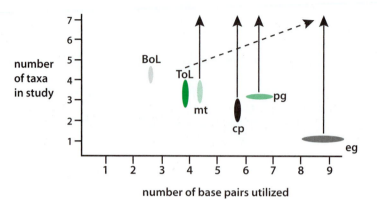

are similar to the database categories discussed in Chapter 5. The home page lists the genomes that have been deposited into GenBank cellular organisms (chromosomal genomes), mitochondria and plastids (organelle genomes like chloroplast genomes), or viruses (viral genomes). Information for each of these genomes is accessed by clicking on the name of the genome of interest. For example, under "GOLD Genome Projects Database," a list of the organisms with which genome projects are associated is provided: eukaryotic, fungi, insects, mammals, microbial, and plants. Bacterial genome projects can be accessed by clicking on "Microbes" (Figure 16.2). There are several other genome-specific options in this part of the homepage. The "Organelles" database lists all of the completed mitochondrial and other plastid genomes (**Figure 16.3**).

Any sequenced and published genome can be downloaded from the database. Some unpublished genomes can also be downloaded because the researchers producing the sequence have uploaded their data before publication. We demonstrate in the Web Feature for this chapter how to download a viral (HPV), a mitochondrial (marsupial mitochondrial genomics for *Monodelphis domestica*), and a bacterial (*Haemophilus ducreyi*) genome from NCBI.

Comparative Genomics Approaches

There are several basic approaches to comparative genomics. For the purposes of this book, these approaches are categorized as either simple "counting" analyses, in which the numbers and kinds of genes are compared, or more complicated phylogenomic comparisons, in which tree-building approaches are used.

Venn diagrams, EDGAR, and Sungear visualize the overlap of genes from two or more genomes

One initial way to compare genomes is to construct a Venn diagram that shows the overlap of genes from two or more genomes. Two genomes are compared to each other by use of BLAST to determine the presence or absence of genes in the compared genomes overlap for specific genes. Each genome is then represented as a circle in a diagram. For example, the number of genes that overlap in two genomes is found in the region where the two circular diagrams representing the genomes overlap (**Figure 16.4**).

This method, however, is difficult to apply when more than three organisms are compared because visualizing a large number of comparisons is difficult. The efficient database framework for comparative genome analysis using BLAST score ratios (EDGAR) is one of these sites (see Web Feature for this chapter for a demonstration of EDGAR). Another Web-based program that organizes comparative genomic data and is useful for more than three taxa is Sungear, which takes

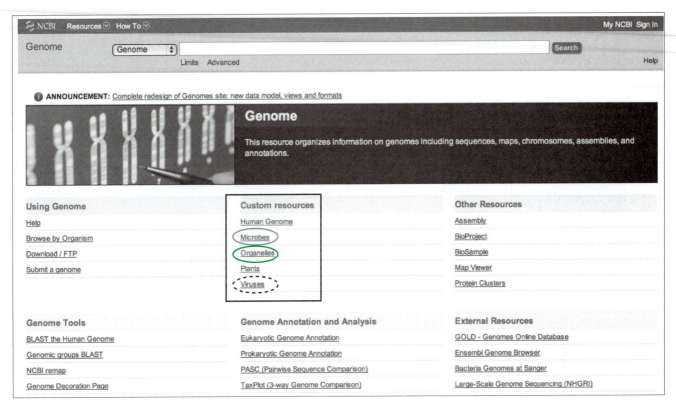

Figure 16.2 Screenshot of the NCBI Genomic Biology home page. The large box indicates where the Custom Genome Resources can be found. Microbial genome data (gray oval) and viral genome data (dashed black oval) can be accessed from these sites. Organelle genomes (green oval), such as chloroplast and mitochondrial genomes, can be located here. (Courtesy of the National Center for Biotechnology Information.)

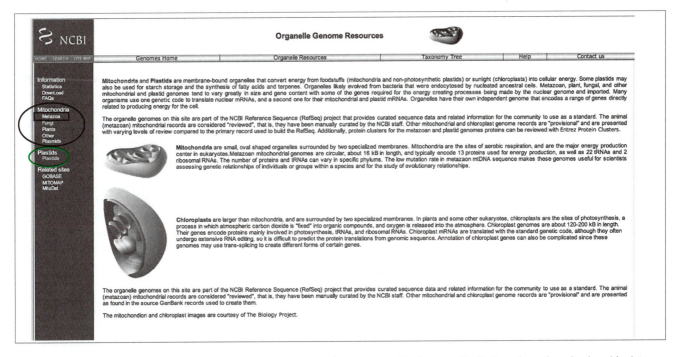

Figure 16.3 Screenshot of the Organelle Genome Resources home page. The large oval indicates where the mitochondria data are stored and the small rectangle within this oval indicates where metazoa mtDNA genomes are stored. The small oval identifies the site in which plastid genomes, such as chloroplast genomes, are located. (Courtesy of the National Center for Biotechnology Information.)

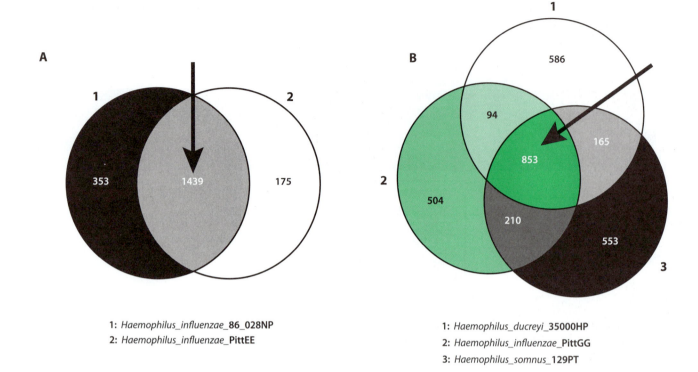

A

1: *Haemophilus_influenzae_***86_028NP**
2: *Haemophilus_influenzae_***PittEE**

B

1: *Haemophilus_ducreyi_***35000HP**
2: *Haemophilus_influenzae_***PittGG**
3: *Haemophilus_somnus_***129PT**

Figure 16.4 Venn diagrams obtained from the EDGAR Web site. A: comparison of gene overlap in the genomes of *Haemophilus influenzae* strains. B: comparison of gene overlap in the genomes of three different *Haemophilus* species. Arrows point to the region of the Venn diagram where the full overlap occurs. (Adapted from J. Blom, S. Albaum, D. Doppmeier, A. Pühler, F.-J. Vorhöter, M. Zakrzewski, A. Goesmann, *BMC Bioinformatics* 10:154, 2009. Courtesy of CeBiTec, Bielefeld University.)

biological information and organizes it in a geometric shape. We describe the use of this approach below.

The input into Sungear is the list of genes that are present in a particular genome, which the program then uses to create an overlap diagram. If there are *N* taxa in the analysis, then the output of Sungear will show an *N*-pointed symmetrical trapezoid, where the vertices of the trapezoid represent the species in the comparison. Each of the white dots in the Sungear output represents a set of genes that are present in the taxa that occupy that specific area of the plot. The general idea of a Sungear plot is that the dots in the center of the diagram (the space of the plot that is occupied by all species in the analysis) are found in all of the taxa. Where the dots are placed depends on which species the genes are in. So, for instance, if a dot cluster appears in the very center of the plot in a Sungear analysis, this means that all species have the genes represented by the white dot plot. If a white dot cluster appears on the outer ring of the plot next to a species name, this means that the genes represented by this dot are unique to that species. The output for a Sungear analysis for 10 eukaryotic species is shown in **Figure 16.5**.

Generating Genome Content Phylogenies

Information derived from whole genomes has been used in a unique way, based on the fact that a particular kind of gene can be certified as being present or absent in a fully sequenced genome. Since the annotations of the genomes are presumed to be complete, any gene that is not annotated in a particular species is not found in that species. This fact allows one to construct matrices for species with whole genomes that indicate the presence or absence of genes in their genomes. Such matrices are also known as genome content matrices and are important for phylogenomic studies because the presence/absence information can be used to generate phylogenies. In addition, the presence/absence (genome content) approach in conjunction with gene ontology can determine which functional classes of genes are present or absent in the genomes of organisms. The basic approach used in constructing presence/absence matrices for whole genome information involves the use of an approach called single linkage clustering, which we discuss below.

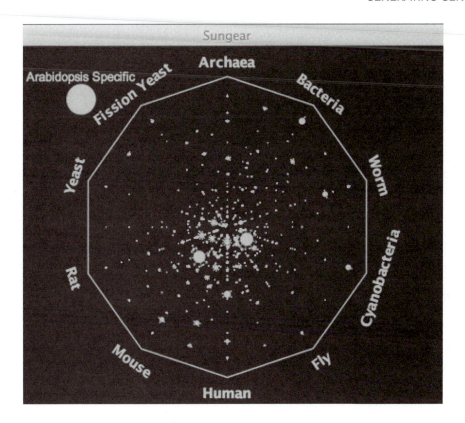

Figure 16.5 Screenshot of Sungear Web site. For this example, the input into Sungear was the list of genes present in 10 eukaryotic species, which the program then used to create an overlap diagram. The vertices of the trapezoid correspond to the species in the comparison, and each white dot represents a set of genes present in the taxa that occupy that specific area of the plot. If a dot cluster appears in the very center of the plot, this means that all species have those genes. If a dot cluster appears on the outer ring of the plot next to a species name, this means that the genes are unique to that species. (Image generated using Sungear.)

Clusters of orthologous groups is a method that enables identification of orthologs of genes across multiple species

The most well-established method for determining presence/absence-based phylogenies is clusters of orthologous groups (COGs); which was developed by Eugene Koonin's group at NIH. The basic goal of this technique is to identify orthologs of genes across multiple species. As discussed in Chapter 3, the ortholog of a gene in species X is the gene in species Y most related to that gene from species X through speciation. For well-annotated genes such as hemoglobin, this is an easy task, but for the vast majority of genes that are not well annotated or shared across species, the task is more difficult.

In order to identify COGs, an all-against-all BLAST search is performed for the genes being studied. The translated gene sequence is used since there is more conservation of amino acid sequence at the sequence level than DNA (see Chapter 3). After the BLAST results are obtained, cross-species reciprocal best hits are determined. Any proteins that are top hits for each other in the corresponding genomes are included in a COG. As a trivial example, take genomes A, B, and C which have genes 1, 2, and 3, respectively. Between genomes A and B, genes 1 and 2 are reciprocal top BLAST hits for each other. This means that gene 1 has a higher similarity to gene 2 than to any other gene in genome B, while simultaneously gene 2 has a higher similarity to gene 1 than to any other gene in genome A. Also, genes 2 and 3 are reciprocal best hits across genomes B and C, and genes 1 and 3 are reciprocal best hits across genomes A and C. This set of reciprocal best hits forms a COG triangle, which is the most basic COG element (**Figure 16.6A**). This triangle can be expanded by the inclusion of any other genes that are reciprocal best hits with any of the proteins in the triangle, regardless of whether they have any relationship to the other proteins in the triangle. For example, take genomes D and E, having genes 4 and 5. Genes 3 and 5 are a set of best hits, along with genes 1 and 4, and no other matches are found between these genes. This addition

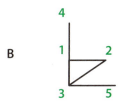

Figure 16.6 Steps to establish clusters of orthologous groups. A: clusters of orthologous group (COG) relationships for three taxa. B:new COG relationships after addition of two more taxa.

A

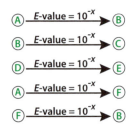

B

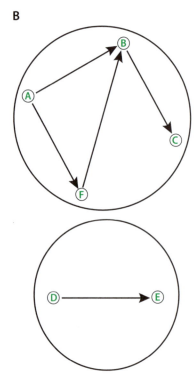

Figure 16.7 Basic steps of single linkage cluster approach for determining a gene presence/absence matrix. There are six species in the comparison (A–F). The *E*-values for comparing a putative homolog for several pairwise comparisons are shown in panel A, and the single linkage clusters based on these *E*-values are shown in panel B. In this example, A, B, F, and C cluster together and D and E cluster together

results in the new COG shown in Figure 16.6B. These COGs can become very large and jumbled as genomes are added to the analysis. A key point to this analysis is that all of the genes within a COG are grouped together, whether or not there are reciprocal best hits between them.

Single linkage clustering compares genes in a cross-species context based on sequence

One method for determining the presence and absence of genes in species is to use a single linkage clustering approach where the transitive property of gene similarity is utilized. As an example, take six genes from different species labeled A–F. Each of these genes is compared against the others by use of BLAST, and the *E*-values of the matches are determined. These matches are then filtered to include only those pairs of genes that match each other below a certain *E*-value threshold and are hence significant. In this example, there are six matches found that exceed the threshold (**Figure 16.7A**). On the basis of these matches, the genes are clustered into groups. Here are the steps of the clustering in this example as we proceed down the list of matches:

- Genes A and B are put into group I, as they are the first match.

- Gene C is added to the group because it has a match to A, which is already in the group.

- A new group (II) is started for genes D and E, since they do not have any matches in group I.

- Gene F is added to group I because it matches gene A.

- Nothing needs to be done for the match of genes B and F, since they are already grouped together.

The result of this clustering is the two groups of proteins shown in Figure 16.7B. One important result of clustering performed in this manner is that each group represents closely related genes in a cross-species context without regard to gene name. The grouping is based only upon the sequence, not the gene names.

In this example, we did not indicate a certain *E*-value threshold, but in practice researchers use multiple different *E*-values as their cutoff for a positive match and then examine the quality and robustness of the patterns obtained to determine the optimum threshold. In general, when closely related species are being examined, a more stringent threshold (for example, a very small *E*-value) needs to be utilized in order to distinguish the fine-grained differences in gene content between the species. For more distant species, a less stringent threshold (for example, a relatively large *E*-value) can be used.

A presence/absence matrix is constructed via single linkage clustering

Constructing a matrix representing presence/absence information for genes is somewhat complex but has been automated by several laboratories. One whole genome is BLASTed against another, a procedure known as a genome-by-genome BLAST. The end product of this BLAST is a list of genes that have significant hits across the two genomes. This procedure is in turn conducted for all pairwise comparisons of species that one wants to place in a gene presence/absence matrix. This step results in pairwise lists of genes that are similar, based on the BLAST scores of each comparison. The third step is to use approaches that can sift through the lists of species-based pairwise BLASTs and indicate which genes are present and which are absent in all of the species. The most common procedure for establishing these results for all species is called single linkage clustering, as discussed above (see Figure 16.7).

Constructing a presence/absence matrix for five species requires running 10 pairwise whole genome BLASTs for each gene in these genomes. Genes that appear in single species can be scored as present only in the species in which they originate. These are genes that will have *no* significant BLAST hit to any other species in any of the pairwise whole genome BLASTs. A highly conserved type of gene that is present in all genomes shows a significant BLAST score for that gene in each pairwise comparison. In a matrix of gene presence/absence, this gene would be scored present in all species in the study. Any gene that does not fall into one of the categories described above needs to be examined by the single linkage clustering approach.

As an example of this approach, we compare five species with small genomes that contain 10 genes in each genome. We begin by comparing species A with B and then move through the rest of the species. If A and B are BLASTed against each other, resulting in 10 significant hits, all 10 genes are present in A and B. If only nine genes show significant hits, there are nine genes in both genomes, with one gene present in A that is not in B and one gene present in B that is not in A. The number of genes that occur in the two genomes together is now 11 instead of 10. Conducting all the possible pairwise BLASTs for these hypothetical genomes yields the results in **Table 16.1**. The single linkage clustering algorithm described above is then used to construct a presence/absence matrix.

Each comparison value is an *E*-value. An *E*-value is essentially the probability that the similarity observed is detected as a result of luck (see Chapter 4 for detailed explanation of this value). An *E*-value of e^{-100} means that there is a $1/e^{100}$ probability that the result is random (a very small probability). Therefore, if we use an *E*-value of e^{-5} as a cutoff for significance, the 1s in Table 16.1 should be considered nonsignificant hits, while the 10s and 100s are significant. The single linkage clustering procedure for each of the 10 genes from species A as a starting point is shown in **Figure 16.8**. We go through each gene and construct the final presence/absence matrix in the following.

For gene 1, there are significant hits for all comparisons. All five species are connected by lines in the single linkage cluster diagram for this gene, signifying that they are in the same cluster and that the gene is present in all species. In a presence/absence matrix, presence is represented by a 1 and absence is represented by a 0. The first row in the presence/absence matrix is

```
         A    B    C    D    E
gene 1   1    1    1    1    1
```

Table 16.1. Example of *E*-values for five species (A, B, C, D, and E) for 10 hypothetical genes (all values in the table are negative powers of e).

	AvB	AvC	BvC	AvD	BvD	CvD	AvE	BvE	CvE	DvE
Gene 1	100	100	100	100	100	100	100	100	100	100
Gene 2	10	10	1	100	1	100	100	1	100	100
Gene 3	100	100	100	100	100	100	1	1	1	1
Gene 4	1	1	100	100	1	1	1	1	1	1
Gene 5	10	10	10	1	1	1	1	1	1	100
Gene 6	100	1	1	1	1	10	1	1	10	10
Gene 7	1	100	100	100	1	100	100	1	100	100
Gene 8	100	100	100	10	10	10	10	10	10	10
Gene 9	100	100	100	1	1	1	1	1	1	1
Gene 10	1	1	100	100	100	100	100	100	100	100

For gene 2, species B has insignificant hits to all other species for this gene. This gene is present in species A, C, D, and E and absent in B. Therefore, a new gene (gene 11), which is present only in species B, needs to be created to accommodate species B. The presence/absence matrix is revised as follows (completes row 2 and reflects the creation of gene 11):

	A	B	C	D	E
gene 1	1	1	1	1	1
gene 2	1	0	1	1	1
gene 11	0	1	0	0	0

Figure 16.8 Single linkage clustering. The 10 genes are listed in Table 16.1; e⁻⁵ was used as a cutoff. The cutoff value refers to *E*-values that are significant when they are smaller than the cutoff and insignificant when they are larger than the cutoff. Here, any *E*-value smaller than e⁻⁵ is considered significant and any value greater than e⁻⁵ is considered nonsignificant.

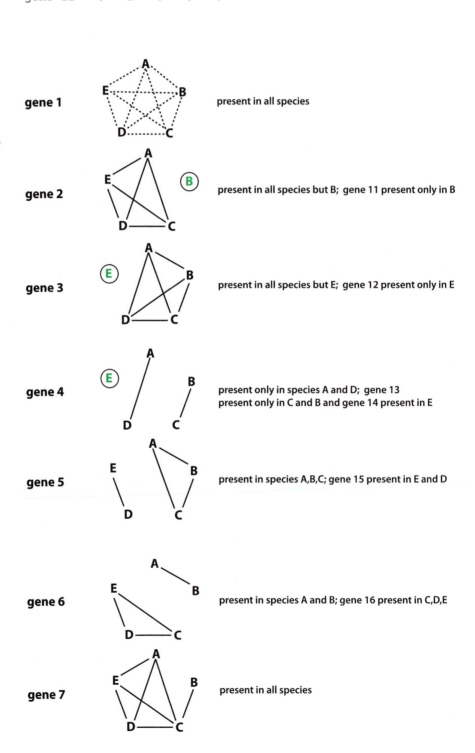

gene 1 — present in all species

gene 2 — present in all species but B; gene 11 present only in B

gene 3 — present in all species but E; gene 12 present only in E

gene 4 — present only in species A and D; gene 13 present only in C and B and gene 14 present in E

gene 5 — present in species A,B,C; gene 15 present in E and D

gene 6 — present in species A and B; gene 16 present in C,D,E

gene 7 — present in all species

Figure 16.8 Single linkage clustering (continued).

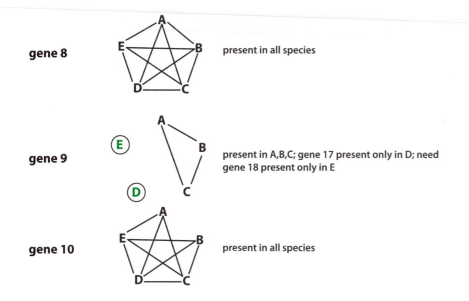

gene 8   present in all species

gene 9   present in A,B,C; gene 17 present only in D; need gene 18 present only in E

gene 10   present in all species

For gene 3, species E has insignificant hits to all other species for this gene, which is present in species A, B, C, and D and absent in E. A new gene (gene 12) is created to accommodate species E, with the resulting matrix shown below.

	A	B	C	D	E
gene 1	1	1	1	1	1
gene 2	1	0	1	1	1
gene 3	1	1	1	1	0
gene 11	0	1	0	0	0
gene 12	0	0	0	0	1

Gene 4 has significant hits for A and D, but none from A or D to any other species. This indicates a unique linkage of A and D as indicated by the line between them in Figure 16.8. It also has significant linkage between species B and C. Species E is off by itself. These results indicate that two new genes need to be created. One row depicts the presence of one of the new genes (gene 13) in B and C, and the second new gene (gene 14) is present only in E, as shown below.

	A	B	C	D	E
gene 1	1	1	1	1	1
gene 2	1	0	1	1	1
gene 3	1	1	1	1	0
gene 4	1	0	0	1	0
gene 11	0	1	0	0	0
gene 12	0	0	0	0	1
gene 13	0	1	1	0	0
gene 14	0	0	0	0	1

For gene 5, linkages occur for A, B, and C; hence, this gene is present in species A, B, and C and absent in D and E. A second linkage for a new gene exists for species E and D, so a new gene (gene 15) is created that is present in species E and D.

For gene 6, linkages exist for species A and B; therefore, this gene is present in species A and B and absent in the others. A second linkage cluster exists for species C, D, and E, so gene 16 is created, which is present in species C, D, and E.

Gene 7 is interesting because there is insignificant linkage between species B and species A and E. However, species B has significant linkage to species C and D. Figure 16.8 shows that all of the species are connected by at least one linkage (the "single" in single linkage); therefore, the gene is considered to be present in all species.

Gene 8 is present in all species.

For gene 9, linkage occurs between species A, B, and C. Species D and E have no hits to each other and hence two new genes (genes 17 and 18) are created to accommodate these nonlinked species.

For gene 10, similar to the case for gene 7, there is no significant hit from species A to species B, but significant hits occur in all other comparisons. This gene is present in all species. The presence/absence matrix shown below incorporates all the information described for genes 5 through 10:

	A	B	C	D	E
gene 1	1	1	1	1	1
gene 2	1	0	1	1	1
gene 3	1	1	1	1	0
gene 4	1	0	0	1	0
gene 5	1	1	1	0	0
gene 6	1	1	0	0	0
gene 7	1	1	1	1	1
gene 8	1	1	1	1	1
gene 9	1	1	1	0	0
gene 10	1	1	1	1	1
gene 11	0	1	0	0	0
gene 12	0	0	0	0	1
gene 13	0	1	1	0	0
gene 14	0	0	0	0	1
gene 15	0	0	0	1	1
gene 16	0	0	1	1	1
gene 17	0	0	0	1	0
gene 18	0	0	0	0	1

All of the species have 10 genes included in the matrix, which is highly dependent on where the significance cutoff is placed. For instance, if we were more stringent about our cutoffs and accepted only an E-value of -50 or less or more as a significant hit, the presence/absence matrix would change. In such a case, gene 1 would lose the A to B to C linkage to other taxa and to themselves (dotted lines in Figure 16.8), drastically changing the interpretation of presence and absence of this gene.

Genome content analysis was first demonstrated for bacterial genomes

The first gene presence/absence studies used the whole genome sequences of bacteria. Snel et al. were the first to propose the use of genome content to reconstruct phylogeny. They used the fully sequenced genomes of 13 unicellular species. A pairwise measure of the number of genes that overlap in two genomes was established by use of E-value cutoffs of e = 0.01 and 0.001. These pairwise gene overlap data were then transformed into a distance matrix and the matrix was then analyzed by the neighbor joining (NJ) algorithm. Their results suggested that the approach could easily recover the phylogeny of the 13 taxa because their tree was highly congruent with a phylogeny for the same species constructed by use of the standard 16S rDNA gene (**Figure 16.9**). Almost simultaneously, Fitz-Gibbon and House used 11 fully sequenced genomes and a modified single linkage cluster approach to generate a presence/absence matrix for the gene content analysis of these 11 species. This study constructed matrices by use of several cutoffs and compared the results for these different cutoffs, as well as using the presence/absence matrix for both parsimony and distance analysis. They also partitioned their gene presence/absence matrix into "operational" and "informational" genes. The informational genes are those involved in transmission of information in the cell, such as genes involved in transcription, DNA replication, and translation. Operational genes are all other kinds of genes. While there are subtle differences among the trees generated by the different ways the authors analyzed the data, the result of their study was that the presence/absence data, just as with the pairwise distance approach, were in close congruence with the 16S rDNA tree.

Genome content analysis is currently performed by computer software packages

Since these seminal studies, several researchers have examined the potential for genome content to recover phylogeny. These subsequent studies have resulted in refinement of the approach and the development of several software packages that accomplish genome content analysis. In addition, researchers have realized that the order of genes in a genome can also be used to gain even more information for whole genome content phylogenies. The first extension of the approach was to apply it to smaller genomes than the fully sequenced unicellular genomes, such as chloroplast or large viral genomes. Other researchers noted that the way the presence/absence matrix was constructed resulted in consistent biases that needed to be corrected. Subsequently, several research groups all developed a set of approaches called conditional reconstruction (CR) approaches that refined the way matrices were built for genome content studies. One of the specific problems that CR attempted to correct was the problem that small genomes cause in presence absence studies. Small genomes accrue an inordinate number of "absent" scores in the presence/absence matrix, and this will cause what some authors have called "small genome attraction" in genome content trees. This phenomenon occurs because the absent scores in the small genome taxa make such taxa look similar, and hence they are attracted to each other in phylogenetic analysis. Gu and Zhang used an "extended gene content" scoring approach to refine the presence/absence matrix scores. They recognized that "the status of a gene family in a given genome could be absence, presence as single copy, or presence as

Figure 16.9 Examples of phylogenetic trees. A: genome content tree. B: 16S rDNA tree. (Adapted from B. Snell, P. Bork, M.A. Huynen, Nat. Genet. 21:108–110, 1999. Courtesy of Macmillan.)

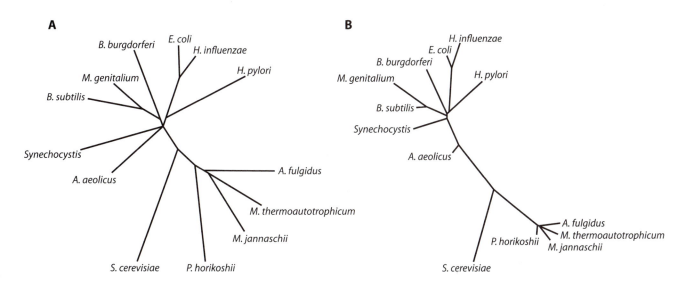

duplicates" and all of these could be used as character states for estimating distance matrices for phylogenetic analysis.

This last example points to the need for the development of models for how the presence/absence characters evolve so that likelihood approaches might be used to estimate phylogeny from genome content. In this context, researchers have developed models for the evolution of genome content that allow for likelihood approaches to be applied. Another way to circumvent the problem first realized by Gu and Zhang is to focus on the presence and absence of protein domains. This approach uses the establishment of the presence or absence of protein domain architecture as a phylogenetic tool for over 174 complete genomes. Because some proteins have similar protein domains, the number of characters is reduced in such an analysis, but the problem of orthology is reduced. As mentioned before, Fitz-Gibbon and House explored the impact of the significance cutoff for the establishment of presence of a gene in their seminal study. Subsequent studies examined this phenomenon in more detail. Specifically, changing the E-value changes the presence/absence scores, which in turn change distance measures and presence/absence matrices. Changing matrices result in changing tree topologies. In general, as the E-value cutoff is made more stringent (that is, as the cutoff approaches 0.0), the tree topologies become less resolved, because the number of genes where presence/absence can be recovered is reduced. On the other hand, very relaxed E-values (values closer to 1.0) result in phylogenies with obviously incorrect topologies, because the orthology statements established by these lax cutoffs are suspect. Overall, though, genome content analysis is a valuable approach in phylogenomics. It tends to recover relationships that are more important to the interior of phylogenetic trees, and perhaps it should be used in conjunction with other kinds of genome-level data such as sequence data from well-established orthologs.

Summary

- The Genome Biology Website, which can be accessed via the National Center for Biotechnical Information (NCBI) home page, offers several resources by which various genomes can be analyzed.

- The overlap of genes from two or more genomes can be visualized by use of Venn diagrams or programs such as EDGAR and Sungear.

- Identified orthologs of genes across multiple species can be grouped into clusters of orthologous groups (COGs).

- Single linkage clustering compares genes in a cross-species context, based only upon sequence, and allows the construction of a presence/absence matrix.

- Genome content analysis was first demonstrated for bacterial genomes and can now be performed by computer software packages and has been extended to include many eukaryotes.

Discussion Questions

1. The following is a table of *E*-values for four species. Generate a gene presence/absence matrix for these four species by use of a similarity cutoff of e^{-5}.

	AvB	AvC	BvC	AvD	BvD	CvD
gene 1	100	1	1	1	1	1
gene 2	1	1	1	1	1	100
gene 3	100	100	100	100	100	100
gene 4	1	1	1	10	1	100
gene 5	100	100	100	1	1	1
gene 6	100	1	1	1	1	100
gene 7	1	100	100	100	1	100
gene 8	100	100	100	10	10	10
gene 9	1	1	10	1	10	10
gene 10	1	1	100	100	100	100

The cutoff value refers to *E*-values that are significant when they are smaller than the cutoff and insignificant when they are larger than the cutoff. For this problem, any *E*-value smaller than e^{-5} is considered significant and any value greater than e^{-5} is considered not significant.

Draw the single linkage clusters for all of the genes.

2. Generate a presence/absence matrix for the *E*-values for the four species in Question 1, but this time use a similarity cutoff of e^{-20}. Draw the single linkage clusters of all the genes.

Further Reading

Fitz-Gibbon ST & House CH (1999) Whole genome-based phylogenetic analysis of free-living microorganisms. *Nucleic Acids Res.* 27, 4218–4222.

Gu X & Zhang H (2004) Genome phylogenetic analysis based on extended gene contents. *Mol. Biol. Evol.* 21, 1401–1408.

Huson DH & Steel M (2004) Phylogenetic trees based on gene content. *Bioinformatics* 20, 2044–2049.

Lake JA & Rivera MC (2004) Deriving the genomic tree of life in the presence of horizontal gene transfer: conditioned reconstruction. *Mol. Biol. Evol.* 21(4), 681–690.

Lienau EK, DeSalle R, Rosenfeld JA & Planet PJ (2006) Reciprocal illumination in the gene content tree of life. *Syst. Biol.* 55, 441–453.

Medini D, Donati C, Tettelin H et al. (2005) The microbial pan-genome. *Curr. Opin. Genet. Dev.* 15, 589–594.

Snel B, Bork P & Huynen MA (1999) Genome phylogeny based on gene content. *Nat. Genet.* 21, 108–110.

Snel B, Huynen MA & Dutilh BE (2005) Genome trees and the nature of genome evolution. *Annu. Rev. Microbiol.* 59, 191–209.

Tatusov RL, Fedorova ND, Jackson JD et al. (2003) The COG database: an updated version includes eukaryotes. *BMC Bioinf.* 4, 41.

Wolf YI, Rogozin IB, Grishin NV & Koonin EV (2002) Genome trees and the tree of life. *Trends Genet.* 18, 472–479.

Phylogenomic Perspective of Biological Diversity: Tree of Life, DNA Barcoding, and Metagenomics

In Chapters 8 and 9, we looked at very basic phylogenomic analysis for small data sets. Remember that, in phylogenetic analysis, exact solutions for three species involve evaluating only three trees. However, study systems in modern evolutionary biology involve hundreds of species. When 50 terminals are involved, there are 2.75×10^{76} trees that we need to evaluate to get an exact solution. Even after adding only 70 taxa to an analysis, the number of trees that would need to be examined and evaluated would be 10^{117} or more than the number of particles in the universe! As we discussed in Chapters 8 and 9, these problems, and indeed the phylogenetic analysis problem, are solved by using heuristic approaches. Since the Tree of Life project proposes to establish a hierarchical branching diagram for all of the named species on the planet, this involves a tree with over 1.7 million terminals. The number of trees to examine for this project is staggeringly large, and so the tree of life project relies on the application of shortcut methods and innovative computing to generate a tree. With the development of massively parallel sequencing methods and the promise of the "$1000 genome," it is not outrageous to assume that future phylogenetic Tree of Life studies will involve high-throughput genomics approaches. In this chapter we will discuss the dynamics of utilizing such approaches in phylogenetics. While the current application of whole genome information to phylogenetics is somewhat limited because of a dearth of data, we feel it important to demonstrate, with the few model systems that have been developed to date, that the phylogenomic approach is viable for the Tree of Life. In the meantime, other approaches have been developed to assist in identifying the diversity of life on the planet. We will also discuss two of these approaches: shallow genome sequencing, to generate phylogenetic hypotheses, and DNA barcoding, to generate an identification system for organisms based on DNA sequences. Another modern high-throughput project to characterize biodiversity concerns the analysis of sequence information for at least 10 individuals from each of the 1.7 million named species on this planet. This project proposes to use a standardized DNA sequence as a barcode to identify species. The large number of individuals examined for each species takes into account the variation within species. Such an approach introduces novel analytical problems to understanding large amounts of sequence information that we will discuss. A subset of the barcode of life problem is the identification of microbial species in metagenomic studies, and we discuss this aspect of modern high-throughput analysis to characterize biodiversity.

A Hierarchy of Sequencing Approaches for Biodiversity Studies

High-throughput sequencing has been used in many different ways to obtain sequences for biodiversity studies. Shallow genome sequencing begins with the selection of a small number of genes that are presumed to have interesting phylogenetic characteristics. These genes are then sequenced in as many taxa as is monetarily feasible. For the purposes of this chapter, we will examine two of the major phylogenetic questions in the tree of life that have been addressed in this way. These systems involve the relationships of plants and the relationships of animals. DNA barcoding, which we discuss at the end of this chapter, is a highly reduced form of this targeted gene approach (a single gene is the focus of analysis). Another approach to obtaining large amounts of sequence information for phylogenetic analysis involves expressed sequence tag (EST) analysis, where expressed genes are utilized to amass a collection of gene sequences for analysis. Finally there is the approach of sequencing whole genomes for phylogenetic analysis. We have already discussed how whole genomes can be used to generate presence/absence matrices in Chapter 16, and now we will delve into how the sequences of all of the genes in a genome can be used to generate phylogenetic hypotheses. Fortunately there are some excellent examples of this approach in the literature for us to examine in detail.

In shallow targeted sequencing, a small number of interesting genes are sequenced in as many taxa as feasible

So far the genes that have aided in the phylogenetic approach are conserved genes such as the 18S and 28S ribosomal RNA genes both for plants and animals, genes like the mitochondrial cytochrome oxidase I gene, the *rbcL* gene for plants, and the 16s rDNA gene for bacteria. Other genes have been targeted in the last decade, and substantial databases are being developed to incorporate the targeted gene approach. In a *tour de force* study, Goloboff et al. gathered DNA sequence information for over 70,000 taxa for 13 different genes. **Table 17.1** shows the list of genes used.

Note that the majority of genes are from chloroplast or mitochondrial genomes and that the information is found in both DNA sequence format (four bases and gaps) and amino acid format (20 amino acids and gaps). More importantly, note that no single gene covers all taxa, meaning that the matrix has a large number of missing entries in it. In fact, 92% of the entries in the matrix are missing. The reason there are so many missing entries is that researchers working on different organisms use different genes as sources of information. Sometimes this is unavoidable. For instance, most researchers studying plants use chloroplast genes that do not exist in animals, simply because animals do not have chloroplasts. In the study of Goloboff et al., the large amount of missing data also stems from the

Table 17.1 DNA sequence information for selected genes

Gene	Fragments	Taxa	Characters	Scope	Type	Genome
LSU-rRNA	11	11,700–1267	312–115	global	DNA	nuclear
MatK	1	11,855	792	Embryophyta	DNA	plastid
NdhF	1	4864	1209	Embryophyta	DNA	plastid
RbcL	1	13,043	43	Embryophyta	DNA	plastid
COXI	1	7310	1296	Metazoa	protein	mitochonrial
COXII	1	8315	437	Metazoa	protein	mitochonrial
COXIII	1	2309	272	Metazoa	protein	mitochonrial
CytB	1	13,766	337	Chordata	protein	mitochonrial
NDI	1	4123	349	Metazoa	protein	mitochonrial
SSU	6	20,462–19,336	293–26	global	DNA	nuclear
SSU	1	1314	464	Hexapoda	DNA	mitochonrial
RNA PII	2	869–333	515–203	Fungi	DNA	mitochonrial
LSU	1	752	314	Ascomycota	protein	nuclear

Source: P.A. Goloboff, S.A. Catalano, J.M. Mirande et al., *Cladistics* 25:211–230, 2009.

fact that these studies were all done independently over a long period of time in different labs. Again, because there is no standard gene that has been established for phylogenetic studies, some studies will use markers that other studies avoid.

The analysis by Goloboff et al. could be accomplished because of overlap of some genes over *all taxa*, which link the matrix throughout the taxonomic sample. For instance, all plants might have the 18s rDNA gene sequenced, while 90% of animals do not. The 10% of animals that do have this marker sequenced will allow for the linking of plants and animals in a concatenated matrix. The mega tree generated by their approach is shown in **Figure 17.1**. In general, groupings in this tree follow well-accepted taxonomic and systematic lines established via anatomy and other characters over the last century.

Goloboff et al. attempted to include as many taxa as possible and to that end did not focus on maximizing the genes used. One way to increase the data set size for a particular phylogenetic question is to target specific genes, by use of conserved polymerase chain reaction (PCR) primers and Sanger sequencing, to generate a broader array of genes for the full taxonomic sampling of organisms in a study using standard Sanger sequencing. The targeted primer approach is

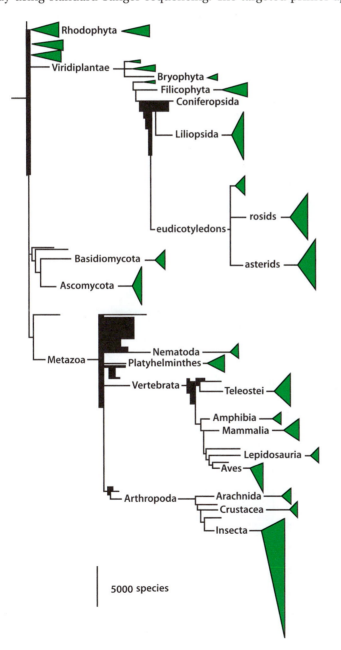

Figure 17.1 Goloboff's supertree.
This tree is based on analysis of a large concatenated matrix with over 70,000 terminals. The triangles represent many species circumscribed by the higher taxonomic name; the number of species in each triangle can be estimated from the bar at the bottom, which represents 5000 species at terminals. (Adapted from P.A. Goloboff, S.A. Catalano, J.M. Mirande et al., *Cladistics* 25:211–230, 2009.)

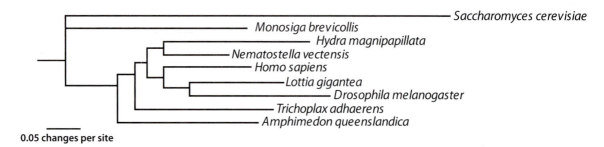

Figure 17.2 Metazoan phylogeny. After the sequencing of the Placozoa genome, this phylogeny was produced by analyzing over 100 genes from the indicated taxa. (Adapted from M. Srivastava, E. Begovic, J. Chapman et al., *Nature* 454:955–960, 2008.)

laborious, as it involves designing and testing primers on a wide variety of species and any primer designed should be somewhat "universal" (meaning it should be functional in a wide range of species). Other approaches attempt to limit taxonomic sampling in order to provide a matrix with as few missing data as possible. In the realm of animal or metazoan phylogenetics, this has been done by use of whole genome sequences and targeted shallow sequencing. Upon completion of sequencing of the *Trichoplax adhaerens* genome (phylum Placozoa), a phylogeny was generated by using seven metazoan representatives and two outgroups for over 100 targeted genes (**Figure 17.2**). Similarly, two research groups used targeted shallow sequence sampling to generate phylogenies for the Metazoa (**Figure 17.3**). One study utilized 128 genes and the other study used nearly 50. Note that the topologies of the two studies have many similarities except at the

Figure 17.3 Two phylogenetic hypotheses. The hypotheses were generated for larger taxon sampling of animals via a targeted gene sequencing approach. The two panels represent different studies: one study used approximately 50 genes, and the other utilized 128 genes. Vertical bars represent higher taxonomic levels. Key: BIL, Bilateria; DIP, diploblasts; E, Ecdysozoa; L, Lophotrochozoa; D, Deuterostomia; C, Cnidaria; white triangle, Porifera; asterisk, Placozoa; black dot, Ctenophora.

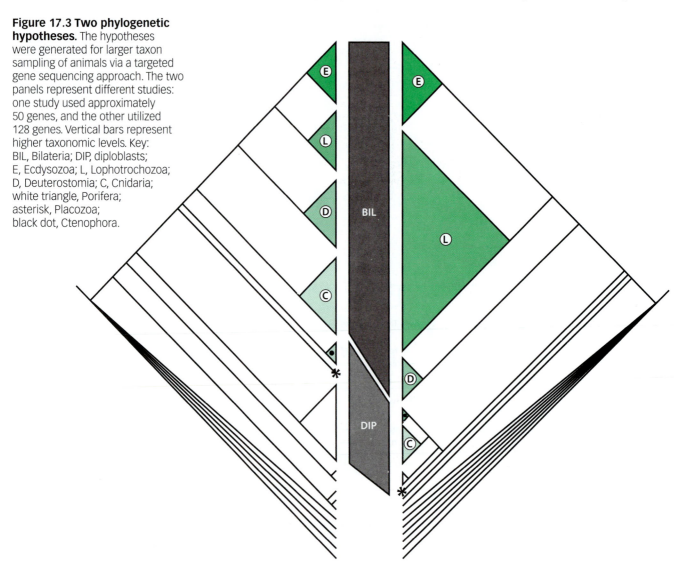

base of the Metazoa, where one study suggests sponges (Porifera) are the most basal animal and the other study suggests *Trichoplax* (Placozoa) as the most basal diploblastic animal. We will discuss these differences below in more detail.

For plants, targeted sequencing of the chloroplast genome has progressed impressively over the last decade. Currently, over 70 plant chloroplast genomes have been fully sequenced, and the phylogenetic analysis of these sequences has been discussed in two studies (**Figure 17.4**). One study used 61 genes from whole chloroplast genomes for over 40 taxa; the other used 81 genes from over 60 taxa to generate these trees. The broad overlap of almost complete agreement of these studies is in direct contrast to the shallow targeted metazoan studies cited above.

Expressed sequence tag approaches use expressed genes to amass a collection of gene sequences for analysis

The EST approach was explained in Chapter 2, but briefly, it involves the isolation of mRNA from tissues. The tissues can be either specific tissues, which might

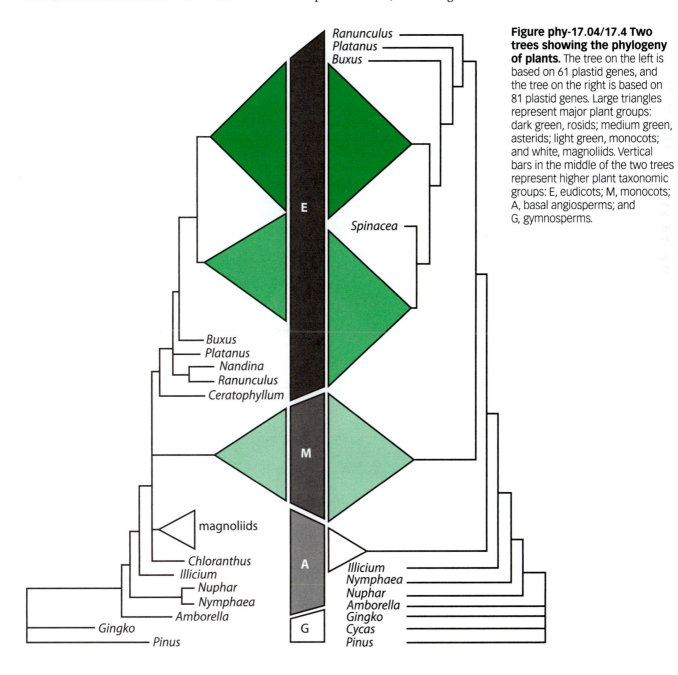

Figure phy-17.04/17.4 Two trees showing the phylogeny of plants. The tree on the left is based on 61 plastid genes, and the tree on the right is based on 81 plastid genes. Large triangles represent major plant groups: dark green, rosids; medium green, asterids; light green, monocots; and white, magnoliids. Vertical bars in the middle of the two trees represent higher plant taxonomic groups: E, eudicots; M, monocots; A, basal angiosperms; and G, gymnosperms.

ensure that the same genes are being expressed across a large range of taxa, or simply chosen as a result of convenience. For instance, if gonadal tissue is chosen as the target tissue for isolating RNA over a broad range of animals, then the same basic genes might be expressed in the gonads of individuals from different taxa and the overlap of gene sequences obtained might be enhanced. Once RNA is isolated, cDNA is made and then the cDNAs are either cloned and sequenced via Sanger sequencing (choosing, say, 5000 clones to in some cases over 100,000 clones for sequencing) or are sequenced using the next-generation approaches we discussed in Chapter 2. Once the sequences are obtained, "unigenes" are constructed by assembling the various sequences into short contigs that represent the mRNAs in the initial pool of messenger RNAs.

Both animal and plant systems have been approached by EST techniques. The "hit or miss" nature of EST libraries is an encumbrance of the approach, but as we have seen in the targeted approaches, missing data are difficult to avoid. The problem is that overlap of gene sequences from EST libraries diminishes as more and more taxa are added to an analysis. For instance, for two taxa the overlap of genes might be 50%. But adding a third taxon might decrease the overall overlap down to 25% (**Figure 17.5**), and so on as more and more taxa are added. This does not mean that missing data decrease at this drastic a rate, though the lack of overall overlap when adding more and more taxa is augmented by increased partial overlap of pairs of taxa (Figure 17.5).

Dunn et al. generated EST libraries for over 25 animals and combined those data with existing data for over 50 animals to generate a broadly taxonomically sampled metazoan phylogenetic matrix. The phylogeny generated from this matrix showed the same monophyly of the major animal groups: Ecdysozoa, Lophotrochozoa, and Deuterostomia. Like an earlier study (Figure 17.3, right panel), Cnidaria were the closest relative of Bilateria, but the most basal metazoans were ctenophores. While the taxonomic sampling for this study is broad, there is a considerable amount of missing data, as demonstrated in **Figure 17.6**. Likewise, large-scale plant phylogenomic studies generate matrices made up mostly of EST information mined from the EST database.

One group of researchers constructed a matrix from 136 plant species and obtained 18,896 gene partitions (orthologs) with over 2.9 million characters. The tree they generated has many of the major groups that the previous studies have (Figure 17.4) and will not be discussed here. The matrix of another group, Lee et al., has 22,833 sets of orthologs from 150 plant species. These species, belonging to 101 different genera, represent a broad taxonomic range of angiosperms and also deeply cover extant gymnosperms. To reduce the size of the data set for some of the

Figure 17.5 The number of overlapping genes declines with the addition of more terminals. While the full overlap region decreases in size, more genes with partial overlap can be added to the matrix.

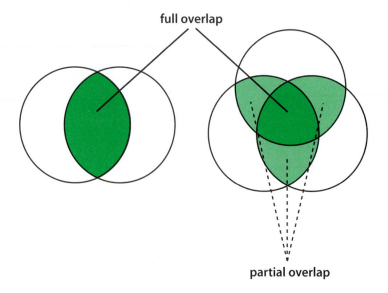

full overlap

partial overlap

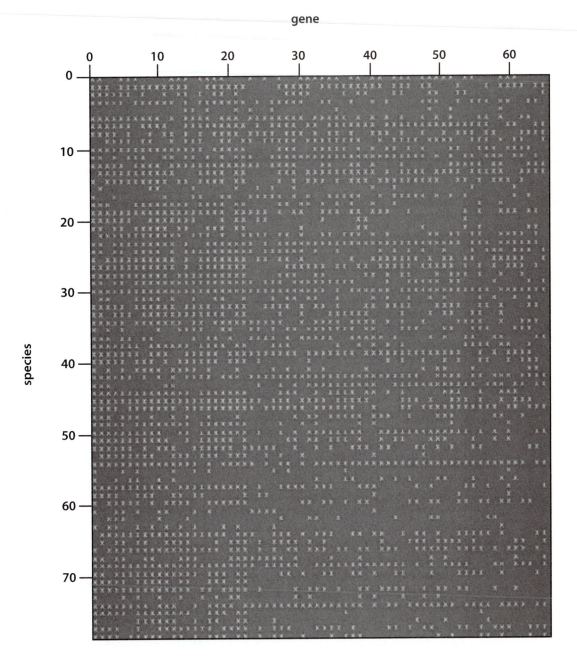

gene

species

Figure 17.6 A broadly taxonomically sampled metazoan phylogenetic matrix. Information content in the matrix generated by Dunne et al., by the expressed sequence tag approach, was analyzed by Siddall. An X in the matrix indicates missing data. Species are listed on the y-axis and genes on the x-axis. Species correspond to those analyzed by Siddall (Courtesy of Mark Siddall).

analyses and to decrease the impact of missing data, Lee et al. removed ortholog partitions with the most missing data, thereby producing a matrix that included only genes with at least 30% representation across all genera (**Figure 17.7**). The median number of gene partitions in which a taxon is represented is 2071. The analysis shown in Figure 17.7 has 101 taxa (genera), derived from 2970 gene partitions and 1,660,883 characters. The average number of genera represented in each gene partition is 41 (40.6%) in this matrix, and hence a relatively good estimate of the amount of missing data is about 60%. The relationships seen in this tree are similar in many respects to the two previous plant phylogenies shown in Figure 17.4, indicating that the nuclear DNA inferences are in broad agreement with chloroplast inferences. The EST analyses support the gymnosperms as a monophyletic group (all descendents coming from a single common ancestor to the exclusion of other taxa) as in all molecular data sets except the ones based upon *rbcL*. By contrast, most morphological analyses and some molecular analyses retrieve gymnosperms as paraphyletic (an arrangement where the group in question is basal and has a single common ancestor, but there are other taxa in

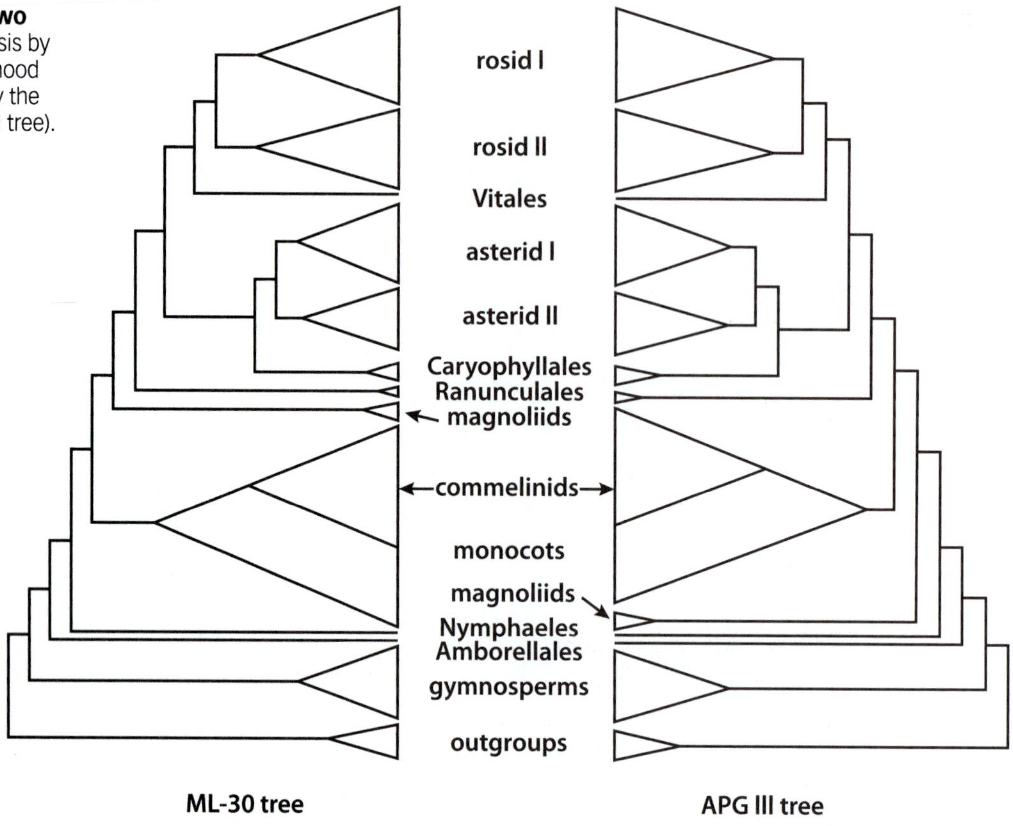

Figure 17.7 Comparison of two trees. (Left) Full plant EST analysis by Lee et al., using maximum likelihood (ML-30 tree). (Right) Tree used by the Angiosperm Plant Group (AGP III tree).

rosid I

rosid II

Vitales

asterid I

asterid II

Caryophyllales
Ranunculales
← magnoliids

←commelinids→

monocots

magnoliids

Nymphaeles
Amborellales

gymnosperms

outgroups

ML-30 tree   APG III tree

the analysis that share that same common ancestor in derived positions in the tree), as opposed to most molecular data sets that retrieve a monophyletic gymnosperm group. The differences between the topologies obtained with the different molecular data sets involve the placement of the gnetophytes. Of the varied topologies with respect to those derived from molecular data sets, the analyses of Lee et al. conclusively support a monophyletic gymnosperm group, with the gnetophytes as a sister group to all other gymnosperms.

The basic topology of the angiosperm tree used by the Angiosperm Phylogeny Group (APG; Figure 17.7) is generally supported but with some important changes. Lee et al. retrieved the same topology of major groups on the backbone of the tree starting with Amborellales followed by Nymphaeles, magnoliids, Ranunculales, Caryophyllales, asterids, and rosids. One difference from the chloroplast data sets (tree comparison not shown) is the placement of the monocots between Nymphaeles and magnoliids by Lee et al., in contrast to the monocots being placed between magnoliids and ranunculids by APG. The very few discrepancies between the tree of Lee et al. and the APG working hypothesis are at nodes with lower support and are likely the consequence of ambiguous orthology due to incomplete genomes.

Yeast and *Drosophila* represent examples of whole genome approaches

While the number of whole genome sequences has grown rapidly in the last 5 years, the only taxonomic group with sampling greater than 100 is microbes. In this part of this chapter we will examine two specific examples of whole genome phylogenomics: the first real phylogenomics study of eukaryotes using yeast (8 taxa), and a study of *Drosophila* relationships (12 taxa). Bacterial phylogenomics will be discussed in the following section. Before looking at these examples, we first need to make it clear that establishing orthology of genes is absolutely critical for these genome-level studies. The subject of orthology has been discussed in detail in Chapter 5.

Rokas et al. presented a phylogenomic analysis of several yeast species in the first eukaryotic genome-level study. They constructed their matrix based on whole genome sequences of eight yeast species and selected 106 genes that they suggested did not suffer orthology problems. They determined the orthology of their genes using sequence similarity and chromosomal position (synteny) of the genes in *Saccharomyces cerevisiae*. The phylogenomic analysis of all of the data combined together for these taxa gave a fully resolved and very strongly supported hypothesis, with all nodes showing 100% bootstraps for maximum likelihood (ML) and maximum parsimony (MP). Rokas et al. pointed out that while the overall hypothesis generated by concatenating all the genes together into a single matrix was well supported, individual gene analysis resulted in a potpourri of trees. In fact, 12 different trees are generated by single-gene analysis, with the concatenated tree being the most frequently obtained at 55%. They made suggestions from these analyses about the optimal number of genes in an analysis and the optimal amount of sequence for an analysis to be robust. This paper generated a great deal of enthusiasm and response from the systematics community. We will discuss one of these problems here.

Rokas et al. showed that their concatenated tree (CT) was at odds with one relationship in particular. As we will see, the same phenomenon happens with the *Drosophila* example, where two hypotheses are preferred by most of the data, with one of the hypotheses being in higher percentage than the other. In the case of yeast, the CT hypothesis is supported in about 55% of the genes used in the study and the alternative hypothesis (AH), which involves the switching of two taxa, is supported in 35% of the cases. These two topologies are shown in **Figure 17.8**. The sister relationships of the two taxa (labeled Skud and Sbay) are different in the two topologies. One could argue that these differences are real and that there are two different evolutionary histories being expressed in the 55% of genes that give the CT and the 35% of genes that give AH. There are three main reasons why the data, when examined at a gene by gene level, might show different evolutionary histories for genes. First, horizontal transfer of a specific gene will often produce a radically different history for that gene than other genes that do not experience horizontal transfer. Second, in eukaryotes, genes sort in different patterns due to biparental reproduction and random assortment. This sorting often leads to different lineage-specific patterns and is observed as a phenomenon called lineage sorting (see Chapter 15). In this case, two genes that have undergone different sorting processes will have different histories. The third reason is an artifact of rooting trees for closely related species that we discuss below.

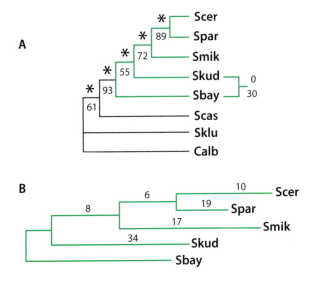

Figure 17.8 Results of reconsidering the rooting of the yeast tree. A: total evidence or concatenated tree from the study of Rokas et al., with the number of genes supporting each node. An asterisk on the nodes signifies that there is a 100% MP and ML bootstrap value for that node. The five taxa that are stable in the phylogeny are marked by the green branches. The branching on the right of the tree indicates the "spurious" sister group relationship of Skud and Sbay. Since the Sbay and Skud node is not in the concatenated tree, it has a bootstrap of 0, but it is supported in 30 of the 106 genes used in the study. B: tree showing the number of genes that support each of the possible rooting points.

Gatesy et al. examined the possibility that rooting of trees with closely related taxa might result in spurious gene trees by use of the yeast data set. They suggest that a better way to look at the 106 yeast gene trees is to realize that, for the five ingroup taxa in the analysis and the concatenated tree topology, there are seven different branches where an outgroup can root the five taxa (Figure 17.8). Five of these "roots" will land on branches that unite Skud and Sbay and only two roots are on branches that will result in the CT topology. It can be seen in Figure 17.8 that the outgroup lands in a fairly random fashion on all seven branches. Part of the reason for this is the taxonomic sampling used by Rokas et al. All the outgroups outside of the five taxa shown in Figure 17.8 have incredibly long branches relative to the ingroups. In fact, when Gatesy et al. analyze the data using only the five ingroups, all 106 genes give the same topology, suggesting that any anomalous topologies resulting from adding taxa in the study of Rokas et al. are caused by random rooting on the seven branches in the five-taxon topology. We will return to this problem in the next section.

In 2000, the fruit fly *Drosophila melanogaster* was the first multicellular eukaryotic genome to be sequenced. After that time, researchers realized that a collection of closely related species whose relationship and time of divergence were well-known would be a very important asset. To that end, they sequenced the complete genomes of 11 additional *Drosophila* species. These 12 genomes have been analyzed together to gain further understanding of *Drosophila* genetics and general evolutionary trends. At this point, the 12 *Drosophila* genomes are the largest resource of genomes for closely related taxa.

Several studies have analyzed the 12 genomes at the sequence level, and one study analyzed gene presence/absence patterns by use of the whole genome data. Each of these analyses settles on the same overall topology no matter the tree building approach or models imposed on the analysis or data set (presence/absence or amino acids or nucleic acids), and each study reports extremely high support for the relationships. An interesting result with respect to the relationships of *D. yakuba*, *D. erecta*, and *D. melanogaster* was pointed out in Chapter 11 that is very similar to the result discussed above for yeast.

Specific Problems Caused by Bacterial Phylogenomics

At the time of the writing of this book, several thousands of bacterial, archaeal, and eukaryotic genomes have been sequenced. Several studies have attempted to accomplish phylogenetic analysis of subsamples of this large number of taxa. There are two ways that these analyses can proceed. The simple gene presence/absence states of these genomes can be used to generate phylogenomic hypotheses, or the actual residue sequences (usually amino acids) can be used to generate phylogenetic hypotheses.

There are two problems we need to address here. The first is very simple: does a bifurcating tree represent the history of life when it comes to bacteria? Many scientists suggest that, because of rampant horizontal gene transfer (HGT), vertical history in the bacterial part of the tree of life is obscured. HGT is the process by which genes are passed between different bacteria in a non-parent–child relationship. The most common method for HGT is the formation of a pilus that allows for the transfer of genetic material from one bacterium to another; this is thought to be common among bacteria.

The history of life in this context is suggested to be of a web or a rake (imagine a leaf rake with prongs all coming out of a large base) with very little resolution in the tree. Researchers holding this view are called microbialists (not to be confused with microbiologists) and they more or less deny that treelike structure exists for organisms undergoing large amounts of HGT. Other systematists have argued that

HGT, while affecting vertical signal in the tree of life, does not destroy it and that if enough information is gathered, a treelike structure can be obtained. This brings us to the second problem in phylogenomics context: if we do assume there is a treelike structure to the history of life on our planet, do we use all genes in an analysis? The argument is that if some genes have been horizontally transferred, then vertical history should be lost in those genes and they should be excluded from attempting to infer phylogeny. These two problems have raised a large amount of controversy in the phylogenetics community with respect to reconstructing the tree of life. What can be done to address these two problems?

Does a tree of life really exist for bacteria?

Some of the arguments for a tree of life (ToL) are philosophical, and others are pragmatic and empirical. There are three conceivable tests for whether or not a vertical tree of life exists before the idea should be thrown out. First, we should attempt to build trees with data from this problematic part of the ToL. If an unresolved tree results, then the first test is not passed and we have to assume that a vertical or bifurcating ToL does not exist. Many microbialists have simply not even tried this approach and, from first principles, refrain from constructing a tree at this level. The second test is valid only if the first is passed: it involves assessing the robustness of a vertical tree if so obtained. If the vertical tree is not robust (does not have high bootstrap values or large Bayesian posteriors) then the microbialists are correct and a vertical ToL does not exist or is extremely weakly supported. If it is weakly supported in a scientific context, this is bad news for a vertical tree, as any subsequent work will more or less result in unstable inferences at this level. The third test is only relevant if the first two are passed: it involves asking how real are the results of a robust vertical tree. While there is no way to tell if a particular tree topology is true, we can compare the tree to already assessed knowledge about the classification and relationships of bacteria based on other kinds of information. This approach is somewhat circular but unavoidable. However, it is also logical to suggest that if there are glaring inconsistencies between a vetical tree of life generated from sequences and existing classifications, then one (or both) is wrong. Since it is the ToL that is being tested, we must logically reject the ToL if there are glaring inconsistencies.

Using both presence/absence and amino acid sequences to approach these three tests, Lienau et al. have consistently generated a ToL hypothesis that is well-resolved and robust. The tree is the result of analysis of a concatenated matrix with both amino acid sequences and presence/absence characters (over 2.5 million characters). Note that all nodes in their tree are resolved and all nodes have strong bootstrap proportions (via MP analysis). Hence, the ToL passes two of the three tests relevant to the acceptance of a vertical ToL. The third test can be assessed by examining the monophyly of major well-accepted groups such as the proteobacteria, the firmicutes, and other well-accepted classifications for bacteria. In nearly all cases, the ToL agrees with accepted classification. Lienau et al. suggest, therefore, that a ToL exists and can be constructed from genome-level information.

Can all genes be used in a concatenated analysis?

At the risk of putting the cart before the horse, we can now address whether or not we need to consider if genes undergoing large amounts of HGT should be restricted from use in phylogenetic analysis. The alternative to concatenating all the information into a single matrix and using all information for generating a ToL is to use what are called "core" genes to construct phylogenies. Core genes are very similar to targeted shallow analyses, but instead of obtaining genes opportunistically, only those genes that appear to be resilient to HGT are used to generate phylogenies. Such core genes have been discussed by many microbial phylogeneticists and are usually genes involved in the translational and transcriptional

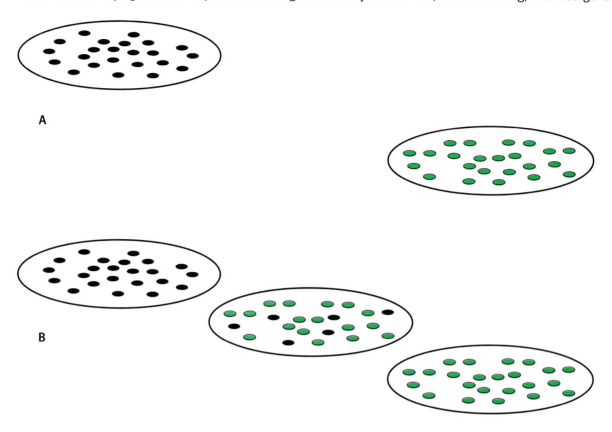

Figure 17.11 Surveying too few populations can lead to overdiagnosis. A: two hypothetical populations that appear to be diagnostically different. B: adding a third intermediate population with polymorphism can destroy the diagnosis.

different. However, if a third middle population of similar individuals is discovered and examined, then the two outside populations have to be aggregated together (Figure 17.11B).

Is there enough information in a single gene to do DNA barcoding?

A major problem that a lot of scientists saw with the early formulation of DNA barcoding was that a single gene contains only a finite amount of information. After all, the *COXI* gene is only a little over 1000 base pairs long. Is this enough information to provide diagnostics for 1.7 million species? Hebert pointed out that if only 15 polymorphic sites exist in *COXI* and each of the four bases can be substituted in these 15 sites, then there are 4^{15} = 1 billion possible combinations of bases for these 15 sites. This is about 50 times the information needed to barcode 1.7 million species. Since there are many more than 15 sites that vary over the region that the DNA barcoding community has chosen to use in *COXI*, this calculation is probably an underestimate of the number of potential "barcodes" that exist in animals. Another way to demonstrate the potential for a single gene to hold enough information for barcoding is to consider the situation in **Figure 17.12**. This diagram shows two populations (one above the line and one below) for a short stretch of DNA sequence. The shaded columns represent various kinds of pure diagnostics in these sequences for these two hypothetical populations. The dark green shading in the first and last positions shows two single-site diagnostics. The last position shows an interesting case where neither species is fixed, but because the polymorphism in the top population is different from the one in the bottom, the polymorphic state becomes diagnostic. The medium green shading shows a really interesting case where two sites can be combined to become diagnostics. Note that in this example for both sites, the bottom population has a G in it that is not in the top. These kinds of sites are called "private sites" because the G is private to the bottom population. These private sites can be combined so

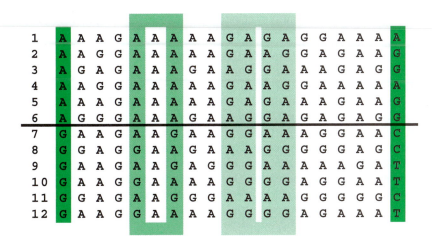

Figure 17.12 Two hypothetical populations undergoing population aggregation analysis. The horizontal line demarcates population 1 from population 2. The dark green columns indicate single pure diagnostic characters. The medium green column depicts a complex diagnostic pairing. The light green columns show a polymorphic combination of four positions that together are purely diagnostic. (Adapted from R. DeSalle, M.G. Egan, M. Siddall, *Philos. Trans. R. Soc. B* 360:1905–1916, 2005.)

that the top population is diagnosed by being AA and the bottom population by being AG or GA at these two sites. Finally even wholly polymorphic sites might be useful in detecting diagnostics. The light green shading shows four fully polymorphic sites. However, when all four are examined simultaneously, it is clear that the top populations can be diagnosed by having being GAGA, GAAG, AGAG, or AGGA and the bottom population for these four sites is diagnostic with GGAA, AAGG, AAAA, or GGGG. The potential for using these complex pure diagnostics has so far been untapped by DNA barcoding.

Potential new species are flagged by DNA barcoding

One operational aspect of classical taxonomy is the process of erecting a novel species existence hypothesis. Such hypotheses are tested by use of character-state data, and diagnostics are established from the character-state data. One of the really interesting aspects of DNA barcoding approaches is that they allow for very specific marking or flagging of entities for further tests of species status. This approach has been articulated with moths from Costa Rica. For example, in this approach a set of animal specimens from known species are sequenced for *COXI*. The barcodes are then used in a phylogenetic analysis with other specimens. The accepted taxonomy is then overlaid on the phylogenetic result, and if there is a good correspondence of the tree topology with the accepted taxonomy, then the barcodes are established as identifiers for the species in the analysis. If there are anomalous parts of the topology—for instance, if a previously described single species appears polyphyletic with respect to other known species or if two clades in a phylogenetic tree that represent individuals from the same species have unusually large genetic distances between them—then the anomalous entities can be "flagged" for future taxonomic work.

Metagenomics

Another aspect of comparative biology that has benefit from advances in high-throughput sequencing is the field of microbial environmental biology, which requires obtaining samples and identifying the microbes that live in these samples. Prior to the development of high-throughput approaches, environmental microbiology relied principally on culturing microbes from samples. However, since many of the microbial species that exist on the planet are difficult to culture, a large proportion of microbial diversity was being missed in these kinds of studies. The approach is now known as "metagenomics". If genomics, as we have defined it throughout this book, refers to the study and characterization of the single genomes of organisms, then the "meta" in metagenomics simply refers to a large mixture of genomes being analyzed. Because the environmental samples mentioned above contain many species, any DNA from the sample will contain the genomes of many species; hence it would be considered a metagenomic sample.

High-throughput DNA sequencing approaches provide the tools to identify the majority of species of microbe in a sample without culturing the individual species. The procedure is outlined in **Figure 17.13**. For the last several decades, microbiologists have used the 16S ribosomal RNA gene as a species identification and phylogenetic tool. The 16S rRNA gene is about 1800 base pairs long and codes for a small RNA that makes up part of the small ribosomal subunit. It was chosen as a marker for bacterial species because it is present in every bacterial species examined to date and its rate of evolution appears to be appropriate for making inferences about bacterial relationships. Over 900,000 bacterial sequences exist in the Ribosomal Database (RDB; see Web Feature for this chapter for a demonstration), and they serve as a major resource for identifying bacteria. Since the 16S rRNA is made up of stretches of highly conserved regions dispersed among rapidly evolving regions, it has become the standard tool for doing bacterial identification by DNA. The highly conserved regions serve as "anchor" regions for primers for PCR analysis, and the variable regions between these primers contain the information needed to identify a bacterial species or strain.

In analyzing microbial species, one of two approaches can be used, based upon the size of the fragment to be analyzed. First, a sample of water or soil is obtained and strained through a filter with pore sizes that retain the microorganisms on the filter. DNA is then isolated from these organisms, together with any components that are larger than the pore size of the filter, including eukaryotes. This process produces a pool of chromosomal DNA from all of the organisms that existed in the original water sample. The chromosomal DNA is then amplified with PCR by use of 16S rRNA primers set between 400 and 1400 base pairs apart, yielding a pool of PCR products for the 16S rRNA gene. To generate large fragments, the PCR pool is cloned into vectors that individualize the PCR products and make them available for analysis. In a typical environmental sample with the Sanger sequencing approach, millions of PCR products are cloned, but only 500–1000 are randomly selected for sequencing. Once the cloned PCR products have been sequenced, these sequences are compared to a database to determine which species are present. These results are tabulated, and an estimate of the diversity of the sample is obtained. One limitation with this approach is the small number of clones that can be sequenced per sample, since the number of sequences (500–1000) is a

Figure 17.13 Flow of an environmental sample study. One method applies amplification of the 16S rDNA gene using the polymerase chain reaction (PCR) and subsequent cloning of the PCR products followed by Sanger sequencing (left) and the other method utilizes PCR amplification of small fragments of the 16S rDNA and subsequent next generation sequencing (NGS; right).

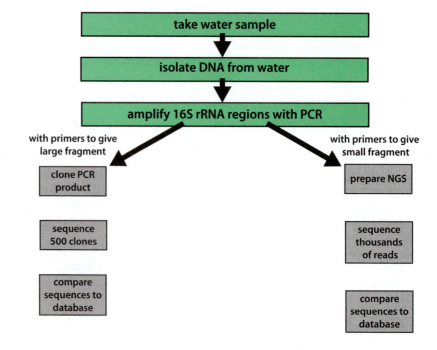

small proportion of the organisms that exist in a given sample. Consequently, new methods are available by which the completeness of a sample, as represented by the sequences obtained via the above procedure, are assessed (see Sidebar 17.3).

To produce small fragments, NGS technologies are used (right side of Figure 17.13). The steps in this method are similar to those described above, up to the PCR analysis. At this point, a primer pair is set between 60 and 200 bases apart. Short primers are used because the read length of NGS approaches technologies is much shorter than standard Sanger sequencing. Two highly variable regions of the rRNA gene, which are situated between highly conserved stretches, are used for NGS (right side of Figure 17.13). The PCR products are prepared for 454 sequencing and run on the 454 sequencer, where as many as tens of thousands of sequences per sample can be obtained. These sequences are then compared to the RDB to determine the species composition of the sample. Given the number of sequences generated, this approach provides a more robust estimation of the diversity of the community.

Upon completion of the sequencing, the resulting 16S rRNA sequences need to be identified. One way in which this identification can be made is to compare the query sequence to a reference sequence (sequences where the species of origin is known) in the database and achieve an exact match. Another method involves building a phylogenetic tree by use of both the query sequence and a set of reference sequences (see Chapter 6). The query sequence will be attached to the tree to species to which it is most closely related and can thus be identified this way. Fortunately, there are nearly 900,000 reference sequences in the database for bacteria, and these sequences have been organized by the Ribosomal Database Project (RDP) at Michigan State University. The student can explore the functions and resources of the RDP. In the context of metagenomics, we will look closely at one function, the "Classifier," in the Web Feature for this chapter.

Sidebar 17.3. Rarefaction curves in metagenomics.

The result of an environmental or human microbiome metagenomic study is a list of species that are present in the original sample. This list will give the frequency with which specific species are found. As part of the experiment, the number of sequences used to generate the list will also be available. With these data, the fraction of the total actual number of species in the sample can be assessed via a rarefaction curve. These curves are generated by plotting the number of species observed as a function of the number of sequences in the sample (**Figure 17.14**). Most rarefaction curves are steep in the initial parts of the curve near the origin of the plot. If a good sampling strategy has been used, the curve will "flatten" out. This is because fewer and fewer new species are being discovered per sequence sample obtained. An environmental sample can be interpreted by examining the rarefaction curve. For instance, the more extreme the slope of the curve, the bigger the contribution of how many sequences are obtained to a full understanding of the sample. In other words, an extreme slope of a rarefaction curve means that a large number of sequences will be needed to complete a survey of an environmental sample. Curves that have very gentle slopes represent samples that require little or no extra work to fully characterize them.

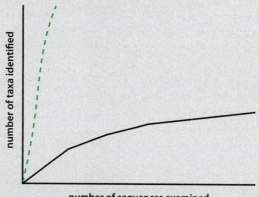

Figure 17.14 Range of rarefaction curves that can be obtained in metagenomic studies. The gently sloping solid line represents an experiment where the number of taxa in the sample will be easily determined. The dashed line represents an experiment where the rarefaction curve is extremely steep and estimation of the number of the species in the sample will be difficult.

Summary

- Sequencing approaches for biodiversity studies include shallow targeted sequencing, expressed sequence tag approaches, and whole genome approaches. Whole genome approaches have been used for yeast, flies, and bacteria.

- Bacterial phylogenomics poses specific problems: whether a tree of life really exists for bacteria, and whether all genes can be used in a concatenated analysis.

- When the number of taxa becomes unmanageable, various supertree approaches are used. Grafting breaks down a problem into a series of smaller problems. Matrix representation allows trees in different studies to be linked together in a supertree matrix that summarizes each subtree as a series of nodes. The divide-and-conquer approach is similar to grafting, but no a priori assumptions about relationships are made.

- Taxonomy and speciation studies both involve the study of species delimitation. DNA taxonomy is significantly different from classical taxonomy.

- Approaches to DNA barcoding may be character-based or distance-based. Calculations indicate that there is sufficient information in a single gene for DNA barcoding. Potential new species can be flagged by this technique.

- Metagenomics uses high-throughput DNA sequencing approaches to identify the majority of microbe species in a sample without culturing individual species.

Discussion Questions

1. Discuss the merits and potential drawbacks of using a supertree approach to construct a tree of life. Do the same for a supermatrix approach.

2. Discuss the potential differences observed in the metazoan tree of life among the studies presented in this chapter. Do the same for the plant studies presented in this chapter. What factors might be involved in generating the differences in topologies among these studies?

3. How might the large amount of missing data that will occur in EST-based phylogenetic studies affect the resolution of a tree? What effect might it have on the accuracy of a tree? How about the robustness associated with the nodes in the tree?

4. Discuss the statement "DNA barcoding does not require a species concept."

5. When a DNA barcode system is constructed for a group of organisms, there will inevitably be specimens that are involved in the analysis that aren't represented in the database. What should be done about these specimens?

Further Reading

Clark AG, Eisen MB, Smith DR et al. [Drosophila 12 Genomes Consortium] (2007) Evolution of genes and genomes on the *Drosophila* phylogeny. Nature 450, 203–218.

DeSalle R, Egan MG & Siddall M (2005) The unholy trinity: taxonomy, species delimitation and DNA barcoding. *Philos Trans R Soc London, B: Biol. Sci.* 360(1462), 1905–1916.

Doolittle WF & Bapteste E (2007) Pattern pluralism and the Tree of Life hypothesis. *Proc. Natl. Acad. Sci. U.S.A.* 104, 2043–2049.

Doyen JT & Slobodchikoff CN (1974) An operational approach to species classification. *Syst. Zool.* 23, 239–247.

Dunn CW, Hejnol A, Matus DQ et al. (2008) Broad phylogenomic sampling improves resolution of the animal tree of life. *Nature* 452, 745–749.

Gatesy J, DeSalle R, & Wahlberg N (2007) How many genes should a systematist sample? Conflicting insights from a phylogenomic matrix characterized by replicated incongruence. *Syst. Biol.* 56(2), 355–363.

Goldstein PZ & DeSalle R (2011) Integrating DNA barcode data and taxonomic practice: determination, discovery, and description. *BioEssays* 33(2), 135–147.

Goloboff PA, Catalano SA, Mirande JM et al. (2009) Phylogenetic analysis of 73,060 taxa corroborates major eukaryotic groups. *Cladistics* 25, 211–230.

Hebert PDN, Ratnasingham S & deWaard JR (2003) Barcoding animal life: cytochrome c oxidase subunit 1 divergences among closely related species. *Proc. R. Soc. Lond. B* 270, 596–599.

Lee EK, Cibrian-Jaramillo A, Kolokotronis SO et al. (2011) A functional phylogenomics view of the seed plants. *PLoS Genet.* 7, e1002411.

Lienau EK, DeSalle R, Rosenfeld JA, & Planet PJ (2006) Reciprocal illumination in the gene content tree of life. *Syst. Biol.* 55, 441–453.

Lienau EK, DeSalle R, Allard M et al. (2011) The mega-matrix tree of life: using genome-scale horizontal gene transfer and sequence evolution data as information about the vertical history of life. *Cladistics* 27, 417–427.

Marco D (ed) (2011) Metagenomics: Current Innovations and Future Trends. Caister Academic Press.

Philippe H, Derelle R, Lopez P et al. (2009) Phylogenomics revives traditional views on deep animal relationships. *Curr. Biol.* 19, 706–712

Ranwez V, Berry V, Criscuolo A et al. (2007) PhySIC: a veto supertree method with desirable properties. *Syst. Biol.* 56, 798–817.

Rokas A, Williams BL, King N & Carroll SB (2003) Genome-scale approaches to resolving incongruence in molecular phylogenies. *Nature* 425, 798–804.

Schierwater B, Eitel M, Jakob W et al. (2009) Concatenated analysis sheds light on early metazoan evolution and fuels a modern "urmetazoon" hypothesis. *PLoS Biol.* 7, e1000020.

Srivastava M et al. (2008) The *Trichoplax* genome and the nature of placozoans. *Nature* 454, 955–960.

Microarrays in Evolutionary Studies and Functional Phylogenomics

Functional genomics refers to studying the function of genes on a genome-wide scale. Prior to the generation of whole genome sequences, function was studied on a gene-by-gene or protein-by-protein basis. With the availability of whole genome sequences, functional studies have started to include larger numbers of genes, and a new dimension has been added. This new dimension involves the interaction of genes to produce phenotypes, anatomical trends, behavioral trends, and other kinds of evolutionary phenomena. One of the new workhorses of functional genomics is the microarray. This approach, described in Chapter 2, results in a large amount of information about the transcription of genes from a genome. Approaches have been designed to generate information about the kinds of genes that are active under different experimental conditions. In addition, microarrays can be used to examine the repertoire of genes expressed in different kinds of cells and tissues. Other cleverly designed experiments have examined the genes that are expressed in different developmental stages or life history stages such as queens, workers, and drones in bee populations. These experiments are interesting in an evolutionary context, because they are designed to examine different cells, tissues, developmental stages, and life history types that all emanate from the same starting genome of the target organism being examined. Yet most often radically different suites of genes are expressed in different tissues. Understanding how to manipulate the microarray data in these experiments and interpret the information on thousands of genes is an important aspect of understanding the function of genes. Whereas genomics and proteomics study the sum of genes and proteins, respectively, in an organism, functional genomics makes use of the vast amount of information known about transcription, translation, and protein–protein interactions as a basis of study. Only recently have functional genomicists incorporated phylogenetic methods into the study of function.

Why does phylogenetics have a place in functional genomics? Knowing the phylogeny of organisms when doing functional studies allows us to order evolutionary events and hence reconstruct the evolutionary history of the events involved in the overall function of genes and proteins. The scale of analysis in these studies is very different from most phylogenetic studies and requires that data not only be managed for the various taxa in a study but also for the various nodes in a phylogenetic tree that hypothesizes the relationships of the taxa. Manipulating this large amount of data results in large-scale inferences about the function of genes and proteins. Finally, we can look at functional phylogenomics from a level-based approach. That is, we can examine how transcription proceeds, how proteins are translated, and how proteins interact with each other. This approach leads to an understanding of how cells behave as a result of genomic influence. All of these can be viewed from a phylogenomic perspective.

Transcription-Based Approaches

Microarrays have become extremely important tools for detecting and quantifying the expression of genes in cells. The technique is based on the classic knowledge of DNA–DNA hybridization and the pioneering work of Edwin Southern.

As we discussed in Chapter 2, Southern realized in the 1970s that DNA could be affixed to a surface and used as a means to detect other nucleic acids, because of the natural tendency of single-stranded nucleic acids to anneal to each other if they are complementary. This technique became known as a Southern blot and has been extensively used in biology. Scientists next realized that they could slightly modify this approach to use it to detect and quantify RNA. As a twist on Southern's name, the technique for RNA is known as a northern blot. A northern blot allows a researcher to determine the amount of a particular messenger RNA (mRNA) that was found in cells. For instance, to compare the amount of tubulin mRNA in heart cells and lung cells, one would first prepare an RNA probe sequence based upon the known tubulin sequence and label it with a radioactive marker. The full complement of mRNA from heart cells would be run on an electrophoresis gel to separate it by size, and a second gel would be prepared for the total mRNA from lung cells. The size-separated mRNA would then be transferred ("blotted") from each gel to a nitrocellulose paper. The labeled tubulin probe would then be hybridized to each of the two nitrocellulose-bound mRNA samples. Excess of probe would be used, so that every mRNA molecule on the nitrocellulose would bind to a radioactively labeled probe. The radioactivity of each sample would then be quantified by use of a special film, and the sample with the greater amount of radiation would be the sample with the greater amount of tubulin mRNA. Since the level of mRNA can be roughly correlated to the amount of protein, this approach would allow you to determine whether there is more tubulin in heart cells or lung cells. While this approach was powerful, it was extremely limited and tedious. Each northern blot could only examine the expression level of one gene in one cell type. This limitation has been overcome by the much more powerful microarray and NGS technology.

Microarrays resemble high-throughput Southern and northern blots

A microarray can be thought of as an extremely high-throughput version of a Southern and/or northern blot. For a microarray experiment, first, sequences matching every known coding sequence from a species are bonded to a microscope slide. These can either be from cDNA sequences or, in the case of oligonucleotide arrays such as those produced by Affymetrix, the DNA sequence is printed onto the array by a process akin to that used by an inkjet printer. Once the array is prepared (or in most cases purchased from a company such as Affymetrix), the mRNA from the heart and lung cells is made into copy DNA (cDNA; see Chapter 2) and then labeled with a fluorescent dye. This labeled DNA is then hybridized to the microarray and the amount of fluorescence matching to each sequence on the chip is determined by use of a laser that excites the fluorescent dye. The two different types of microarrays that are generally used differ in their hybridization approach. cDNA microarrays are hybridized simultaneously with two differentially colored mRNA samples. For our example from above, the heart cell mRNA would be given one colored label while the lung cell mRNA would be given a different-colored label. These two labeled sets of mRNA would then be hybridized at the same time, and the difference in fluorescence for each of the colors would be determined. For an oligonucleotide microarray, each sample is labeled and hybridized individually, so the heart and lung cells would be labeled with the same dye, but each would be hybridized to a different array. While this difference matters for some of the primary image analysis, for our purposes it is not significant. The most important point is that either type of microarray will determine the expression level of thousands of genes at the same time. This is clearly a great advance over a northern blot, which could probe only one individual gene at a time. Currently, microarrays have been produced and are available for many different species.

Microarrays are used for class comparison, prediction, and discovery

Before we discuss how an array is analyzed for categorization, we need to discuss the role of arrays in modern diagnostic science. There are three goals that can be addressed by use of arrays. First, to determine which gene expression profiles differ between pre-ordained groups of cell lines or experimental treatments, a class comparison is performed. Second, microarray data are used to predict classification structure. In this approach, there are pre-described groups of data such as cell type or experimental condition, but instead of simply quantifying the differences, the microarray data are used to predict group inclusion. This approach is called class prediction. Finally, data without pre-described groups are analyzed to discover new groups. This approach is called class discovery.

Class comparison of two biological samples is an important task, and it was one of the first main uses of microarrays. By use of a microarray, a researcher was able to take two cultures and determine which genes had differing expression levels. For example, establishing which genes are expressed uniquely in cancer cell lines versus normal cell lines is an important endeavor. This approach uses a variety of statistical approaches that rely on two sample test statistics. The most commonly used is the two-sample t-statistic. If a test statistic is at least as extreme as the observed t-value at a predescribed confidence level, then the difference between the two samples is significant. Because this test statistic is simple and does not always address the realities of array data, other two-sample test statistics such as the paired t-test, Welch's test, signed rank tests, and one-way analysis of variance (ANOVA) are used. All of these tests attempt to impart some degree of statistical significance to the comparison of samples and are incorporated into standard array analysis programs and Websites.

Microarray data can be very useful for *class prediction*, where one wants to predict a disease state on the basis of expression patterns. In this approach, the data at hand are used to develop a set of criteria for inclusion into a group. Data from a number of cell lines or array experiment data points are grouped according to some a priori criterion, such as disease state or experimental parameter (high pH, low temperature, etc). Once the data are placed into groups, the researcher can determine what expression patterns are unique to the imposed groupings. These expression patterns are then established as markers or predictors for the disease state or experimental condition. The problem here is that this sounds a little circular or inductive, and it is. A model can be created to accommodate the process, but sometimes the model will overfit the data, and this will have consequences on the decision rules that are made for subsequent tests. This problem can be alleviated by splitting the raw data into two parts. The first part is used as a *training set* to select the decision rules. The second part of the initial data set is then used as a *validation data set*. Since inclusion of the second data set as members of the same group as the training data set is certain, then the accuracy and precision of the decision rules gleaned from the training set can be assessed.

An example of this technique was the experiment performed by Todd Golub and colleagues in 1999, where they used microarrays to diagnose patients as having either one of two types of leukemia, acute myeloid leukemia (AML) or acute lymphoblastic leukemia (ALL). Rather than requiring extensive histology or expert pathology, the diagnoses were made directly by looking at the gene expression pattern of the two types of cells. Since that time, this technique has been used for hundreds of disease comparisons, and there are some cases where it has become a standard clinical test.

In the *class discovery* approach, no predescribed groups are set. Instead, the data set is used to discover new groups. Clustering can be done by using treatment or the kind of cell line as the element to be clustered. Alternatively, the genes used in

Figure 18.1 Structure and nature of microarray data. A: schematized drawing of compiled array results. For simplicity, only two array intensities appear on this schematic array: intensity of 0.9 = gray and intensity of 0.1 = green. B: transformed intensities that correspond to the array in panel A. C: array intensity data are transformed into a distance or similarity matrix. The numbers in the matrix refer to positions in the matrix, indicating that for 10 experiments there are 45 potential pairwise comparisons. D: data in panels A and B are transformed into a character state matrix. This table is constructed by rounding the values in panel B, so that 0.9 rounds up to 1 and 0.1 rounds down to 0. These rounded data can now be treated in a character-based context. (Adapted with permission from P.J. Planet, R. DeSalle, M. Siddall et al., *Genome Res.* 11:1148–1155, 2001. Courtesy of Cold Spring Harbor Laboratory Press.)

the array experiment themselves can be the object of clustering. There are many ways to cluster microarray data, and these are described in detail in the next section. In order to develop a discussion of the clustering methods, though, we first need to detail how array data are transformed into a format suitable for use in class discovery approaches.

Microarray intensity data are transformed for use in dendrograms and other techniques

The general procedure used to transform array data for use in the construction of dendrograms and other approaches is shown in **Figure 18.1**. Array experiments for several cell types or experimental conditions are performed, and the data are normalized across arrays. This process results in a matrix of spot intensities for each cell line in the experiment. Next, a metric of distance for each cell line to all other cell lines is calculated via a mathematical approach. The most commonly used is Euclidean distance, which works best when the data are flatly distributed in space. This process results in a half-matrix with dimensions corresponding to the number of cell lines being compared (Figure 18.1). At this point, any number of phenetic tree-building approaches can be used to generate a clustering diagram from the cell line similarities.

For class discovery, the normalized expression values are clustered into groups. One of the most popular clustering methods is the *K*-means algorithm, which finds

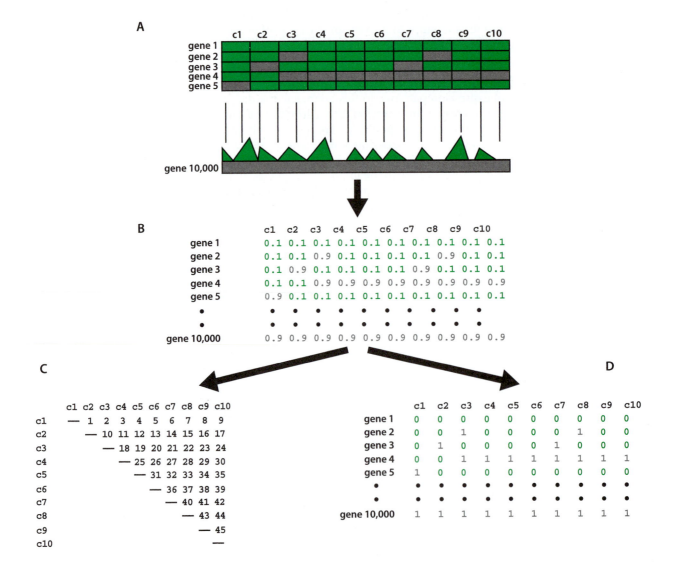

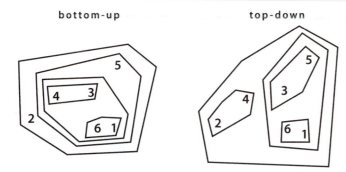

Figure 18.2 Different results from bottom-up versus top-down approaches to clustering. Bottom-up methods successively join nearest observations (least dissimilar) to other observations (or groups of observations). Top-down methods successively divide up observations (or groups of observations) into subgroups. (Adapted with permission from H. Chipman and R. Tibshirani, *Biostatistics* 7:287–301, 2006.)

K clusters to minimize the squared Euclidean distance between each observed distance and its respective mean cluster distance. This algorithm alternates iteratively between two steps: (1) find the observations with the smallest distance to the center of the cluster and assign the observations in that cluster, and (2) once a cluster has been assigned, update each cluster center so that new distances are computed, and repeat step 1. Because the starting point in this procedure affects the final clustering of observations, it is recommended that the clustering be performed with several starting points, much like when searching for trees in tree space. There are two approaches to clustering in this context. Bottom-up methods successively join nearest observations (least dissimilar) to other observations (or groups of observations). Top-down methods successively divide up observations (or groups of observations) into subgroups (**Figure 18.2**). Different results will be attained by the two different approaches, and it appears that top-down methods tend to give smaller numbers of clusters while bottom-up methods result in larger numbers of clusters. For this reason, hybrid methods have also been developed to treat dissimilarity data for microarrays. The most common algorithms for hierarchical clustering are single linkage clustering, complete linkage clustering, average linkage clustering, average group linkage, and Ward's linkage.

An example of microarray data is shown in **Figure 18.3**. This figure is called a heat map, because the darker a particular data point is the higher its value is. For details on how to make heat maps, see the Web Feature for this chapter. In this hypothetical example, three experimental treatments (separated by white lines) have been performed on several cell lines. The data for each cell line examined is displayed in columns in the figure. This figure shows expression data for about 50 genes. The darker the spots in the columns are, the higher the level at which the corresponding gene is expressed. Qualitatively, it can be inferred from this figure that, for these 50 or so genes, treatment C induces gene expression at a higher level than the other two treatments because the columns are in general darker under treatment C. As we mention above, the data in this figure can be displayed as a dendrogram or clustering diagram.

Another interesting approach to analyzing array data concerns transformation of the raw data into discrete state data, as presented in Figure 18.1D. A discrete character state is one that has a fixed number of possibilities, rather than being a continuum of values. The simplest discrete characters are either present or absent (in our example, expressing or not expressing). In the example in Figure 18.1D, we have simplified the approach by scoring the genes as either being "on" (1) or "off" (0). Note that instead of compressing the initial intensity data into a dissimilarity metric for pairs of entries, each gene retains information for each cell line or experimental condition. The matrix can then be analyzed by parsimony or Bayesian approaches where parsimony is assumed as a model. In addition to the simple scoring of the initial intensity data as present or absent (1 or 0), other scoring methods for array data are possible (**Table 18.1**). There are a number of these approaches that are called "binning;" they attempt to establish breaks in the data where bin boundaries can be placed. In the first column of Table 18.1, note

treatments

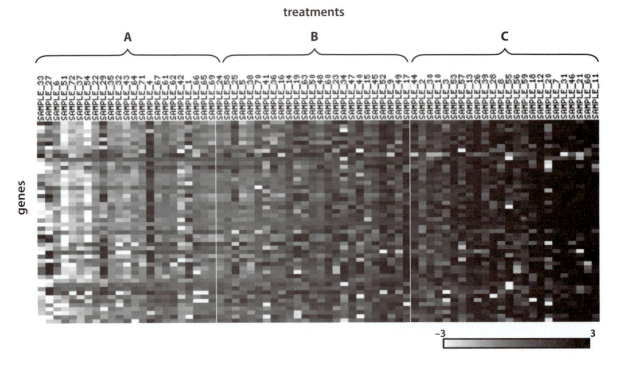

Figure 18.3 Construction of dendrograms to represent the results of a microarray experiment. Hypothetical example of a microarray with several cell lines (in columns) and three treatments (A, B, and C, separated by white lines). Expression profiles for genes (rows) are represented by the heat map. The darker the entry, the higher the level of gene expression for the gene and corresponding cell line. The scale bar at the bottom shows a relative scale of expression from −3 to +3.

that the "intensity" data are continuous from 1.0 to 2.0. One way to score these is to round the intensity values. When this is done, values 1.1, 1.2, 1.3, and 1.4 are rounded down to 1 and values 1.5, 1.6, 1.7, 1.8, 1.9, and 2.0 are rounded up to 2. If we simply use these rounded scores, we have a two-state character. However, there are some obvious drawbacks to this approach. First the two values that are

Table 18.1 Examples of binning microarray data.

Raw	Rounded	Binning 1[a]	Binning 2[b]	Binning 3[c]
1.1	1	0	0	0
1.2	1	0	0	0
1.3	1	0	?	0
1.4	1	0	?	?
1.5	2	?	?	1
1.6	2	?	?	1
1.7	2	1	?	1
1.8	2	1	?	?
1.9	2	1	1	2
2.0	2	1	1	2

Hypothetical raw data are shown in the first column and the results of rounding are shown in the second column. Three kinds of binning discussed in the text are presented in the next three columns. Question marks (?) represent ambiguous scoring, implemented as discussed.
[a]Binning 1: 1–1.4 = 0, 1.7–2.0 = 1. [b]Binning 2: 1–1.2 = 0, 1.9–2.0 = 1. [c]Binning 3: 1–1.3 = 0, 1.5–1.7 = 1, 1.9–2.1 = 2.

Source: Adapted with permission from P.J. Planet, R. DeSalle, M. Siddall M et al., *Genome Res.* 11:1148-1155, 2001. Courtesy of Cold Spring Harbor Laboratory Press.

most distant from each other with respect to initial values (1.5 and 2.0) are scored the same, and two of the values that are closest to each other (1.4 and 1.5) are scored differently. Binning attempts to correct for these problems when array data are used. One way to bin is to establish breaks in the data based on statistical distributions and to score intermediate values as missing. Because these values could be placed either in the bin above or in the bin below (with respect to value), this approach seems reasonable, as any value that is ambiguous should be scored as such, and missing (scored as a "?") is a good way to do this (see columns 2–4 in Table 18.1). With these methods of binning and converting raw array data to matrices, any of the phylogenetic tree-building methods can be used to construct a tree to depict relationships.

Next-generation approaches are applied to transcription analysis

Next-generation sequencing, using massively parallelized sequencing approaches, promises to augment the array approach. One approach that is particularly promising is the serial analysis of gene expression (SAGE). In this approach, messenger RNA is isolated from a target cell or tissue and cDNA is prepared. The cDNA is then sheared to a specific length and prepped so that only 3′ ends of the copied mRNA are present. This DNA is then used as a template for polymerase chain reaction (PCR) that extends pairs of short single-stranded tagged primers, producing a single-stranded library of all transcripts in the starting cells or tissue. All of the amplification is accomplished on the surface of a glass flow cell, and then the DNA is sequenced via the Illumina approach. The end product is millions of 36-base-pair reads of the 3′ end of transcripts in the target cell. These million reads then are mined to determine which genes are expressed in the target cell. The bioinformatics step is perhaps the most difficult but involves the many techniques we have discussed already. This approach will detect very rarely expressed genes and will also be useful for determining the quantity of transcript made in a cell.

There are variations of this approach, such as the ChIP-chip and ChIP-seq approaches, that allow for examination of transcription factors. ChIP-chip stands for chromatin immunoprecipitation (ChIP) plus microarray technology (chip). ChIP-seq stands for chromatin immunoprecipitation plus sequencing technology (seq). The goal of these approaches is to determine which transcription factors and other proteins interact with DNA involved in gene expression, such as histones. A biological question is posed: for example, do transcription factors bind to different sites in cancer cells than in normal cells? Two cell preparations are then prepared: one for the cancer cells and one for the normal cells. Next, the cells or tissues are treated with formaldehyde to cross-link the DNA and any proteins that are bound to it. This protein–DNA complex is then sheared to produce small fragments. If the biological question concerns, say, the role of transforming growth factor-β (TGF-β) in cancer cells, then an antibody to TGF-β is used to precipitate out all of the small fragments with that protein bound to them. The precipitate will contain all of the regions of the genome where TGF-β was bound before the fixation step with formaldehyde. The cross-linkages are reversed and the DNA that had TGF-β bound to it is recovered. In ChIP-chip methods, the DNA is used as probe on a microarray chip system that tells the experimenter which genes are interacting with TGF-β. In the ChIP-seq approach, the small DNA fragments are sequenced by a high-throughput method such as Illumina. Such experiments result in millions of 36-base-pair reads that had TGF-β attached. Bioinformatics methods are used to determine the genomic locations that are the source of the reads. These data are computationally intense to analyze. Several programs have been developed to deal with the computational problems involved in using these approaches (see Table 7.1).

Microarrays are useful in evolutionary and phylogenomic studies

One of the earliest studies to use the microarray approach in evolutionary studies examined the difference in transcription profiles of young workers and queens in bee colonies. The young stage is important because young bees have the potential to differentiate into workers and queens. The study characterized gene expression in these organisms by an array approach and found the expression patterns of the three categories of honeybees. Another honeybee microarray study demonstrated gene expression differences in individual honeybee brains as a product of behavior. In a carefully controlled experiment, the researchers examined brains of "young" nurse (YN) bees and compared them to the brains of "old" forager (OF) bees. They examined a total of 60 brains from these two experimental classes and could show that 39% of the 5500 genes in their microarray showed expression-level differences between the YN and OF bees. The researchers of this study showed that only a small number of gene expression profiles were needed to predict with 100% certainty the behavioral state (YN or OF) of the bees.

One recent study in plants examined the expression profile differences of weedy and wild populations of the sunflower *Helianthus annuus*. This is an interesting question because the weedy populations have adapted to novel ecological niches, and any gene expression differences might be involved in the adaptation. Researchers examined both wild and weedy populations from Kansas, weedy from Indiana, weedy from California, and weedy from Utah. Their results indicate that four genes can discriminate the Kansas wild population from the weedy sunflowers from all locations. They also examined a Utah wild population, and found that seven genes are differentially expressed between that wild population and the weedy populations.

Primate brains have also been examined by the microarray approach. One of the most interesting questions in all of comparative biology is how the human brain evolved. The interesting result in these studies concerns the up-regulation of genes in human brains versus other primates (**Figure 18.4**). Three studies were conducted and the percentage of genes with increased expression in the cerebral cortex (the most "human" part of the brain) was determined. In all three studies, there are significant proportions of overexpressed genes in the human cerebral cortex relative to the chimpanzee cortex.

An example of next-generation sequencing techniques can be found again in Hymenoptera, this time in wasps. Massively parallel sequencing using 454 approaches of wasp (*Polistes metricus*) brains from four different reproductive

Figure 18.4 Problems that occur when arrays are used across species boundaries. A: a single nucleotide difference between human and chimpanzee is shown, with the effect of the single nucleotide polymorphism on hybridization efficiency. Small squares to the right of the sequences show the patterns of spot intensity produced by using different probes. In these squares the intensity for a perfect match (PM) and a mismatch (MM) are shown for the different mRNAs hybridized to a microarray. B: results of three microarray studies that examined expression differences between chimpanzee and human brain tissues. The *y*-axis shows relative expression levels and the human and chimpanzee results are depicted in different colors. (Adapted with permission from T.M. Preuss, M. Cáceres, M.C. Oldham, & D.H. Geschwind, *Nat. Rev. Genet.*.5:850–860, 2004. Courtesy of Nature Publishing Group.)

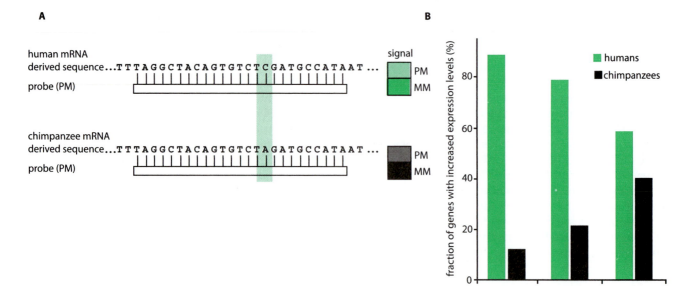

and provisioning (procuring food for workers and gyne) types of wasp was done. In this species, females that establish new colonies are called foundresses. Foundresses are both reproductive and provide maternal care. After a first generation is reared, the females develop into workers and successful foundresses become queens. At this point the workers provide maternal care only and the queen is reproductive only. A fourth category called gynes develop late; they are basically freeloaders because they neither provide care nor are reproductive. However, gynes survive over the winter and become the next generation's foundresses. In this study nearly 400,000 sequence reads were obtained for these four types of *Polistes*. When the representation of genes being expressed in the brains of these wasps is examined in detail, it is found that expression patterns from the next-generation sequencing approach reveal differences among the four types of wasp. A linear discriminate analysis was accomplished that simply shows correlation of the expression of specific genes with each other for the four kinds of wasps. The hierarchical clustering analysis (**Figure 18.5**) indicates the gyne is the most divergent from the other three categories, with very distinctive gene expression patterns. Workers and foundresses have very similar expression patterns, except for a group of four genes where the foundresses and workers are very different. Queens lie intermediate between the other types (Figure 18.5). The differentially expressed genes that are significant in this study appear to be insulin-related genes. This result is significant because it implies that the evolution of this system (eusociality) involved nutritional pathways and reproductive pathways.

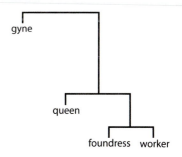

Figure 18.5 Next-generation sequencing technology approach reveals differences among four types of wasp. The dendrogram was generated from analyzing 454 transcriptome data for the four kinds of wasps. (Adapted with permission from A.L. Toth, K. Varala, T.C. Newman et al., *Science* 318:441–444, 2007. Courtesy of the American Association for the Advancement of Science.)

Corrections are needed in cross-species microarray studies

Cross-species questions can be approached by use of microarrays. However, recent work raises many technical questions about cross-array and cross-species comparisons. Two problems occur with respect to comparisons across arrays. First, the conditions for arrays can vary from experiment to experiment. Some of this problem can be alleviated by use of standards. The second problem involves the use of, say, a human array and challenging it with chimpanzee cDNA. Because sequence divergence has occurred between humans and chimpanzees, the annealing of chimp cDNA to the array will not reflect accurate expression patterns. Corrections that use mismatch oligonucleotides on microarrays can be implemented to make comparisons across species valid (Figure 18.4A).

Protein–Protein Interactions

The next level of analysis concerns understanding the interaction of proteins with each other. We have already discussed yeast two-hybrid assays for establishing databases for protein interactions. In this section, we examine how these data are analyzed to give protein interaction networks.

Various approaches are used to examine protein–protein interactions

Perhaps the simplest way to analyze protein interactions is through the use of graph theory. In this approach, individual proteins are represented as graph vertices that are connected by edges validated by experimental evidence (such as two-hybrid data) as interacting. In other words, each protein is represented as a point in space, connected by lines to all other proteins with which it interacts. This approach results in diagrams that show all pairwise interactions of the proteins examined (**Figure 18.6**). Some of the more interesting results from these kinds of studies show asymmetries with respect to highly connected nodes versus rarely connected nodes. One study that used this approach of constructing interaction diagrams demonstrated that proteins at highly connected nodes, when altered, are more prone to result in lethality than proteins that are rarely connected, suggesting that highly connected nodes might be more highly interdependent on

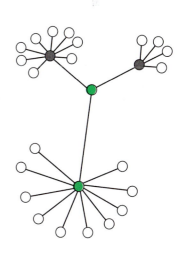

Figure 18.6 Display of hypothetical pairwise protein interactions. Each sphere represents a protein that has been tested for interaction with the other proteins by two-hybrid assays. Two proteins (green circles) have many interactions with other proteins and are considered hubs of interaction. Other proteins either interact singly with the hub proteins (white circles) or are smaller hubs themselves (gray circles).

each other or phylogenetically correlated (that is, they have evolved in concert, much like hosts and parasites do). Nodes that have few connections might be new pathways or expendable. Taking the approach one step further, proteins with many connections might be disease genes, and those with few connections probably are not. The Web Feature for this chapter demonstrates how these interaction profiles can be used to construct protein interaction networks.

Mendelian phenotypes in humans and model organisms are studied by Web-based approaches

Recently the research communities that study humans, worms, and flies have banded together to start ENCODE (the ENCyclopedia Of DNA Elements; http://www.genome.gov/ENCODE/). ENCODE was initiated by the National Human Genome Research Institute (NHGRI) in 2003, "with the goal of defining the functional elements in the human genome." Later, the model organism ENCODE project began in 2007 and is called modENCODE (where "mod" stands for model; http://www.modencode.org/). The ENCODE project Websites are a treasure trove of information for understanding "phenotype" at the molecular control level.

While the end goal of much genomics work is to understand phenotype, this is a very difficult problem. There are a few Web-based approaches to this problem. One is the Online Mendelian Inheritance in Man database (OMIM; http://www.ncbi.nlm.nih.gov/omim/). This database has existed since the 1960s and contains a collection of genetic disorders in man. A keyword for a disease or genetic disorder such as "hemophilia" can be typed into the query box on the Website, and the database can be searched for information on this disorder and others with similar names. Accession numbers and profiles for proteins associated with the disease can be obtained, and the list of items that can be displayed for the disease can also be obtained via this Website. Alternatively, the entire report for the disorder can be downloaded. It is best to explore this site by reading through the very detailed descriptions for OMIM. Other Websites that explore phenotype are very specific to model organisms (Table 18.2).

Phylogenetic Approaches to Understanding Function

Now that we have reviewed some of the classical comparative ways of examining functional data, we can examine how phylogenomic approaches are used to understand function. The real utility of phylogeny in understanding function is in the structure of the phylogenetic tree that can guide the behavior of genes and proteins through time. Because the tree will imply a sequence of divergence of species and also implies common ancestors, using trees to infer function can offer a new dimension in functional studies: the dimension of time that involves common ancestry.

Table 18.2. Organismal databases.

Database	URL
FlyBase	http://flybase.org/
WormBase	http://www.wormbase.org/
Saccharomyces Genome Data Base (SGB)	http://www.yeastgenome.org/
Mouse Genome Informatics (MGI)	http://www.informatics.jax.org/
The Arabidopsis Information Resource (TAIR)	http://www.arabidopsis.org/index.jsp

Phylogenomic gene partitioning is used to explore function

The potential to uncover insights into gene function by simultaneous character analysis of large multigene-to-genome phylogenies has been poorly explored. Characters from different data sets are traditionally used to produce a phylogeny by total evidence approaches, conditional combination of data sets, and taxonomic congruence. Gatesy et al. point out that there is little consensus on how conflict among data sets is quantified in traditional and even more recent methods. They propose a set of phylogenetic metrics that measure the congruence of gene partitions and individual characters in a phylogenetic analysis, focusing on variations of methods to assess support or conflict for a particular node. This approach is related to, but in many ways very different from the approaches we described that detect branch specific differences in natural selection that we discussed in Chapter 14.

Congruence measures of character evolution such as consistency, degree of support, and hidden support as described by Gatesy et al. are useful in mining genomes for patterns of protein function. In this discussion we are mostly concerned with variants of Bremer support and their use in assessment of the overall contribution (positive, negative, or neutral) of a particular gene to the various nodes or branches in a phylogenetic hypothesis.

The review by Gatesy et al. came at a time when systematists were using at most 10 gene partitions and perhaps a single morphological partition in their analyses. With the onslaught of genome-level sequencing and large EST studies in recent years, the number of gene partitions and ways of partitioning phylogenetic information has exploded. Measures of branch support can be used to identify proteins and characters that may have functional significance in the evolution of organisms. Functional significance in this context is based on the degree of support each gene gives to the concatenated hypothesis obtained from phylogenomic trees constructed from large genome studies. Specifically, we will examine the role of support and function using the *Drosophila* 12 Genomes database and a large plant EST database that have recently been used in the literature.

Individual tree statistics such as the total number of characters, the number of phylogenetically informative characters (PI), the consistency index (CI), retention index (RI), rescaled consistency index (RCI), and variations of traditional Bremer support (BS) are easily computed (see Chapter 9 and Sidebar 18.1). The latter measures (BS variants) are the most important measures of tree and branch support for the purposes of our discussion. They measure the stability of a group (clade) by quantifying the difference in character steps (tree length) between a tree containing a group of interest and a similar tree where this group is absent. High positive BS values would then reflect the stability or robustness of the group in question. Modified elaborations of Bremer support, such as partitioned branch support (PBS) and partitioned hidden branch support (PHBS), apply Bremer support metrics to trees constructed from combining data from various sources (for example, morphological and DNA; mitochondrial and nuclear DNA; or genes/proteins from different functional categories), whereby the contribution of particular/individual data sets (partitions) can be evaluated to measure the stability of relationships in the context of simultaneous analysis of concatenated data sets. In the case of phylogenomics, the partitions are the various genes in the genome and higher partitions such as gene ontology (GO) categories or other classification of genes.

By definition, for a particular combined data set, a particular node, and a particular data partition, PBS is the minimum number of character steps for that partition on the shortest topologies for the combined data set that do not contain that node, minus the minimum number of character steps for that partition on the shortest topologies for the combined data set that do contain that node. PHBS is the difference between PBS and BS for that data partition. Values for these metrics

Sidebar 18.1. Calculating Bremer support and related metrics.

Bremer support. Bremer support tells the researcher how many characters support a particular node in a parsimony tree. There are several ways to look at Bremer support. One is to simply calculate the total support a data set has for a parsimony-based hypothesis. If multiple genes or partitions exist in a data set, another way to look at Bremer support is to calculate the support each partition lends to the concatenated tree. The former calculation is called Bremer or branch support (BS), and the latter calculation is called partitioned Bremer or branch support (PBS).

The hypothetical DNA sequence matrix in **Figure 18.7** is used to demonstrate calculations for BS and PBS. In this matrix there is information for three "genes" or partitions (x, y, and z). There are three ingroup taxa (a, b, and c) with an outgroup (o), and DNA characters A and T. The three possible trees for the three ingroup taxa are shown. Calculating BS requires that one consider trees *without* certain nodes, which we will call here "antipartition" trees. In the example:

- Partition x supports the tree on the left, so it is the partition (PT) tree for x.
- Partition y supports the tree in the middle, so it is the PT tree for y.
- Partition z supports the tree on the right, so it is the PT tree for z.
- The middle and right trees are at odds with partition x, so they are anti(PT) trees for x.
- The left and right trees are at odds with partition y, so they are anti(PT) trees for y.
- The middle and left trees are at odds with partition z, so they are anti(PT) trees for z.

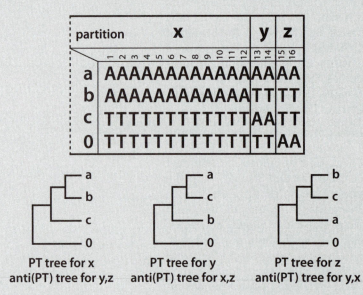

Figure 18.7 Calculating Bremer support. A hypothetical matrix is shown at the top. In this matrix there are three partitions (x, y, and z). There are three ingroup taxa (a, b, and c) with an outgroup (o), and DNA characters A and T. The three possible trees for the three ingroup taxa are shown. Partition (PT) and antipartition (anti(PT)) trees are described in the text. At the bottom, the number of steps for PT and anti(PT) trees for each partition is given.

partition	tree length PTx	tree length PTy	tree length PTz	length anti(PT)
x	12	24	24	(24 + 24)/2 = 24
y	4	2	4	(4 + 4)/2 = 4
z	4	4	2	(4 + 4)/2 = 4
x + y + z	(12 + 4 + 4) = 20	(24 + 2 + 4) = 30	(24 + 4 + 2) = 30	

These names are summarized in the middle panel of Figure 18.7. Tree lengths for the different partitions on their PT and anti(PT) trees are shown in the bottom panel. Bremer support (BS) for each partition for the single relevant node

Sidebar 18.1. Calculating Bremer support and related metrics. (continued)

in the tree is defined as steps for the partition on the anti(PT) tree minus steps for the partition on the PT tree. If more than one tree exists, then the steps from the multiple trees are averaged. The following shows the calculation of BS for each partition.

Partition	Anti(PT)	PT	BS on partition's best tree
x	(24 + 24)/2 = 24	12	24 – 12 = +12
y	(4 + 4)/2 = 4	2	4 – 2 = +2
z	(4 + 4)/2 = 4	2	4 – 2 = +2

For partition x, the number of steps on its best tree (the PT tree) is 12. The number of steps for partition x in anti(PT) trees has to be calculated as the average for the two anti(PT) trees: (24 + 24)/2 = 24. The BS for this partition for the relevant node is then the difference of the two: 24 − 12 = +12, which is the BS for partition x on its best tree. Similar calculations are made for the other two partitions, leading to BS = +2 each for partitions y and z as support on their best trees.

Now that we have the BS values for each partition on their best trees, we can determine the overall BS for the best tree of the combined or concatenated matrix (CM). The total BS for the CM tree is given by the difference between support for the anti-CM trees minus support for the CM tree. These calculations are shown below, and the BS for the CM tree is 10.

Anti(CM)	CM	Anti(CM) – CM = BS
24 + 3 + 3 = 30	12 + 4 + 4 = 20	30 – 20 = +10

Partitioned Bremer support. We can also calculate the support that exists for each partition on the combined or concatenated matrix (CM matrix) best tree. To do this, we need to know which tree is the best tree for the concatenated data set. It should be obvious that the best tree for partition x is the best tree for the concatenated matrix. So the partition x tree becomes the CM tree, and the trees for partitions y and z become anti-CM trees.

The hypothetical matrix in **Figure 18.8** is used to demonstrate calculations for partitioned Bremer support (PBS). In the matrix there are three partitions (x, y, and z). There are three ingroup taxa (a, b, and c) with an outgroup (o), and DNA characters A and T. The three possible trees for the three ingroup taxa are shown. Calculating PBS also requires that one consider trees *without* certain nodes, which we will call here "anti" trees. The concatenated matrix trees for the combined data set are shown in the middle panel of Figure 18.8. Since the tree on the right is the best tree for the combined matrix, it is called the CM tree, and the other two trees are called anti(CM) trees. PBS is calculated as steps for the partition on the anti(CM) tree minus steps for the partition on the CM tree. The number of steps is averaged if more than one tree exists. The following shows these calculations for each of the three partitions.

Partition	Anti(CM)	CM	PBS
x	(24 + 24)/2 = 24	12	24 – 12 = +12
y	(4 + 2)/2 = 3	4	3 – 4 = −1
z	(4 + 2)/2 = 3	4	3 – 4 = −1

These results for this hypothetical data set indicate that partition x imparts all of the positive support for the node in the tree, and partitions y and z do not support the node.

Hidden support measures. Now that we have information for all three partitions in our hypothetical data set, we can start to quantify the impact of combining data. To do this, we use hidden support measures. The hidden Bremer support (HBS) for a data set for a tree is given by the difference between BS for that node in the combined analysis and the sum of BS values for that node from each data partition.

Sidebar 18.1. Calculating Bremer support and related metrics. (continued)

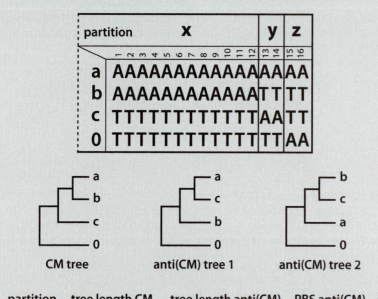

Figure 18.8 Calculating partitioned Bremer support. A hypothetical matrix is shown at the top. In this matrix there are three partitions (x, y, and z). There are three ingroup taxa (a, b, and c) with an outgroup (o), and DNA characters A and T. The three possible trees for the three ingroup taxa are shown. Concatenated matrix (CM) and anticoncatenated matrix (anti(CM)) trees are described in the text. At the bottom, the number of steps for each tree for the partitions is given.

partition	tree length CM	tree length anti(CM)	PBS anti(CM) – CM
x	12	(24 + 24)/2 = 24	24 - 12 = 12
y	4	(4 + 2)/2 = 3	3 - 4 = -1
z	4	(4 + 2)/2 = 3	3 - 4 = -1

For the examples in Figures 18.7 and 18.8, the BS for the node in the combined data set is +10. The sum of BS values for the node from each partition is 12 + 2 + 2 = +16. HBS is therefore 10 − 16 = −6, indicating that there is negative hidden support of −6 for this node given the partitions. Combining data in this case does not reveal hidden support, which makes sense because there is only a single node in the analysis and there are four characters that conflict with the combined analysis tree. We can also calculate the hidden support for each partition (PHBS). This is given by the difference between PBS for the partition at the node for the combined data set and the BS value for that node for that data partition. For partition x, PBS for the node in the tree is +12 and BS for the node is +12, so PHBS is 12 − 12 = 0. This makes sense because all the support for the tree comes from this partition and none of it is hidden. For partition y, PBS is −1 and BS is 2. PHBS for partition y is therefore (−1 − 2) = −3. Likewise, for partition z, PHBS is (−1 − 2) = −3. This means that, even though these two partitions only conflict at two characters, the hidden support is actually a −3. Note that the sum of the three PHBS for the three partitions is 0 + (−3) + (−3) = −6, which is simply the HBS for the entire data set.

can be positive, zero, or negative, and the values indicate the direction of support for the overall concatenated hypothesis: positive lends support, zero is neutral, and negative gives conflicting support. For the purposes of this chapter, we will discuss only the impact of positive values for these branch support measures. The calculations for BS, PBS, HBS, and PHBS, as described in Sidebar 18.1, are summarized in **Figure 18.9**.

A gene presence/absence matrix was employed to compare 12 *Drosophila* genomes

The *Drosophila* 12 Genomes Consortium released the genomes of 12 species of flies in 2007. This data set has turned out to be one of the most important whole genome data sets. The well-accepted phylogenetic hypothesis, both from current

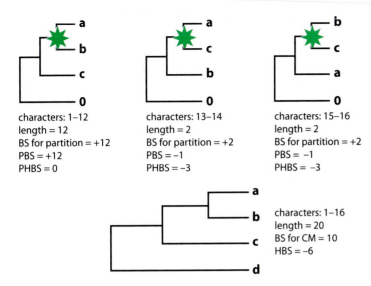

Figure 18.9 Summary of calculated support results. Hypothetical matrices in Figures 18.7 and 18.8 were used as an example, and the calculations are detailed in Sidebar 18.1. The asterisk indicates the node in the trees for which the support values are calculated. BS, Bremer support; PBS, partitioned Bremer support; HBS, hidden Bremer support; PHBS, partitioned hidden Bremer support; CM, concatenated matrix.

knowledge of the flies in this group and from the best tree obtained by any phylogenetic approach, is illustrated in **Figure 18.10**. The tree is extremely robust, with traditional bootstrap measures at 100% and Bayes probabilities at 1.0 for every node in the tree. This data set was analyzed by a gene presence/absence study in order to gain further understanding of the relationships between the species. The goal was, first, to generate a phylogeny based on the presence and absence of genes for this group of flies, and second, to see if some categories of genes were overrepresented at nodes, giving strong support to those nodes. Not surprisingly, the presence/absence matrix, made up of 0s and 1s for over 14,000 genes, generates the well corroborated tree in Figure 18.10 with very high support at all nodes in the tree. The next step was to use gene ontology (GO) categories to group genes into larger partitions. This is easily accomplished by use of the GO Website we discussed in Chapter 10. Each of the single GO categories can be partitioned:

```
#NEXUS
Begin data;
        Dimensions ntax=12 nchar=14908;
        Format datatype=protein symbols="01" gap=- interleave;
        Matrix
Drosophila_grimshawi        1110100111111111111001110111010111111110110111110 ...ᵃ
Drosophila_ananassae        1110110111111111111111111101110111111111110111111100 ...ᵃ
Drosophila_erecta           1111111111111111111111111111111111111111111111111111 ...ᵃ
Drosophila_melanogaster     1111111111111111111111111111111111111111111111111111 ...ᵃ
Drosophila_mojavensis       1110100101111111111001101011101011111111101101111110 ...ᵃ
Drosophila_persimilis       0110100101111111010101111101110001010101101101111110 ...ᵃ
Drosophila_pseudoobscura    0000100110111111010101110100011000110111110000111110 ...ᵃ
Drosophila_sechellia        1111111111111111111111111110111111111111111111111111 ...ᵃ
Drosophila_simulans         0111110111111111111111111111111111100111111111111111 ...ᵃ
Drosophila_virilis          1110100111111111110011101110101111111110110111110 ...ᵃ
Drosophila_willistoni       1110100111111110110011101110011111111110110111110 ...ᵃ
Drosophila_yakuba           1111111101111111111111111110111111111111111111111111 ...ᵃ
ᵃAnd on for the number of orthologs in data set
;
end;
```

```
begin sets;
charset GO2=4149;
charset GO3=631 141;
charset GO9=7314;
charset GO10=6879 9186;
charset GO12=2211 7881;
charset GO14=2466 3023 2413;
charset GO15=176;
charset GO22=01733 2222 2983 6486 7462 4463...
   (etc for the 210 columns that fit this GO term);
charset GO26=6916;
charset GO30=5689 5961 8240 1418 339 1113;
charset GO38=6320;
charset GO39=9371 9372;
etc. for 4000 GO categories.
```

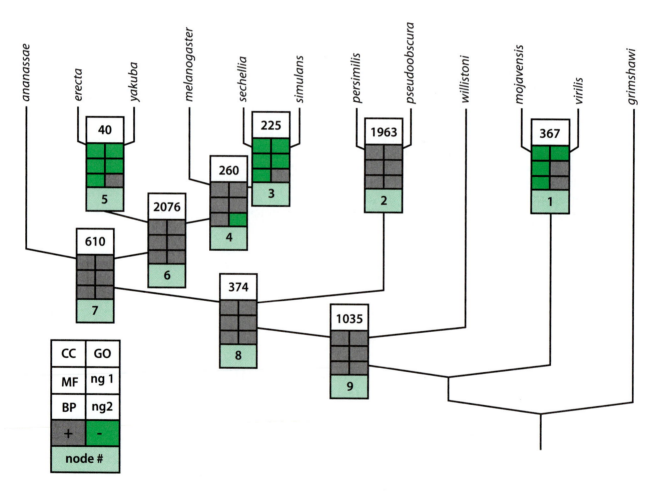

Figure 18.10 Phylogenetic hypothesis for genomes of 12 *Drosophila* species. This hypothesis is based on the gene presence/ absence matrix from Rosenfeld et al. All nodes are supported by 100% bootstrap and 100% jackknife values and posterior Bayesian probabilities of 1.0, except for the node defining the sister pair *D. erecta* + *D. yakuba*. (See Chapters 9 and 10 for how phylogenetic trees are constructed and how the support methods work.) The boxes on the nodes represent partitioned support measures. The numbers in the white boxes represent the total Bremer or branch support measure (BS). The six smaller boxes below each white box indicate whether the partitioned Bremer support is positive or negative for the corresponding partition. The number in the light green box at the bottom is the node number. For example, for the node uniting *D. persimilis* and *D. pseudoobscura* (node 2 in the light green box), the branch support is 1963, and all partitions are positive for partitioned Bremer supports. GO, gene ontology partitions: CC, cellular component; MF, molecular function; BP, biological process; ng1 and ng2, categories of genes without GO terms. (Adapted with permission from J.R. Rosenfeld, R. DeSalle, E.K. Lee & P. O'Grady, *Fly* 2:291–299, 2008. Courtesy of Landes Bioscience.)

In this table the matrix of 0s and 1s appears at the top in NEXUS format (the format used in PAUP). Each column of the matrix is given a character number. So the first column of 0s and 1s is character number 1, the second column is character number 2, and so on for the 14,908 genes in the matrix.

So, for instance, the GO category with ID number GO:0000030 (GO30 in the graphic) has characters (columns) 5689, 5961, 8240, 1418, 339, and 1113 in it. If this functional category is important at a node in the phylogeny, then the genes represented by these columns should show positive support measures at that node.

With each gene tagged in this way, it is a simple matter of partitioning further into higher or more restrictive categories. In a similar fashion we can also break the partitions into smaller functional categories, using what the GO system calls "goslim" categories. These categories are simply more restricted groups of genes based on more restricted functions such as growth, binding, or metabolism, to name a few.

charset 1_Cyto=446 1143 1239 3050 3350 10191 825 1082 1084 1158 3544 4125 1165 ...

charset 2_DNAS=504 1450 1496 1972 2097 3420 4210 4515 7687 210 1119 1376 405 851 ...

charset 3_APOP=1598 1739 2145 2438 2688 4940 8781 10240 1493 1602 2372 4951 10217 ...

charset 4_CelC=1211 2064 6053 8829 1942 1771 2058 3350 1140 1531 411 1248 218 599 ...

charset 8_CCOM=1703 3736 4172 4333 5150 5291 3784 4068 4641 8779 4726 7818 2342 ...

charset 9_CADH=274 279 534 677 711 731 741 814 923 951 1276 1351 1462 1863 1917 ...

In this example, genes that are important in apoptosis (3_APOP in the graphic) can be grouped by listing them in the partition. Characters 1598, 1739, 2145, 2438, 2688, 4940, 8781, 10240, 1493, 1602, 2372, 4951, 10217, and so on are now genes that can tell us something about apoptosis at the various nodes in the tree.

Finally, for the GO root categories molecular function (MF), cellular component (CC), and biological process (BP), we can create partitions by developing a list of all of the genes in these categories.

charset component = GO811 GO785 GO786 GO776 GO791 GO805 GO792 GO790 GO796 GO439 ...
charset function = GO339 GO403 GO739 GO823 GO774 GO309 GO900 GO703 GO1591 GO3677 ...
charset process = GO902 GO732 GO705 GO723 GO338 GO289 GO710 GO578 GO819 GO731 ....

Here, since the root categories are composed of less inclusive GO categories already defined, the root categories can now be described by the more specific categories. So, for instance, the CC root category (charset component in the graphic) is composed of the genes in partitions GO811, GO785, GO786, GO776, GO791, GO805, GO792, GO790, GO796, GO439, and so on.

Once the partitioning into functional GO categories is accomplished, some very interesting approaches can be taken to determine if one category behaves differently than another with respect to the degree of support the categories lend to the phylogenetic hypothesis. We will examine here the dynamics of support for the three root categories CC, MF and BP on the phylogeny in Figure 18.10. The genes in each of these three categories can be examined for overrepresentation at the nine nodes that exist in the tree (see the numbered nodes). Heat maps, which illustrate the degree of support the three categories lend to each node, are shown in Figure 18.11. Light green indicates high support and dark green indicates low support, with black representing intermediate or equivalent support. Genes belonging to the BP root category are underrepresented by PBS values at almost all nodes, suggesting that genes involved in biological processes are least likely to be impacted by or respond to branching events in species divergence and in divergence of major groups of *Drosophila*. In addition, it appears that CC category

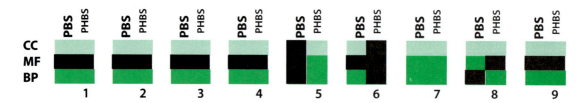

Figure 18.11 Heat map for *Drosophila* 12-genome tree. The number of statistically significant overrepresented genes rendering support at each of the nine nodes for the three GO root categories (CC, cellular component; MF, molecular function; and BP, biological process) is depicted. Two kinds of support are shown: partitioned Bremer support (PBS) and positive partitioned hidden Bremer support (PHBS). Dark green indicates a low number of genes, light green indicates a high number of genes, and black represents an intermediate level of support. (Adapted with permission from J.R. Rosenfeld, E.K. Lee, P. O'Grady & R. DeSalle, *Fly* 2:291–299, 2008. Courtesy of Landes Bioscience.)

genes are generally most affected by major divergence events that are deep in the phylogeny. Also, for almost all nodes, CC support > MF support > BP support. Node 7, which supports the *D. melanogaster* species group, is the only exception to this rule. At this node, MF genes and BP genes are equal in their contribution of support. This approach, when used with more strict GO categories, can give even more precise information on what kinds of genes and hence what kinds of functions are important in the divergence of these flies.

Expressed sequence tags and phylogeny can be used to study plant function

The previous example demonstrates the utility of partitioning presence/absence data for functional studies. More often, though, phylogenomic matrices are composed of amino acid or DNA sequences of large numbers of genes. The structure of these matrices allows determination of the degree of support for single genes and clustering and partitioning can proceed differently for these kinds of studies. One study compiled a large phylogenetic matrix of EST sequence information for 16 plant species. The matrix consists of over 2200 genes, all of which have been clearly annotated and given GO category identifiers. The 16 taxa cover a wide range of seed plants in two major clades, angiosperms and gymnosperms, as well as a seed-free group used as an outgroup. By use of the same partitioning methods described above, critical nodes in the tree can be examined for the kinds of genes that impact the divergence of seed plants. This is accomplished by determining the PBS for each gene in the analysis for each of the nodes in the tree. Lists of genes are then created that show positive support for each of the nodes. These genes can also be ranked according to the magnitude of the partitioned support. Next the lists are examined for overrepresentation of GO categories by use of a Z-score which is a statistic similar to the E-values we discussed in the context of BLAST searches in Chapters 4 and 5. The Z-scores are compared to the distribution of that GO term in the *Arabidopsis* genome as a null distribution. Generally, a score of $Z = \pm 7$ is considered to be a threshold of significance, although when dealing with limited random subsets (such as orthologous partitions derived from EST libraries), the information given by a negative score (underrepresentation) is practically null.

In this analysis, genes in the GO category representing ovule development (enclosed in a circle in **Figure 18.12**) are the only genes overrepresented at node 3 that rightfully is the node defining seed plants (angiosperms + gymnosperms). This is a completely expected result and validates the approach to a certain extent. This study also found that genes involved in post-transcriptional regulation by small RNAs are highly overrepresented functional categories in the phylogeny (shown in rectangles in Figure 18.12). Post-transcriptional gene silencing and mismatch repair received high Z-values at several nodes, for PHBS and/or PBS. The functional role of highly conserved micro-RNAs (miRNAs) and small interfering RNAs (siRNAs) is well-known. Mutations in conserved small RNA pathways, for instance, are important for developmental phenotypes in different tissues. Therefore, overrepresented siRNAs and miRNAs that provide positive support for a particular node are interesting, as they may have a novel or specific function. The specific molecular changes of these genes can also be examined at the sequence level.

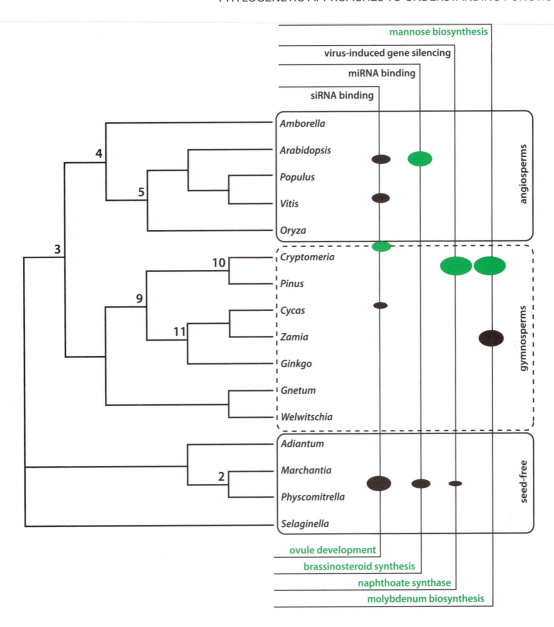

Figure 18.12 Tree resulting from a large phylogenetic matrix of EST sequence information for 16 plant species. The matrix consists of over 2200 genes, all of which have been clearly annotated and given GO category identifiers. The 16 taxa cover a wide range of seed plants in two major clades, angiosperms and gymnosperms, as well as a seed-free group. Genes with positive partitioned Bremer support that are overrepresented across the phylogeny are shown as ovals at the relevant node; the size of the oval corresponds to the statistical significance of the inference. Green ovals and green labels correspond to categories that are found only in gene clusters unique to that particular node. Black ovals and labels represent genes that are found across the node and thus can be shared with other nodes in the phylogeny. (Adapted with permission from A. Cibrián-Jaramillo, E. de la Torre- Bárcena, E.K. Lee et.al., *Genome Biol. Evol.* 2:225–239, 2010. Courtesy of Oxford University Press.)

Gene clustering in *Caenorhabditis elegans* was determined from RNA interference phenotype

Recently, a set of genes enriched in the ovary of the nematode *C. elegans* has been constructed. Most of these genes have also been analyzed for their impact on embryonic phenotype. This latter characterization can be done by RNA interference experiments (RNAi). Such experiments essentially knock down a gene in an embryo, and then the embryo can be examined for the phenotypic effect. Because the function of the gene used in the RNAi experiment is known and annotated, its GO category is also known, and hence the overrepresentation of genes in certain categories can be correlated with phenotypes. Most of the RNAi experiments result in embryonic lethality with very discrete phenotypes up to a certain developmental stage. Piano et al. have used this approach by creating RNAi knockdowns for single genes. They film the development of worms with the knockdown phenotype and from the developmental data are able to classify the different genes into different phenotypes. In these experiments, 47 RNAi phenotypes are classified for 161 genes, from which a matrix with presence/absence data is generated. For instance, a gene called embryonic-27 (emb27) creates a profile of 47 0s and 1s representing the kinds of phenotypes it produces. This matrix of 0s and 1s can then be used to generate a tree that depicts how the genes relate to each other in the context of phenotype, where phenotypes are at the tips of the tree. The resulting clusters of genes are termed "phenoclusters." This kind of analysis can determine which GO categories are overrepresented in certain phenoclusters. For instance, in the study by Piano et al., one of the major phenoclusters shows overrepresentation of chromatin and chromosome structure genes. It is very possible that the genes in this phenocluster are controlled by chromatin-regulating processes. Another phenocluster shows overrepresentation of protein synthesis and signal transduction genes. This result might indicate that these processes as important in regulating the genes in this phenotypic cluster.

Careful annotation of genes into GO categories and detailed phylogenetic analysis can be combined to reach some interesting conclusions. As suggested by Piano et al., these studies clearly show that "phenoclusters correlate well with sequence-based functional predictions and thus may be useful in predicting functions of uncharacterized genes." More studies like the seed plant study and the *C. elegans* phenocluster study need to be done to further determine the correlation of function and phylogeny.

Summary

- Microarrays allow researchers to detect and quantify the expression of genes in cells in a high-throughput manner. Next-generation sequencing, using massively parallelized sequencing approaches, augments the array approach.

- Microarrays are used as tools to determine the differences in gene expression profiles, to predict classification structure, and to discover and analyze new groups.

- Raw intensity data from microarrays can be transformed by binning and conversion to matrices. Tree-building methods can then be used to depict relationships.

- Microarrays have been used to examine transcription profiles of bees at different life stages; adaptation of sunflowers to novel ecological niches; evolution of primate brains; and gene expression patterns among different types of wasps.

- Functionality, such as protein–protein interaction and determination of phenotype, is studied in classical comparative ways with the aid of various Web-based resources.

- Phylogenetic approaches offer a new dimension in functional studies: the dimension of time that involves common ancestry. The structure of a phylogenetic tree can explain the behavior of genes and proteins through time.

- Phylogenomic gene partitioning is used to explore function in various organisms. Calculations of Bremer support and its variations are important for interpreting the stability of groups (clades) in such studies.

- A gene presence/absence study provided information on what kinds of genes, and hence what kinds of functions, are important in the divergence between 12 *Drosophila* species.

- A large phylogenetic matrix of sequence information from many plant species was analyzed to determine the kinds of genes that impacted the divergence of seed plants.

Discussion Questions

1. Suppose that you are interested in the possible correlation of a gene's function with an important evolutionary trend. You suspect that a particular gene (candidate gene) is involved, so you examine it for positive Darwinian selection at the node where the trend is initiated. You find no detectable dN/dS at the node in the phylogeny of the organisms you are studying. On the other hand, you examine the partitioned support at the same node, and it is significantly large. How do you explain this result? Discuss other potential scenarios that might occur when comparing natural selection for a gene at a node in a phylogeny and the value of node support that the gene renders to the node.

2. Compare the quality of data obtained from a microarray for a specific evolutionary question versus next generation approaches for obtaining transcription information.

3. Almost all microarray studies to date have used dendrograms to display the results of the studies. Discuss the ramifications of using dendrograms in microarray studies. Are there better approaches to building trees from microarray data? What is the major difference between using trees in microarray studies versus phylogenetic studies?

4. Calculate BS, PBS, HBS, and PHBS for the following matrix with three partitions:

Taxon	Gene 1	Gene 2	Gene 3
A	GGAAGGGG	GGGGGAA	GGGGGGG
B	GGAAAAGG	GAAGGAA	GGGGGGG
C	AAGGAAAA	AAAAAGG	AAAAAAA
O	AAGGGGAA	AGGAAGG	AAAAAAA

Further Reading

Buck MJ & Lieb JD (2004) ChIP-chip: considerations for the design, analysis, and application of genome-wide chromatin immunoprecipitation experiments. *Genomics* 83, 349–360.

Cibrián-Jaramillo A, De la Torre-Bárcena E, Lee EK et al. (2010) Using phylogenomic patterns and gene ontology to identify proteins of importance in plant evolution. *Genome Biol. Evol.* 2, 225–239.

Gatesy J, O'Grady P & Baker RH (1999) Corroboration among data sets in simultaneous analysis: hidden support for phylogenetic relationships among higher level artiodactyl taxa. *Cladistics* 15, 271–313.

Hegde P, Qi R, Abernathy K et al. (2000) A concise guide to cDNA microarray analysis. *BioTechniques* 29, 548/endash562.

Lai Z, Kane NC, Zou Y & Rieseberg LH (2008) Natural variation in gene expression between wild and weedy populations of *Helianthus annuus*. *Genetics* 179, 1881–1890.

Piano F, Schetter AJ, Morton DG et al. (2002) Gene clustering based on RNAi phenotypes of ovary-enriched genes in *C. elegans*. *Curr. Biol.* 12, 1959–1964.

Planet PJ, DeSalle R, Siddall M et al. (2001) Systematic analysis of DNA microarray data: ordering and interpreting patterns of gene expression. *Genome Res.* 11, 1149–1155.

Preuss TM, Cáceres M, Oldham MC, & Geschwind DH (2004) Human brain evolution: Insights from microarrays. *Nat. Rev. Genet.* 5, 850–860.

Rosenfeld JR, DeSalle R, Lee EK & O'Grady P (2008) Using whole genome presence/absence data to untangle function in 12 Drosophila genomes. *Fly* 2:291–299.

Schena M, Heller RA, Theriault TP et al. (1998) Microarrays: biotechnology's discovery platform for functional genomics. *Trends Biotechnol.* 16, 301–306.

Whitfield CW, Cziko A-M & Robinson GE (2003) Gene expression profiles in the brain predict behavior in individual honey bees. *Science* 302, 296–299.

Young K (1998) Yeast two-hybrid: so many interactions, (in) so little time. *Biol. Reprod.* 58, 302–311.

INDEX